New Methods of Competing in the Global Marketplace

Series on Resource Management

Rightsizing Inventory
by Joseph L. Aiello
ISBN: 0-8493-8515-6

Integral Logistics Management: Operations and Supply Chain Management in Comprehensive Value-Added Networks, Third Edition
by Paul Schönsleben
ISBN: 1-4200-5194-6

Supply Chain Cost Control Using Activity-Based Management
Sameer Kumar and Matthew Zander
ISBN: 0-8493-8215-7

Financial Models and Tools for Managing Lean Manufacturing
Sameer Kumar and David Meade
ISBN: 0-8493-9185-7

RFID in the Supply Chain
Judith M. Myerson
ISBN: 0-8493-3018-1

Handbook of Supply Chain Management, Second Edition
by James B. Ayers
ISBN: 0-8493-3160-9

The Portal to Lean Production: Principles & Practices for Doing More With Less
by John Nicholas and Avi Soni
ISBN: 0-8493-5031-X

Supply Market Intelligence: A Managerial Handbook for Building Sourcing Strategies
by Robert B. Handfield
ISBN: 0-8493-2789-X

The Small Manufacturer's Toolkit: A Guide to Selecting the Techniques and Systems to Help You Win
by Steve Novak
ISBN: 0-8493-2883-7

Velocity Management in Logistics and Distribution: Lessons from the Military to Secure the Speed of Business
by Joseph L. Walden
ISBN: 0-8493-2859-4

Supply Chain for Liquids: Out of the Box Approaches to Liquid Logistics
by Wally Klatch
ISBN: 0-8493-2853-5

Supply Chain Architecture: A Blueprint for Networking the Flow of Material, Information, and Cash
by William T. Walker
ISBN: 1-57444-357-7

ERP: Tools, Techniques, and Applications for Integrating the Supply Chain
by Carol A. Ptak with Eli Schragenheim
ISBN: 1-57444-358-5

Introduction to e-Supply Chain Management: Engaging Technology to Build Market-Winning Business Partnerships
by David C. Ross
ISBN: 1-57444-324-0

Supply Chain Networks and Business Process Orientation
by Kevin P. McCormack and William C. Johnson with William T. Walker
ISBN: 1-57444-327-5

Collaborative Manufacturing: Using Real-Time Information to Support the Supply Chain
by Michael McClellan
ISBN: 1-57444-341-0

The Supply Chain Manager's Problem-Solver: Maximizing the Value of Collaboration and Technology
by Charles C. Poirier
ISBN: 1-57444-335-6

Lean Performance ERP Project Management: Implementing the Virtual Lean Enterprise, Second Edition
by Brian J. Carroll
ISBN: 0-8493-0532-2

Integrated Learning for ERP Success: A Learning Requirements Planning Approach
by Karl M. Kapp, with William F. Latham and Hester N. Ford-Latham
ISBN: 1-57444-296-1

Basics of Supply Chain Management
by Lawrence D. Fredendall and Ed Hill
ISBN: 1-57444-120-5

Lean Manufacturing: Tools, Techniques, and How to Use Them
by William M. Feld
ISBN: 1-57444-297-X

Back to Basics: Your Guide to Manufacturing Excellence
by Steven A. Melnyk and R.T. Chris Christensen
ISBN: 1-57444-279-1

Enterprise Resource Planning and Beyond: Integrating Your Entire Organization
by Gary A. Langenwalter
ISBN: 1-57444-260-0

New Methods of Competing in the Global Marketplace

Critical Success Factors from Service and Manufacturing

Richard E. Crandall
William R. Crandall

CRC Press
Taylor & Francis Group
Boca Raton   London   New York

CRC Press is an imprint of the
Taylor & Francis Group, an **informa** business

AN AUERBACH BOOK

CRC Press
Taylor & Francis Group
6000 Broken Sound Parkway NW, Suite 300
Boca Raton, FL 33487-2742

© 2008 by Taylor & Francis Group, LLC
CRC Press is an imprint of Taylor & Francis Group, an Informa business

No claim to original U.S. Government works
Printed in the United States of America on acid-free paper
10 9 8 7 6 5 4 3 2 1

International Standard Book Number-13: 978-1-4200-5126-1 (Hardcover)

This book contains information obtained from authentic and highly regarded sources Reasonable efforts have been made to publish reliable data and information, but the author and publisher cannot assume responsibility for the validity of all materials or the consequences of their use. The Authors and Publishers have attempted to trace the copyright holders of all material reproduced in this publication and apologize to copyright holders if permission to publish in this form has not been obtained. If any copyright material has not been acknowledged please write and let us know so we may rectify in any future reprint

Except as permitted under U.S. Copyright Law, no part of this book may be reprinted, reproduced, transmitted, or utilized in any form by any electronic, mechanical, or other means, now known or hereafter invented, including photocopying, microfilming, and recording, or in any information storage or retrieval system, without written permission from the publishers.

For permission to photocopy or use material electronically from this work, please access www.copyright.com (http://www.copyright.com/) or contact the Copyright Clearance Center, Inc. (CCC) 222 Rosewood Drive, Danvers, MA 01923, 978-750-8400. CCC is a not-for-profit organization that provides licenses and registration for a variety of users. For organizations that have been granted a photocopy license by the CCC, a separate system of payment has been arranged.

Trademark Notice: Product or corporate names may be trademarks or registered trademarks, and are used only for identification and explanation without intent to infringe.

Library of Congress Cataloging-in-Publication Data

Crandall, Richard E., 1930-
New methods of competing in the global marketplace : critical success factors from service and manufacturing / Richard E. Crandall and William Crandall.
p. cm.
Includes bibliographical references and index.
ISBN 978-1-4200-5126-1 (alk. paper)
1. Management. 2. Manufacturing industries--Management. 3. Service industries--Management. I. Crandall, William, 1956- II. Title.

HD31.C687 2008
658--dc22   2008005944

Visit the Taylor & Francis Web site at
http://www.taylorandfrancis.com

and the CRC Press Web site at
http://www.crcpress.com

Dedication

This book is dedicated to Jean and Sue for their encouragement and patience not only during its preparation, but also for all of the other years we have been together.

Contents

Foreword xxi
Acknowledgments xxiii
Introduction xxvii
About the Authors xxxi

1   The Vanishing Manufacturing/Services Boundary 1
    Differences between Manufacturing and Service 2
    Forces That Are Eliminating the Boundary 4
        The Economy Is Going through a Natural Evolution 4
        There Is a Need to Identify Critical Success Factors 4
        Supply Chains Are Expanding to Integrate Manufacturing and Service .. 5
        It Is Easier to Blend Tasks, Resources, and Techniques into Programs 5
        Service Businesses Are Looking for Ways to Operate More Efficiently 5
            Health Care Providers Are Looking for Ways to Cut Costs and
            Improve Quality 6
            Banks Are Facing Competition from Lower-Cost Banks 6
            Investment Bankers Are Facing Competition from Commercial
            Banks and Discount Brokers 6
            The Retail Industry Is Moving in Two Strategic Directions
            — Low Cost and Focus 7
        Manufacturing Is Adding Services to Be More Customer Focused 7
            Post-Sales Services 8
            Financing of Purchases 8
            Online Purchasing 9
            Customer Input to Product Design 9
            Enhanced Customer Relationship Management 9
        Technology Development 9
        A Movement towards a Process Perspective 11
        Outsourcing Focuses More Attention on the Need for Good Service 11

Changing Personal and Organizational Relationships 12
The Need to Integrate Companies into Supply Chains 12
Services Become More Relevant to All Types of Customers 13
Continuous Improvements Continue to Blur the Boundary
between Manufacturing and Services 13
    There Is a Movement toward an Industry Focus and away from
    a Functional Focus 13
    The Mass Customization Concept Requires a Blending of
    Manufacturing and Service 13
The Vanishing Manufacturing/Services Boundary 15
    Phase 1: Separated Disciplines 15
    Phase 2: Internal Improvements in Costs and Quality 15
    Phase 3: Customer Service Improvements 17
    Phase 4: Integrated Product and Service Functions 17
Summary 18
References 18

2   Critical Success Factors and Strategic Planning 19
What Are Critical Success Factors? 21
The Evolution of CSFs in the United States 22
Other Changes during a Country's Economic Life Cycle 26
The Need to Be Effective 29
A Hierarchy of the Planning Process 30
    The Strategic Planning Process 31
    The Business Planning Process 32
    Project and Program Plans 33
    CSFs Plan 33
A Hierarchy of Critical Success Factors 33
The Role of CSFs in Strategic Planning 34
    What Is Strategic Planning? 35
    Background of CSFs in Strategic Planning 36
    Characteristics of Strategic CSFs 37
    The Temporal Nature of CSFs 39
        Start-Up 40
        Growth 40
        Maturity 41
        Decline 41
The Role of CSFs in Operational Planning 41
    Positioning Strategy 43
    Manufacturing Flows 44
    Manufacturing Strategies 45
Role of CSFs in Selecting Management Programs 45
Performance Measurement and CSFs 47

Contents ■ ix

	Summary	48
	References	48
3	**The ITO Model**	**51**
	The ITO Model	53
	Introduction to Models	55
	Types of Models	55
	Why Are Models Used?	56
	Benefits of Using Models	57
	The Basic ITO Model — Inputs, Transformation, and Outputs	58
	The General Systems Model	58
	The Basic Model Extended	61
	Components of the Model	62
	Customers	62
	Outputs	63
	Transformation Process	64
	Inputs	64
	Extending the Basic ITO Model into Supply Chain Configurations	65
	A Look at the Supply Chain	67
	The Closed and Open Systems Model	67
	Closed Systems	68
	Open Systems	68
	Closed-System Strategy	70
	Open-System Strategy	71
	Feedback	73
	The Fine Art of Building Relationships	73
	The Concept of Reverse Logistics	76
	Why the Interest in Reverse Logistics?	76
	Benefits of Reverse Logistics	77
	Barriers to Reverse Logistics Implementation	78
	System Design and Implementation	79
	Summary	80
	References	81
4	**The Role of Management Programs in Continuous Improvement**	**83**
	What Are Management Programs?	85
	Normal or Sustaining Day-to-Day Operations	87
	Problem-Solving Activities	88
	Improvement Programs	88
	Management Program Life Cycles	92
	Life-Cycle Stages	92
	At the End of the Life Cycle	94
	Implications of Program Life Cycles for Management	94

Why Are Management Programs Important? 96
Where Do Management Programs Come From? 97
   Overview of Employee Management Theories.................................98
   Scientific Management..98
   Human Relations Management 99
   Administrative Management...100
   Systems Theory..102
   Contingency Theory..103
Why Are Some Programs Successful and Some Not?.............................105
   Failure to Match Program with Need..107
   Implementing the Program Correctly..109
      Planning and Preparation 109
      Execution and Evaluation 110
Future of Management Programs 111
   Management for the 21st Century...112
Summary 114
References...114

5   How Manufacturing Techniques Are Being Adapted to Service Operations...117
Introduction 118
Description of Manufacturing Process Types..120
Product–Process Relationship 122
Service Industry Classifications...122
   The Chase Model 122
   Schmenner Model 123
Comparison of Manufacturing and Services...123
Manufacturing Objectives 125
   Reduce Product Costs 125
   Reduce Inventories 125
   Increase Resource Utilization 126
   Improve Quality..127
   Reduce Response Time 128
   Reduce Product Development Time..129
Service Objectives...130
Programs That Work in Services...131
   JIT and Lean...132
   TQM and Six Sigma 135
Programs More Difficult to Adapt to Service Operations........................138
   Product Costing 139
   Activity-Based Costing..139
   Materials Requirements Planning 139
   Enterprise Resources Planning 140

Contents ▪ xi

    Performance Measures 140
    Automation 140
    Resource Utilization 141
Keys to Extending Manufacturing Techniques to Services 141
Conclusions 142
References.. 143

6 How Service Techniques Are Being Extended to Manufacturing 147
Introduction 149
    The Rise of Services as a Part of the Economy 150
    The Swing of Power from Manufacturing to Retail 150
    Maturity of the Customer as a Shopper 151
    Increasing Complexity of the Marketplace 151
    Need for Manufacturing Company to Add Services....................... 151
    Move from Product-Centric to Customer-Centric.......................... 153
    Services as a Separate New Business Segment................................. 156
    Economic Growth and the Need for Added Services...................... 156
    The Movement from Make-to-Stock (MTS) to Make-to-Order
    (MTO) 157
    The Movement toward Mass Customization 157
What Are Services? 158
    Distributive Services... 161
    Personal Services 161
    Self-Service... 161
    Producer (Business) Services... 161
    Social Services 161
Knowledge Transfer from Services to Manufacturing............................. 162
    Areas of Manufacturing Expertise.. 162
    Areas of Service Expertise... 163
Examples of Programs Developed in Services 163
    Customer Relationship Management (CRM) 164
        Definitions.. 165
        Background 166
        What Does It Do? 166
        Benefits... 167
        Problems 168
        Relation to the Supply Chain.. 168
        CRM's Future... 168
    Response Time Reduction... 169
        Quick Response Systems (QRS) 169
        Continuous Replenishment Programs (CRP) 170
        Efficient Consumer Response (ECR) 170
        Vendor-Managed Inventory (VMI) 171

Sales and Operations Planning (S&OP) 171
Collaborative Planning, Forecasting, and Replenishment (CPFR).. 171
Supply Chain Management (SCM) 172
Present Status 172
Future 174
Flexibility 174
Product and Service Flexibility: Mass Customization 174
Horizontal Communication and Organization Structure 175
Demand Management 178
Location near the Market versus Lowest Cost 179
Evolution from Job Specialization to Self-Directed Teams.............. 180
Replacement of Inventory with Information 180
Inter-Organizational Communications 180
Traditional EDI 180
Internet EDI 181
Other Service Developments 184
Working in an Open-System Environment..................................... 184
Product Development as a Result of Customer Inputs 184
Quality as Customers' Perceptions, Not Just Conformance to
Specification 184
Managing the Customer Encounter... 185
Nonquantitative Performance Measurement................................... 185
The Use of Business Intelligence 185
Summary 186
Differences between Manufacturing and Services 186
From Services to Manufacturing 187
Collaborative Efforts 188
Conclusion 189
References 189

7 The Role of Technology in Continuous Improvement 195
Definitions.. 199
The Role of Technology in Continuous Improvement................................ 200
Technology and the Infrastructure 201
Technology and Organizational Culture 202
Technology Transfer...205
Technology for Process Improvement 205
Agriculture, Mining, Construction, and Manufacturing: Goods
Producers 206
Services 209
Self-Service..211
E-Business..212
Technology for Resource Enhancement 214

Contents ■ xiii

    Human Resources 214
        As an Aid to the Employee 215
        As a Substitute for the Employee 216
        As an Integral Part of the Process 218
    Equipment 219
        Enhance the Performance of Equipment 219
        Provide a Source of Performance Information 219
    Facilities 220
        Design 220
        Location 221
        Layout 222
    Information Technology (IT)...222
    Integrated Systems..224
        Enterprise Resource Planning (ERP) Systems............................224
        Inter-Organizational Systems (IOS)..224
        Service-Oriented Architecture (SOA) 224
        Structure, Trust, and Collaboration..225
Criteria Used in Decision Making 225
    Strategic Needs versus Short-Term Needs..225
    Behavioral versus Scientific Management Issues..................................226
    Costs versus Benefits of Added Technology...227
Steps in Adding Technology to the Process... 228
    Step 1. Communicate, Communicate, Communicate....................... 228
    Step 2. Identify Needs and Opportunities.. 228
    Step 3. Evaluate Alternatives and Select the Optimum Alternative......229
    Step 4. Educate and Orient 229
    Step 5. Develop the Implementation Plan 230
    Step 6. Implement the Technological Changes.................................. 230
    Step 7. Evaluate Results, Redefine Needs, and Redefine Additional
    Increments..231
Future Considerations for Technology 231
    Customer Acceptance...232
    Work Force Acceptance..233
    Economic Feasibility 233
    Technical Feasibility... 234
Summary 235
References..237

8 The Role of Infrastructure in Continuous Improvement 241
What Is Infrastructure?.. 242
Strategies...243
    Corporate Strategy 243
    Business Strategy... 244

The Four Classical Management Functions 245
    Planning..246
    Organizing..246
    Directing...247
    Controlling...248
Organization Structure 248
    Functional...249
    Product 249
    Geographic...250
Alternate Organizational Structures 251
    The Matrix Organization 251
    The Horizontal Organization 252
    The Virtual Organization 253
Trends in Organizational Structures...254
    Moving from Centralization to Decentralization 254
    Moving from Vertical Structures to Horizontal Structures..................255
    Moving from Autocratic Managers to More Empowered Employees...255
    Moving from Job Specialization to Higher Skill Variety 256
    Moving from Line Managers to Self-Directed Work Teams................257
    Moving from Specialized Departments to Cross-Functional Teams....258
    Moving from Top-Down to Multi-Directional Communications 259
    Moving from Rigid Policies and Procedures to More Flexibility 259
    Moving from Mechanistic Structures to Organic Structures.............. 260
The Role of the Internet in Changing Organizational Structure 261
The Integration of Knowledge Management into Organizational Structure..263
    Data, Information, Knowledge 263
    Why Knowledge Is Not Transferred..266
Does Your Business Need a Change in Its Infrastructure? 267
Notes 268

9 Understanding Organizational Culture — the Elusive Key to Change 271
Introduction 273
What Is Organizational Culture?..275
Why Is Organizational Culture so Important? 276
    Organizational Culture Gives the Company an Identity..............276
    Organizational Culture Helps Employees Make Sense of Things........ 277
    Organizational Culture Enables Employees to Be Committed to the Company 279
    Organizational Culture Helps Add Stability to the Company............281
What Are the Components of Organizational Culture?......................282
    The Components of Culture..282

Values 283
Artifacts: The Display of Organizational Culture 283
   Stories 284
   Language 284
   Symbols, Ceremonies, and Rituals..................................... 286
   Identifiable Value Systems and Behavioral Norms 287
   The Physical Surroundings Characterizing the Particular Culture. 288
   Organizational Rewards and Reward Systems 290
What Types of Organizational Culture Are There? 291
   Types of Cultures 291
   Aligning Culture and Strategy 292
Changing Organizational Culture 293
   State What You Want the Culture to Be 293
   Promote an Ethical Culture 294
   Hire the Kind of People You Want to See Perpetuate Your Desired Culture...296
   Take Care of Your Employees, and They Will Take Care of Your Customers 297
   Implement the Strategy Proposed..298
   Change the Physical Cultural Artifacts 299
   Unlock the Power of Language and Phrases 300
   Tell Stories...301
   Recognize Employees Formally 302
Conclusion...303
Notes 303

10 Integrated Supply Chains from Dream to Reality 307
Introduction 308
   Present Status of Supply Chains 308
   Background of Supply Chains 310
      Material Flows 310
      Organization Hierarchy..310
      Systems Theory 312
   Role of the Supply Chain 312
Setting the Stage..313
Supply Chain Models 317
   Evolution of Supply Chain Models...317
   A Comprehensive Supply Chain Model..320
   The World of Lean Production 322
      Product 324
      Purchasing Process 324
      Production Process 324
      Delivery Process..324

Demand Variation 325
　Steps to Achieve a Lean and Agile Supply Chain 325
　　Commitment 325
　　Concept 326
　　Configuration 327
　　　Marketing 327
　　　Purchasing 328
　　　Manufacturing 328
　　　Distribution 329
　　　Finance and Accounting 329
　　　Top Management 330
　　Communication 331
　　Culture 331
　　Customization 332
　　The Integrated Supply Chain 333
　　Coordination 334
　　Collaboration 334
　Steps in the Change Process 335
　　Investigate 335
　　Initiate 336
　　Invigorate 336
　　Implement 336
　　Integrate 336
　　Institutionalize 337
　　Innovate 337
　A Look Ahead 337
　　Complexity 338
　　Clairvoyance 339
　Conclusions 341
　References 342

11　The Role of Services to Complement the Supply Chain 345
　Introduction 346
　　Producer Services versus Consumer Services 347
　What Are Producer Services? 350
　　Why Are Producer Services Important? 352
　　The Role of Outsourcing 352
　　The Role of Producer Services 353
　　Classes of Producer Services 355
　　Reasons to Acquire Producer Services 355
　　　Gain a Cost Advantage 355
　　　Remove Noncore Types of Activities 356
　　　Supplement Internal Staffs with Added Expertise 356

       Provide Flexible Capacity or to Avoid Overload on Key
          Departments 358
       Acquire Expertise Not Available Internally 358
       Assist in Strategic Planning 359
       Comply with Regulatory or Otherwise Mandatory Requirements . 360
       Manage Major Programs 361
   Steps to Interfacing Business with Producer Services 361
       Recognize the Role of Producer Services in the Company 362
       Develop a Strategy for Each Producer Services Area 362
       Select Suppliers to Provide a Continuing Relationship 362
       Prepare a Plan That Interfaces with the Supply Chain Plan 362
       Implement the Service — Fit It to the Situation 363
       Measure the Progress 363
       Revise and Assimilate 363
   The Future of Producer Services 364
   Producer Services Firms Buy Producer Services 364
What Are Social Services? 365
   Why Are Social Services Important? 365
   The Role of Social Services 365
   Classes of Social Services 366
       Government 366
       Nonprofit Organizations 366
       Quasi-Private Firms 367
   Specific Social Services 367
       Information 367
       Communications Systems 367
       Standards 367
       Education and Training 368
       Location Incentives 368
       Support Infrastructure 368
       Protection 368
       Health Care 368
   Steps to Interfacing Business with Social Services 369
   The Future of Social Services 370
       Level of Regulation 370
       Level of Privatization 370
       Level of Effectiveness and Efficiency 371
       Emphasis on Business-Related Topics 371
       Specific Issues to Resolve 371
What Are Consumer Services? 371
   Roles of Consumer Services 372
   Future of Consumer Services 372
Integrated Service Package 373

The Need for an Integrated Service Package 373
Steps in Developing the Integrated Service Package 373
   Recognize the Need to Change 373
   Identify the Service Package Components 374
   Develop Objectives for the Service Package 374
   Develop Strategies for the Service Package 374
   Develop Objectives for Each Component 375
   Develop Strategies for Each Component 375
   Implement the Strategic Plan 375
   Evaluate the Results 375
   Revise as Needed 376
Summary 376
References 376

12 The Future of Improvement Programs 379
Introduction 380
The Background to Improvement Programs 380
   The Vanishing Manufacturing/Services Boundary 380
   The Foundation Topics 380
   Knowledge Transfer across the Manufacturing/Services Boundary 381
   Agents of Change 382
   Integration of Related Entities 382
Future Areas of Emphasis 383
   Services Will Continue to Increase as a Critical Success Factor
   (CSF) in Business 384
   Continuing Need to Integrate the Product and Service Bundle 384
   Continuing Increase in Globalization 384
   Outsourcing Will Become a More Focused Activity 384
   Increased Need for Project Management Competencies 385
   Decision Making Will Deal with a Blend of Hard and Soft Variables ... 385
   Decisions Will Become More Complex 385
Future of Improvement Programs 385
   Improvement Programs Will Become Increasingly Important 386
   Improvement Programs Will Be Customized to Specific Applications ... 386
   Improvement Programs Will Be Better Managed 386
   Improvement Programs Will Become a Regular Part of Business
   Operations 386
   Improvement Programs Will Lose Their Manufacturing or Service
   Heritage 387
The Drivers of Change 387
   Technology 387
   Structure 387
   Culture 388

Most Likely Future Methodologies 388
   Integrated Supply Chains 388
   Outsourcing 389
   Total Cost of Ownership 389
   Performance Measurement 390
   Project Management 391
   Mass Customization 391
   Virtual Organizations 392
   Information Technology 392
   Environmental Design 393
   Service Sciences 393
   Chaos and Complexity 393
Most Likely Improvement Programs 394
   Cost Reduction 394
   Response Time Reduction 396
   Quality 397
   Customer Service Level 398
   Flexibility 398
   Agility 399
   Compatibility 400
   Integration 400
Industries Most Likely to Stress Continuous Improvement 400
   Health Care: Hospitals and Wellness Centers 400
   Pharmaceuticals 401
   Local Government 402
   Retailing 402
   Education 403
Knowledge Management: Where Does It Fit? 405
   From Data to Information 406
   From Information to Knowledge 408
   From Knowledge to Wisdom 408
Notes 409

Index 413

Foreword

Walk into any fast food–chain franchise in any large city around the world and you will step into a microcosm of preordained high technology, precise processing, and trusting relationships. Take McDonald's. Refrigerated tractor-trailer trucks deliver food supplies to the store every day. Every ingredient is fresh because three separate compartments in the trailer are temperature controlled to three different temperatures, each with high precision. Pre-prepared meals are taken off the shelf when their shelf life exceeds the corporate quality standard. Should a new customer walk into the store at just this moment, the kitchen crew springs into action following their strictly defined process to "cook" the next Big Mac for this new customer. A customer will observe highly disciplined action at each station with timers and alarms used to ensure a repeatable outcome. The meal is delivered quickly with smiles. But wait … this sounds like a manufacturing story.

Is McDonald's a service organization that makes great hamburgers, or a hamburger manufacturer that delivers excellent service? Is FedEx a service organization that makes coordinated shipments, or a shipment manufacturer that offers timely service? Is the Mayo Clinic a service organization that performs surgical procedures, or a human health manufacturer that provides outstanding customer care? Where is the line drawn between service and manufacturing? Clearly the boundary is vanishing before our very eyes! Yet, why should we care?

I asked Dick Crandall the following question as he started writing this manuscript: "If one of your students decided to start her own business and came to you for advice about what kind of business it should be and how it should operate, what would you tell her?"

It is clear as we progress into the new millennium that the old forms of business, particularly manufacturing, are not working. But, no one seems to know exactly what will successfully take their place. In *New Methods of Competing in the Global Marketplace: Critical Success Factors from Service and Manufacturing,* Dick Crandall and his son, Rick, present a very carefully researched volume that brings the reader back to the fundamental principle that any business, whether service or manufacturing, must create value to survive. This book is a timely, in-depth analysis of the best and the worst business practices known today with a strong theme of how to continuously improve on the best. Your time investment to read and absorb this book will help you steer a clear course through the business survival waters your organization may be currently facing. And, you will come out the other side having learned how to create and sustain value within your own environment.

William T. Walker, CFPIM, CIRM, CSCP
Resource Management Series Editor and Supply Chain Architect
Summit, New Jersey

Acknowledgments

We have a number of people to thank for helping us to conceive of and then write this book. Most of the credit for the original idea goes to William T. Walker, or Bill, to most of us who have had the good fortune to be acquainted with him. Bill suggested that there was a need for a book that drew on experience in both manufacturing and service operations. He felt that some of the "Relevant Research" columns that Dick had been doing for the *APICS Magazine* had something to say in this area. That, combined with the practical experience that both authors have in several industries, both manufacturing and service, could provide a foundation for expanding the core idea into a worthwhile book. During our research and writing, Bill has been supportive of our efforts and a wonderful source of ideas about how to make the book more valuable to the reader. We probably wouldn't have tried, and I know we wouldn't have succeeded, without his guidance. Thanks, Bill, for all you've done.

We owe special appreciation to Doug Kelly, Editor-in-Chief, and Jennifer Proctor, Managing Editor, of the *APICS Magazine*. They encouraged Dick to write about topics that fit nicely in the book. We have drawn heavily from the following Relevant Research columns (portions of these articles are copyright APICS 2007, and are used with permission):

- Crandall, R.E., A fresh face for CRM: Looking beyond marketing. *APICS Magazine*, 16, (9), 20, 2006.
- Crandall, R.E., Beating impossible deadlines: A variety of methods can help. *APICS Magazine*, 16, (6), 20, 2006.
- Crandall, R.E., Looking to service industries: New resources for learning a thing or two. *APICS Magazine*, 16, (10), 21, 2006.
- Crandall, R.E., The epic life of ERP: Where enterprise resource planning has been and where it's going. *APICS Magazine*, 16, (2), 17, 2006.
- Crandall, R.E., Let's talk: Better communications through IOS. *APICS Magazine*, 17, (6), 20, 2007.

- Crandall, R.E., Dream or reality? Achieving lean and agile integrated supply chains, *APICS Magazine*, 15, (10), 20, 2005.
- Crandall, R.E., Device or strategy? Exploring which role outsourcing should plan, *APICS Magazine*, 15, (7), 21, 2005.

Thanks, Doug and Jennifer, for your help and encouragement.

We would like to thank all of the companies we worked for during our time in industry prior to joining the academic ranks. Dick spent over 25 years in industry, where he worked for manufacturing companies, most notably Sylvania Electric and TRW in the electronic industries, for service companies in wholesale and retail, and for an international consulting company, where he worked with a variety of manufacturing and service companies. Rick spent over ten years in the food service industry, most of which was with ARAMARK in contract food services. We saw firsthand the differences and similarities between manufacturing and service businesses.

To supplement our own experience, we dug deeply into some of the extensive research that practitioners, consultants, and academics have done. We have reviewed hundreds of articles and dozens of books to uncover the concepts, principles, and techniques that we include in this book. We wish we could have shared more; however, in lieu of that, we have extensive references at the end of each chapter. We hope that you will find them of value and that the information in the book will be helpful to you in your work.

We would like to thank Ray O'Connell, Acquiring Editor, Taylor & Francis Group, who has been the "main man" in this effort. He helped us focus our thinking through a careful evaluation of our proposal and has been a constant source of support, but has not tried to micromanage our efforts. Thanks, Ray, we hope we can do it again sometime.

Jessica Vakili is a project coordinator for the Taylor & Francis Group. More importantly to us, she is the project coordinator for our book. She has been helpful in providing guidance about all of the things we needed to do and has answered any number of questions that we came up with. Jessica, we really did try to give you a "perfect storm" kind of book; however, despite our best efforts, I suspect that we didn't reach the Six Sigma level of performance. Thanks for covering for us, and for helping us through all the pre-production steps in preparing the book for publication.

When we submitted our original typewritten pages of the manuscript to the editor, we wondered how it could ever look as professional as most of the books we read in our daily work. Now we know it takes professionals like our project editor Karen Simon to make that transformation possible. Thanks, Karen, but we think the least you could have done would have been to invite us to come to your office in Boca Raton, Florida, to review the proofs instead of working on them in the middle of winter in the mountains.

We would also like to thank Julia DiCicco, who had the unenviable job of trying to make our words sound enlightened, or at least less obtuse, as well as worrying about grammar, punctuation, pagination, location of exhibits and all those other things that we were very willing to outsource to an expert like her.

We're confident there are a number of other people who helped this book come alive. However, in this age of electronic communication, we don't always have a chance to meet everyone personally. Thank you; we hope you enjoyed it as much as we did.

Finally, we would like to give a special acknowledgment to Jean Crandall, wife of Dick and mother of Rick. She encouraged us, but even more important, she read every word (several times) and offered many helpful suggestions. Despite her best efforts, there may be errors. If you find them, blame us, not her or any of the people mentioned above.

Introduction

Business survival is not mandatory; you have a choice! However, when you make the obvious choice to survive, you must face the reality that you have chosen a most difficult challenge. Because it isn't good enough just to survive; that will carry you only through the short term. To survive in the long term, you must prosper. And to prosper, you must do a lot of things right. One of the things you must do is to continue to improve. This is a book about the programs that help to achieve improvement.

Organization of the Book

Figure I-1 shows how the book is organized. We begin with an introduction to the idea that the boundaries between manufacturing and services are vanishing. Consequently, there is a need to enhance the knowledge transfer between the two areas. For the past century, manufacturing has led the way in developing ways to improve the management of a business, primarily in achieving lower costs and higher quality. However, as the economy and the number of employees continue to move quickly toward the service area, the service industries are becoming the source of a number of new management techniques, especially those that enhance customer service. In this book, we will show how knowledge is flowing freely in both directions.

We arranged the chapters of this book in what we believe to be a logical sequence. We describe some basic concepts in Chapter 2, Chapter 3, and Chapter 4. In Chapter 2, we describe the role of critical success factors in improvement programs and strategic planning. In Chapter 3, we illustrate that an Input–Transformation–Output Model is the basic building block on which global supply chains depend. In Chapter 4, we describe management programs, especially those designed to provide improvements in organizations.

We then illustrate in considerable detail how knowledge, in the form of management programs, flows from manufacturing to services in Chapter 5 and from services to manufacturing in Chapter 6. Not too long ago, it was clear that the information about improvement programs originated in manufacturing. Today,

xxviii ■ *Introduction*

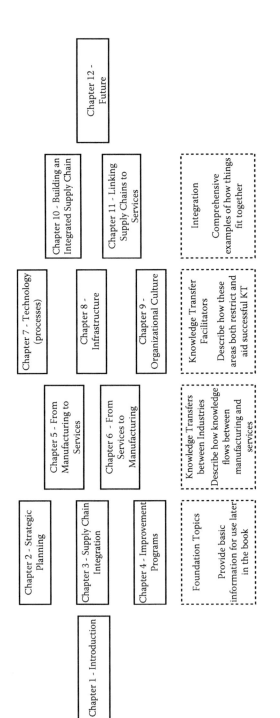

Figure I-1   Organization of chapters

there is an increasing flow from services to manufacturing. Tomorrow, there will be little distinction between manufacturing and services.

In Chapter 7, Chapter 8, and Chapter 9, we describe technology, infrastructure, and culture. Improvement programs depend heavily on new technology. However, technology is only an enabler; it is not sufficient to carry improvement programs to successful conclusions. Although technology is often the driver of change, infrastructure and culture are initially barriers to improvement and have to be changed to become facilitators of improvement.

In Chapter 10, we describe the steps necessary to build an integrated supply chain. It is a journey that many companies have started, but few, if any, have completed to their satisfaction. In Chapter 11, we show how external services are needed to enhance the operation of the entity we describe in Chapter 10. Although there are no simple answers, there are some stimulating challenges ahead.

Finally, in Chapter 12 we show that the past has shaped the present, but the present is just the prologue for what we all hope will be a bright future. However, the future will not be a linear extension of the past. The last chapter is filled with topics that are in your future. You either will use them to your advantage or be victimized by your competitors who do use them.

How to Use This Book

We think you will benefit by reading the entire book. However, if you want to skip around, we tried to make each chapter a stand-alone subject. We suggest that you read Chapter 1 to see what we have included in the book. This is the appetizer. If you find the topics of interest, then move ahead through the succeeding chapters. In these chapters, we dig a little deeper into the topics and provide more substance and illustrations. If you think you don't need that much explanation, look at Chapter 12. See if you are ready for the future. If you are, great! If you decide you may not be, go back to some of the earlier chapters.

You can also get a feel for the chapter content by reading the John Merchant fable that introduces each chapter. John is a retailer struggling to meet the competitive pressures in his business world. He is under enough stress that he dreams at night about business problems and opportunities. He begins … but why don't you read it for yourself?

Who Can Use This Book

We believe this book will be of value to a large audience because it contains a lot of good information that we have assembled from a variety of sources, starting with our own experiences. We brought together a number of important concepts from disparate disciplines. We also tried to strip away the hype and facades of many "hot" topics to provide you with the basic ideas and concepts in a straightforward way.

If you are a manager in a large company, no doubt you are feeling the pressure to acquire more knowledge and apply it in a constructive way. If you are a manager or owner in a small company, you will find a wealth of practical information that you can adapt to your business. This is a book about knowledge transfer and the need to look in new places for it.

In his book *The World Is Flat, A Brief History of the Twenty-First Century* (Farrar, Straus, and Giroux, 2006), Tom Friedman suggests that, in the future, entrepreneurs will play an important role in creating jobs as the global market becomes more fragmented. Large companies may have difficulty in keeping pace with the changes taking place; small companies will have the flexibility to survive. This book provides a picture of what it takes, not only to survive but also to prosper. This is not a book about high-profile success stories in other companies; it is a book filled with ideas about how to make improvements in *your* business. We hope you will find it of value.

About the Authors

Richard E. Crandall is a professor in the College of Business at Appalachian State University, in Boone, North Carolina. He is certified in production and inventory management (CFPIM), integrated resources management (CIRM), and supply chain professional (CSCP) by APICS — The Association for Operations Management. He earned a Ph.D. in production/operations management from the University of South Carolina and is a registered professional engineer and a certified public accountant. Prior to joining ASU, Dick worked as an industrial engineer and in management positions for manufacturing and service companies. He was a consultant with a major consulting firm, installing systems for both operations and financial applications.

William "Rick" Crandall currently serves as professor of management at the University of North Carolina at Pembroke. He earned his Ph.D. in business administration with a focus on management from the University of Memphis. His research interests include topics related to the vanishing service–manufacturing boundaries and crisis management, helping organizations cope with catastrophic events. Prior to entering higher education, Rick worked in management for ARA Services (now ARAMARK), a service management firm based in Philadelphia. (Photo courtesy of Bobby Ayers.)

Chapter 1

The Vanishing Manufacturing/ Services Boundary

John Merchant discovers continuous improvement.

This is a fable. Each chapter begins with a brief scenario involving a fictitious man we call John Merchant. John is facing decisions that are common in the careers of many business owners, top executives, and managers. Each situation that John encounters is meant to serve as an introduction to the topics covered in the chapter.

John Merchant was worried. His retail business was doing well, yet, he had a nagging feeling that he should be doing more to prepare for the increasing competition he faced. He knew he did a good job running the day-to-day operations of the company, and he had been able to correct the problems that periodically popped up before they became a crisis. His business was still profitable although the margins continued to slip a little each year. He had heard that some of his acquaintances in other retail businesses had implemented various improvements that had been very helpful. At the time, he had not paid much attention to their changes; maybe he should have listened more closely.

What could he do? He did not have time to go back to school full time or even take one of the many extended workshops sponsored by some of the professional societies and consulting firms. He could read a book, but he was not sure that would be enough. Maybe that would be a way to start, but what book should he read? There were so many. The more he thought about it, the more he decided that he would find a book that would describe improvement programs and how a company such as his could blend the improvement programs into his business. Maybe then he could get some perspective on what to do next. With that settled, he turned over and drifted into a restless sleep.

During the night, John tossed and turned, periodically waking up to find that his business was still on his mind. He just could not seem to get it out of his mind. During one of his brief periods of sleep, he started to dream. He dreamed he was in a long corridor with a number of storefronts, some of which looked like his own stores. Each had a large display area, but instead of being enclosed with glass they were open, much like a stage. In the first one, he could see people talking, and, because they did not stop when he approached, he decided to listen to what they were saying.

Some of the talking was hard to understand; not that the words were hard to understand, but the concepts seemed different, yet familiar at the same time. There was one conversation that started to repeat itself over and over. John listened closely and realized the people were talking about some kind of boundary that was breaking down. That boundary was not a physical one; it was sort of abstract, but still a real boundary. It was the boundary between manufacturing and services.

The manufacturing/services boundary is vanishing. Because of increased global competition, businesses must find ways to compete on additional critical success factors (CSFs). This quest to compete better has created a somewhat ironic dilemma; manufacturers must become more customer-oriented and service companies must become more effective and efficient. Each side is learning from the other.

Differences between Manufacturing and Service

For a number of years, business scholars have been careful to point out the differences between manufacturing and services. As a result, the implication was that each segment has a different set of goals and means of reaching those objectives. On the contrary, both manufacturing and service companies have the same objective — to obtain customers and provide what those customers want. Their interests are mutually dependent. Over time, businesses have understood the need to view manufacturing and services as a complementary package to be provided to their customers.

We are not trying to say that manufacturing and services are the same. There is certainly evidence that service companies require different perspectives and management techniques from manufacturing firms. This does not mean that the two sectors are incompatible, but it does mean their management requirements are different. However, this observation is true even looking within the manufacturing sector. Management of a job shop is different from the management of a repetitive line flow. Further, line flow management is different from operating a process industry. It is also true for service operations. Managing a fast-food restaurant is different from managing a global management consulting firm.

Let's look at the major differences between manufacturing and service businesses. John Bowen and Robert Ford found a major difference is that manufacturing companies produce a tangible product, and service companies provide an intangible service.[1] Because of the intangibility of their output, Bowen and Ford concluded that service companies show differences in three areas — service organization assessment, service production strategy, and the service production process.

In service organization assessment, the performance measures must include both objective measures of internal performance and subjective measures of customers' perceptions of the quality and value of the service, recognizing that one customer's perception may be slightly different from another customer's. Subjective measures of performance are difficult to develop, but are helpful in providing direction for the business.

In service production strategy, the service product and the service process must be designed to allow customer participation in the design and execution of the service, such as in self-checkout terminals at a grocery store. The service production strategy also has to consider that the consumption of the service may be simultaneous with its production, such as in the attendance at a concert. Because service companies cannot produce inventory like manufacturing companies, they must maintain flexible capacity, usually in the form of the variable staffing of multi-skilled employees.

The service production process is also different. Although technology is widely used in service operations, there is no substitute for an appropriate setting in which knowledgeable and empowered employees can deal with the wide variety of customer encounters. Customers must be attracted, they must be pleased with the initial service experience, and they must be retained, hopefully over a long period of time. Sometimes, customers may experience unsatisfactory service occurrences. When this happens, it is only through the intervention of knowledgeable employees that the customer will remain loyal.

Although there are differences between manufacturing and service objectives, it is becoming more important to combine product-producing activities with service-providing activities to have a successful business. Blending manufacturing and services requires the capability to combine different kinds of businesses. This blending raises the complexity of businesses and increases the need to improve continually.

This book looks at two major themes. The first is that the boundary between manufacturing and service businesses is vanishing and new composite businesses are emerging. The second theme is that these composite businesses continually have to improve their competitive advantage through the introduction, implementation, and assimilation of management improvement programs. The process of achieving this competitive advantage relies on the careful implementation of changes in technology, infrastructure, and organizational culture.

Forces That Are Eliminating the Boundary

What is driving business to recognize the need to eliminate the manufacturing/services boundary? Increased global competition is certainly a major factor. Businesses must find ways to be more competitive, and this requires exploring new avenues for developing a competitive advantage. Usually, this means adding services. In addition, various forces are making it possible to readily combine practices from the manufacturing and services arenas.

The Economy Is Going through a Natural Evolution

One general trend involves the evolution of the economy. Historically, the economy has moved from agriculture to manufacturing and, more recently, to services. This trend is a natural progression and comes about because of the demand for newer services from consumers who can afford to pay for these services. Initially, this demand was somewhat price insensitive; today, consumers are still demanding the service, but at a competitive price. Consequently, there is a need to focus improvement efforts in the areas of controlling costs, which are usually the ones with the highest labor content. Services are segregated into industries, such as health care and financial services. These, in turn, are segmented further into hospitals, extended care facilities, and research institutions. Later in this book, we will discuss the concept of mass customization for both products and services. Businesses that provide improvement programs as a service will also begin to approach mass customization as they adapt improvement programs to specific applications.

There Is a Need to Identify Critical Success Factors

Competition forces businesses to improve. As competition becomes tougher, it is imperative that businesses seek to do those things that are essential to their survival. Companies have found they cannot do everything competently; therefore, they must decide what tasks they "must" do excellently if they are going to compete. In

Chapter 2, we describe the concept of CSFs. A CSF is a strategy, plan, or program that will provide the business with a competitive advantage, for a limited period. However, no matter how innovative and successful a CSF is, it has a limited life; therefore, businesses continually and successfully have to identify and implement new CSFs on a regular basis.

Supply Chains Are Expanding to Integrate Manufacturing and Service

In this book, we are extending the concept of a product–service bundle from a single entity to the entire supply chain. As businesses attempt to link their own organization with other entities, they will interact more formally with other types of businesses. Service businesses become partners with manufacturing companies as they move upstream in their supply chain, and manufacturing companies become dependent on service operations, as they move downstream in their supply chain. This concept will be developed fully in Chapter 3.

It Is Easier to Blend Tasks, Resources, and Techniques into Programs

As part of gaining a competitive advantage, companies are learning to view management concepts and techniques as modular units that can be combined in a variety of programs to meet a specific need. The primary steps include

- Identify basic management theories of scientific, bureaucratic, and human relations management.
- Add systems theory.
- Add contingency theory.
- Blend these two theories with the aforementioned management theories into improvement programs that fit the business need.

We describe this selection and adoption process for improvement programs in Chapter 4.

Service Businesses Are Looking for Ways to Operate More Efficiently

Service businesses are closer to the customer and have always been customer-centric. However, as services become a larger part of the economy and the work force, they have a major impact on the national economy and are being pressured to become more cost and quality conscious. Consequently, they are looking to the manufacturing area for tools and techniques that will enable them to become more effective (do the right things) and efficient (doing things right). In Chapter 5, we describe some of the improvement programs developed in manufacturing and now being implemented successfully in service businesses. The following sections describe examples of ways in which service businesses are seeking efficiencies.

Health Care Providers Are Looking for Ways to Cut Costs and Improve Quality

Rising health care costs are of concern to leaders in business and government as well as the individual consumer. Everyone wants improved health care; however, not everyone is able to pay for it. Health care professionals realize that costs are high and must be reduced; the question is, how? One way of reducing hospital costs is to reduce the length of a patient's stay. This requires not only an improved service process during the patient's stay at the hospital, but also improved management of the pre-admission and post-recovery phases of the patient's experience. One area that is appealing for many health care entities is the use of Total Quality Management (TQM) programs and, more recently, its successor, Six Sigma.[2]

Banks Are Facing Competition from Lower-Cost Banks

Some service businesses have offered increasing levels of services in the interest of attracting and retaining customers. As a result, these service businesses are finding they have provided a greater variety of services than the consumer needs or wants. In an insightful look at this situation, Clayton Christensen, Scott Anthony, and Erik Roth call this "overshooting."[3] They point out that companies innovate faster than customers can change. As a result, products become too good, at least in their variety and complexity. A company that has overshot its market is subject to competition from a competitor that can offer reduced, but still acceptable, services at lower costs. Banking is an industry with this problem. Banks are also trying to reduce costs. They have increased the use of technology to reduce the labor component of their operations. Examples include ATM machines, automated telephone responses, and online bill paying. They have also bought into the need to improve quality; some of the most ardent supporters of the Six Sigma movement include banks.

Investment Bankers Are Facing Competition from Commercial Banks and Discount Brokers

Individual investors have a great deal more information available to them today and often do not need or desire the extensive investment counseling that was once the norm. As a result, low-cost, minimum-level investing is increasing in popularity among investors. The reduced service requirements mean that a greater number and variety of businesses can enter the business of selling stocks. In this case, the business needs to introduce information technology (IT) to improve the flow of information while recognizing that this will reduce the level of personal service offered to the consumer.

The Retail Industry Is Moving in Two Strategic Directions — Low Cost and Focus

The retail industry is moving in two directions. One is toward low cost, broad product lines, lean operations, and minimal customer service, often associated with Wal-Mart. A second direction is toward specialty stores, with a more limited product line and higher quality of service. This trend is luring customers away from intermediate stores that now face the problems of competing in this dichotomous atmosphere. The industry stalwarts — department stores — are under intense competitive pressures and are in the process of consolidating and redirecting their strategies, often by dropping product lines, such as electronics and appliances, which are now dominated by firms like Best Buy and Lowe's. As a corollary, shopping malls are also under competitive pressures. Many consumers no longer have the time to "shop"; they want to "buy." Consequently, they prefer to go to stores with easy access and a new form of shopping area is becoming popular. These take the form of outdoor strip malls with convenient parking, but with larger stores that have a deep product selection. These examples represent why service companies want to provide the "right level of service at a low cost."

Manufacturing Is Adding Services to Be More Customer Focused

It is no longer sufficient to manufacture a low-cost, high-quality product. A complementary service package is necessary. The product is rapidly becoming only the order qualifier; the service that accompanies the product represents the order winner.[4] As a result, manufacturers must focus more attention on the service sectors of their businesses. When they do, they find that some of the techniques they used in manufacturing work in the service sector, such as the organization of work flows or the need for capacity planning. However, they may also find that some functions are different, such as greater difficulty in measuring productivity or the less certain control of quality.

With these realizations, manufacturing managers can acquire a greater appreciation for the services sector and the need to blend it into their total product/service bundle. Even in industries with a heavy emphasis on the product, there is a need for service, such as in the automobile or the computer industries. Figure 1-1 illustrates the overlap between services and manufacturing.

The figure illustrates a continuum with businesses that have a heavy orientation on products shown on the left side of the diagram. Businesses that have a heavy emphasis on services are shown to the right and businesses that sell a blend of product and service are depicted in the center of the figure. Even those businesses that have a heavy product orientation are adding services, such as those described in the following sections.

Post-Sales Services

Consumers are becoming buyers that are more knowledgeable and look beyond just the purchase price of the product. They want to consider the lifetime costs of products and, as a result, are concerned about the reliability and service requirements of products. Businesses are looking beyond the sale of the product and considering the customer's post-sale needs. For example, automobile dealers are putting more emphasis on their service shops and many have systems that enable them to contact the customer, such as with a letter, to remind the customer that his or her car is due for service. These notifications include not only the major repairs or services needed, but also more routine procedures, such as oil changes. Computer sellers are also providing for such services as software upgrades and repairs. One of the more interesting examples of this is the repair facility operated by UPS for Toshiba in Louisville, KY. This outsourcing program makes it possible for the manufacturer to provide faster responses to customers who need it.[5]

Financing of Purchases

Another form of added service is financing of the purchase by the seller. Automobile dealers provide the option of financing the customer's purchase, usually through the dealer's manufacturer. Often, the rates are competitive with other financing sources, such as banks. In some cases, the rates are even lower, as the manufacturer

Figure 1-1 Product service continuum.

uses the financing option as an incentive to the customer. This arrangement also provides a way of maintaining contact with the customer during the lifetime of the purchase. Industrial equipment manufacturers have used this approach as a means of selling their equipment to small businesses. This type of transaction enables firms to buy the equipment if their creditworthiness is not adequate to get a loan from a bank.

Online Purchasing

Consumers can now buy "virtually" anything online. Often, companies offer attractive return policies that reduce the risk of online purchasing. Books and clothing are two areas that have made great progress in attracting buyers. Although major purchases, such as automobiles, are still not widespread on the Internet, consumers can use the manufacturers' Websites to gather information about potential automobile choices before going to a dealer to make the purchase.

Customer Input to Product Design

In the past, manufacturers often operated with the philosophy of producing whatever they could make well, and then trying to find a party that would buy their product. Today, there is more awareness of the necessity to find out first what the customer wants or needs, and then make the product to satisfy that need. This added service may take the form of actually getting input from potential or existing customers through focus groups or surveys. This up-front service will then help to adapt the product to one that is more appealing to the customer.

Enhanced Customer Relationship Management

Customer relationship management (CRM) is becoming more significant for manufacturing businesses. CRM is about retaining customers and many manufacturers are finding that adding post-sales services is a good way to accomplish this goal. The use of CRM and the other examples identified above represent only a few of the ways that manufacturing companies are enhancing their competitiveness by adding services to their product package. We will describe other applications more fully in Chapter 6.

Technology Development

New technology arrives at an almost overwhelming rate. It results in new materials that are lighter, stronger, and more environmentally friendly; equipment that is more reliable and accurate; production facilities that are flexible and pleasant;

information systems that facilitate analysis and communication; and employees that benefit from improved training, availability of tools, and innovative management techniques. Some view technology as the ultimate means of achieving success in organizations. On the other hand, many view technology as a negative force, and are quick to point out the adverse effects of the increased deployment of technology. Despite some resistance, the use of technology, especially in the form of information technology, is expanding rapidly and is forcing change in organizations. We describe these changes more fully in Chapter 7.

Although technology is often a driver of change, it is rarely sufficient to achieve lasting improvement, unless the infrastructure and culture also change within an organization. Figure 1-2 is a hypothetical portrayal of the interaction among technology, infrastructure, and culture during the implementation of a continuous improvement program. Technology is a driver, but may falter unless actively supported by adjustments to the organization's structure and acceptance within the culture of the organization. The infrastructure is important in facilitating communications and collaboration; however, it may be acceptance by the culture that eventually provides the extra support that the program needs to succeed. Eventually, all three change agents should come together to work in synchronization to finish the implementation and assimilate the positive features of the program into the ongoing operations of the company.

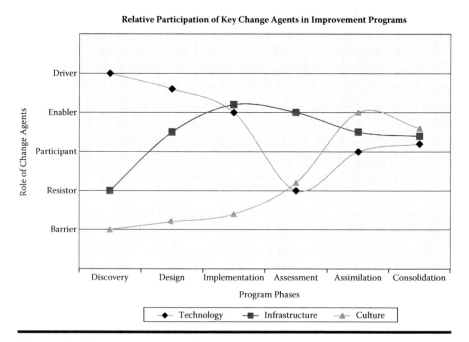

Figure 1-2 Change agents in an improvement program.

A Movement towards a Process Perspective

Organizational concepts are changing and making it easier to blend different functional areas. The rigid, hierarchical line-and-staff structure that has been a mainstay of organizations since the Industrial Revolution is changing. Businesses are considering other ways to restructure to accomplish their objectives. This is especially true when businesses want to implement major strategic changes in their organization. They might use a matrix format to blend functional areas, or a project management approach to achieve a major objective successfully. They may use a virtual organization orientation to enable them to combine and disband groups as needed. Some of the underlying reasons for innovative organization structures include

- Employee work assignments are moving from job specialization to self-directed work teams.
- Communications are moving from paper formats (snail mail) to electronic (e-mail, videoconferencing, and chat formats).
- Organizations are moving from hierarchical (vertical) structures to network (horizontal) formats.
- Businesses are changing their focus from improving transactions to improving processes.

In the past, businesses grouped similar functions together, such as cutting, sewing, and packing in an apparel shop; brake presses, drill presses, and lathes in a machine shop; and accounts receivable, accounts payable, and payroll in an accounting shop. Consequently, the emphasis was on the batch processing of transactions. This grouping worked well, but it was internally oriented and not designed with the customer in mind. As greater awareness of the needs of the customer emerges, businesses are moving toward a process approach that requires the grouping of dissimilar resources to achieve an integrated flow of work. Although the need to change is becoming more obvious, many organizations are not moving as rapidly in changing their organization as they are in implementing new technology. Technology without organizational change is a recipe for disaster. This discussion will be covered more fully in Chapter 8.

Outsourcing Focuses More Attention on the Need for Good Service

Outsourcing requires adjustments in infrastructure and culture as well as technology. Outsourcing itself is a service function and embodies the essence of customer service. Both the manufacturing of goods and the providing of services are being outsourced. The electronic communication capability sometimes makes it easy to outsource services, such as software programming, call center activities, and routine accounting. However, outsourcing triggers the need for organization realignment, also a subject for Chapter 8.

Changing Personal and Organizational Relationships

Another factor that affects the manufacturing/services boundary is the change in the personal and organizational relationships that exist within and outside the organization, including relationships between employees and managers, the company and suppliers, the company and customers, and between the company and its owners or stockholders. Also impacting the manufacturing/services boundary are the relationships between the company and its secondary stakeholders, such as regulatory agencies, service providers, financing institutions, insurers, health care agencies, and professional service firms. Relationships can exist between individuals and between organizations, although the latter is more difficult to establish and manage. Developing effective relationships usually requires changing internal attitudes and corporate cultures, both difficult but not impossible to accomplish.

Workers today possess increased technical knowledge. The result is a more effective person–machine relationship. As their education level increases, employees are able to handle a greater variety and complexity of duties. Many employees want more variety and empowerment, but it is up to the company to create the environment in which this increased knowledge can be used most effectively. These factors are both facilitators and barriers in situations where companies attempt to introduce change, such as in the following:

- Technology, in the form of equipment advances, evolved from hand tools to assist workers, to power tools to assist workers, to automation to replace workers, to interactive equipment to optimize the human–machine interface.
- The work force changed from immigrant workers with little formal education and lacking language skills at the beginning of the 20th century, to college graduates with extensive travel experience and awareness.
- Product lines evolved from companies with a single product line and only domestic locations, to global companies with multiple products and services and locations.

The role of cultures and their potential to hinder or help improvement programs will be covered more fully in Chapter 9.

The Need to Integrate Companies into Supply Chains

Building supply chains requires constructing relationships through communication and cooperation. These are attributes found in service businesses; therefore, building supply chains involves the development of increased service capabilities.

In Chapter 10, we show how this goal requires a careful and comprehensive process that links entities together into an effective and efficient network.

Services Become More Relevant to All Types of Customers

Services are important not only to individual consumers, but to manufacturing and service businesses as well. There are producer services in which one service company provides services to another business. There are consumer services in which service companies provide services to individual consumers. There are also services in which consumers provide services to themselves, or self service. We will describe these various services and how they fit with supply chains in Chapter 11.

Continuous Improvements Continue to Blur the Boundary between Manufacturing and Services

In Chapter 12, we describe the changes that we expect in improvement programs. Some of the changes will be linear and rather obvious extensions of the present. Other changes, however, are apt to be more revolutionary. We provide three examples in the following sections.

There Is a Movement toward an Industry Focus and away from a Functional Focus

With the emphasis today on customers or market segments, it follows that there will be a greater orientation of management improvement programs toward industry applications. Because industries include both manufacturing and services, there will be more of an interest in looking at the similarities along industry lines. This realization will lead to recognizing and exploiting these similarities, and designing improvement programs to fit these areas of focus.

The Mass Customization Concept Requires a Blending of Manufacturing and Service

The mass customization movement is just getting started. Although the concept has been around for over a decade, it is still in the early stages, and companies are still developing ways to best achieve it. David Bowen and William Youngdahl believe that "as the defining characteristics of manufacturing and service technologies continue to blur, we see a 'common industrial paradigm' emerging in which a similar logic is finding appeal in both manufacturing and service sectors."[6] They believe that mass customization provides a way to blend these basic manufacturing and service principles.

Figure 1-3 shows the migration from today's classic manufacturing strategies to that of mass customization. Regardless of a company's present make-to-order strategy, they will be looking for ways to incorporate more customer customization into their processes. It is unlikely that mass customization will become the universal way of doing things, at least not in the immediate future. Rather, it will be a selective strategy for an increasing number of companies.

Now, for a nonlinear example. In recent years, researchers have used chaos theory as an alternate way of thinking about decision areas. Traditionally, managers were trained to think in rational, sequential patterns. This thinking led to the viewpoint that the future forecasts can be based on past data; therefore, businesses can plan by statistically extending the past results. For more than three decades, scientists in fields other than business, such as meteorology, biology, and mathematics, have been developing a theory that addresses nonlinear progressions. They have labeled this field "chaos theory." Managers and forecasters are just beginning to recognize that chaos theory may have applications in the tumultuous environment of the business world. Chapter 12 will offer some additional insights on how chaos theory is being used to solve business problems.

In the previous sections, we described some of the forces that are at work to reduce the differences between manufacturing and services. In the next section, we show how these eroding differences will eventually disappear.

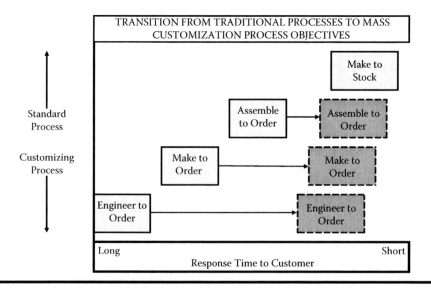

Figure 1-3 Evolution from today's manufacturing strategies to mass customization.

The Vanishing Manufacturing/Services Boundary

In this section, we will describe how manufacturing and services are moving from separate and distinct business entities to a situation in which their mutual interests have caused them to become loosely integrated entities, on their way to becoming a composite entity. We will use the Venn diagram in Figure 1-4 as an illustration of this transition.

Phase 1: Separated Disciplines

At one time, manufacturing and services were separate entities, as shown in the top portion of Figure 1-4. They had contact of course, because manufacturing produced goods that they sold to service organizations in distribution and retail. The manufacturing businesses had two major components — production (of goods) and support services.

Most of the improvement programs in manufacturing organizations focused on the production of goods. Gradually, however, the manufacturing firms began to recognize the need for improvement programs in their support service areas, such as order processing, product development, quality assurance, and distribution. Consequently, these manufacturing organizations developed programs that focused on cost reduction and quality improvement. As an afterthought, they found that some of the programs could be applied to their support service areas as well.

Meanwhile, service businesses were primarily concerned with selling the goods they received from manufacturing. They became proficient at selling, but were not as effective at managing the operations side of their business processes. For a while, they survived and even prospered, as the market demands were growing faster than the services side could keep up with. After a time, however, supply caught up with demand, and these service companies began to recognize their need to become more efficient.

Phase 2: Internal Improvements in Costs and Quality

During the 1970s through the late 1980s, two major shifts in strategy occurred involving manufacturing and service businesses. Manufacturing companies realized they could gain additional cost reductions and quality improvements by adapting some of the same techniques they used in the production areas and applying them to their service areas. They also recognized they should add additional services to supplement their product package. As a result, they began to adapt their continuous improvement programs to service areas.

On the services side, businesses began to feel the pressure of competition, especially from "big box" retailers such as Wal-Mart and Home Depot. Much of the pressure was to reduce costs, and many service companies recognized that manufacturing

The Vanishing Manufacturing and Services Boundary

Manufacturing — We will make it
Services — We will sell it

The boundary is clear
Traditional Interface of Manufacturing and Services

We will help | Interest develops to pursue internal improvements | We will listen

The boundary begins to blur
Increased overlap from mutual interests in internal

Let's work together | Interest in improving customer service and supply chain collaboration | We can help

The boundary further erodes
Increased overlap from mutual interest in customer service

Use a sense and respond strategy to provide an integrated product/service package

The boundary vanishes
Manufacturing and services with congruent interests

Figure 1-4 The vanishing manufacturing/services boundary.

companies were proficient in implementing cost reduction programs. Consequently, these service companies, especially wholesalers and retailers, focused more attention on inventory management, employee scheduling, automatic equipment, and the smooth flow of goods to reduce costs.

Service organizations also became interested in applying the quality improvements techniques used in manufacturing. TQM became a popular topic of conversation among service industries, especially financial institutions and hospitals. Although TQM achieved only moderate success, its lack of success was probably more the result of indifferent implementation than a failure of the quality improvement techniques.

Phase 3: Customer Service Improvements

By the 1990s, businesses of all types recognized the need to consider more closely the role of the customer in their planning. Service companies had always maintained direct contact with customers, but were now faced with the challenge of providing increased facets of customer service, while at the same time reducing the cost and improving the quality of those services. Using the cost reduction and quality improvement techniques developed in manufacturing were no longer a novelty; they were a necessity.

Manufacturing companies also recognized that they could no longer use a make-and-sell strategy where they first decided what to produce and then decided how to sell it to their customers. They had to learn how to find out first what products their customers wanted, and then to decide how best to produce those products. They also found that customers wanted their products complemented with services, so manufacturing companies had to consider more carefully how to bundle services with products. In some cases, the complementary services became as important as the product itself. Thus, manufacturers had to become more customer-oriented and began to learn that service organizations knew a good bit about how to manage the customer.

Most businesses experience this phase today. As service companies continue to learn about cost reduction and quality improvement, manufacturing companies are learning more about the care required for their customers. Stephen Haeckel presents a strong argument that no matter how adaptive an organization is today, it will be inadequate in the future. He advocates that businesses must develop a "sense and respond" approach to be able to cope with the increasing complexity of business and its increasing needs to develop social systems.[7]

Phase 4: Integrated Product and Service Functions

In the future, companies will no longer be classified as manufacturing or service. For example, is IBM a manufacturing company or a service company? What about General Electric? Nike? Dell? Or a host of other companies that have component parts of both manufacturing and service? And what if they do not include needed products or services within their organization? They can simply outsource the creation of those goods or services, or create a virtual organization to satisfy the need.

However, no matter how cleverly they arrange their supply-side constituents, businesses will always face the need to improve continuously. Today's standard of excellence will be tomorrow's minimum acceptable standard. To borrow from Terry Hill, today's order winner will be tomorrow's order qualifier.[8]

That is what we will describe in this book — the design and implementation of continuous improvement programs. We do not have a simple, step-by-easy step process to recommend because effective implementation of continuous improvement programs is not easy. However, we think we provide an array of information about continuous improvement programs that will help managers find a way to cope with the transition to a composite company. In short, this book will help managers deal with the vanishing manufacturing/service boundary.

Summary

As you read further in this book, if you are a manufacturer, look for ways you can extend manufacturing techniques into your service operations and even more important, how you can learn from the service industries. If you are a service operation, look for manufacturing techniques that you can use in your service operations. If you are a nonprofit organization, we will use the term "business" in most cases; however, much of what we describe has application in nonbusiness entities, such as government agencies and other nonprofit organizations.

References

1. Bowen, J. and Ford, R.C., Managing service organizations: Does having a "thing" make a difference? *Journal of Management*, 28, 3, 447, 2002.
2. Snee, R.D. and Hoerl, R.W., *Six Sigma Beyond the Factory Floor,* Pearson Prentice Hall, Upper Saddle River, NJ, 2005.
3. Christensen, C.M., Anthony, S.D., and Roth, E.A., *Seeing What's Next, Using the Theories of Innovation to Predict Industry Change*, Harvard Business School Press, Boston, 2004.
4. Hill, T., *Manufacturing Strategy* (Second Edition), Irwin, Burr Ridge, IL, 1994.
5. www.pressroom.ups.com/pressreleases/archives/archive/0,1363,4421,00.html, Toshiba and UPS join to set new standard for laptop repair, *UPS Newsroom*, April 27, 2004.
6. Bowen, D.E and Youngdahl, W.E., "Lean" service: In defense of a production-line approach, *International Journal of Service Industry Management*, 9, 3, 207, 1998.
7. Haeckel, S.H., *Adaptive Enterprise Creating and Leading Sense-and-Respond Organizations*, Harvard Business School Press, Boston, 1999.
8. Hill, T., *Manufacturing Strategy* (Second Edition), Irwin, Burr Ridge, IL, 1994.

Chapter 2

Critical Success Factors and Strategic Planning

One of the speakers in John Merchant's dream was a dour-looking individual. John heard him saying, "I tell you, the way wealth is distributed in our country is fascinating. Most of it is in the hands of just a few. Why, only about 20 percent of the people control over 80 percent of the wealth. Don't you think it would be better if it were more evenly distributed?" John later learned that the speaker's name was Vilfredo Pareto. Although the other men sitting in the window were sympathetic listeners, they seemed to be concerned with their own thoughts.

The second speaker, who turned out to be a man named Peter Drucker, was eager to voice his opinion. He explained that he had studied business, especially the practice of management. He had observed that managers often became involved in too many different details of running a business in the interest of achieving what he called "efficiency." His conclusion was that first you must be effective in doing the right things, then worry about efficiency, which he described as doing things right. Although John thought this was a somewhat catchy saying, he was not exactly sure what it meant.

The third speaker was a quiet and thoughtful-looking person who answered to the name of Joseph Juran. He also was a student of business, especially in the quality area. He said, "Yes, yes, you are both right. I see that in my work in quality. In reducing defects, the answer to a successful business is to separate the vital few causes of defects from the trivial many. By that, I mean that you should decide the major causes of defects and work on them first to get the greatest reduction in total defects." John seemed to understand this statement the best. In fact, he had seen it in his own business. The problems that emerged in the day-to-day running of the business consistently went back to the same causes. He had become good at finding those causes and addressing them quickly. Still, unexpected problems in quality arose from time to time.

John listened to all of this and thought, "That's nice, but what do the rest of these people have to do with my business?" So he asked that question of the three individuals sitting on the stage. Almost as if one, they turned to him and said with one voice, "Decide on the critical success factors in your business — those things that are essential to the short-term preservation of your business and to its long-term prosperity. Then build your strategic plan around those critical success factors." With that, they turned back to their own conversation.

John thought, "Wow! That sounded really good. But what are critical success factors? And what is a strategic plan? I do some planning now, but it is primarily for the ongoing part of my business. I never thought about including plans for improvement programs. How would I do that? And how is strategic planning done? What does it include? Who does it? Why? How?"

John noticed that there was a pamphlet on the side of the stage in a basket labeled "Take One." The pamphlet was entitled "Critical Success Factors and Strategic Planning." He took one and put it in his briefcase for future reference.

Managerial Comment 2-1. So What Is Corporate Life?

"Corporate life in today's business world is a mix of scarce resources, unrealistic deadlines, instantaneous communication, demanding customers, and fierce competition."[1]

Chapter 1 proposed that the manufacturing/services boundary is vanishing. Why is this happening? Perhaps the best answer is that businesses need to compete more effectively. So how can they more effectively compete? By identifying activities they must accomplish to be successful. Those activities are called critical success factors (CSFs). This chapter describes how businesses can identify CSFs and then incorporate them into their strategic and operational planning and control processes.

What Are Critical Success Factors?

In this section, we show that CSFs are the result of focusing on the important things to accomplish in a business. CSFs help us see that, although an organization can accomplish a vast agenda of activities, not all of those accomplishments are necessary for the success of the business. However, some are critical, and these become the CSFs.

Much of the literature on CSFs takes an information technology perspective. John Rockart is credited with being the first to use the term "critical success factor" in his article about how to improve reporting key information to top executives. He defined CSFs as "…the few key areas where 'things must go right' for the business to flourish…. As a result, the critical success factors are areas of activity that should receive constant and careful attention from management."[2]

Herbert Simon also emphasized the need to separate the important information from the vast amount of information that is generated in a business. See Managerial Comment 2-2.

Managerial Comment 2-2.
The Main Thing

"The main requirement in the design of organizational communication systems is not to reduce scarcity of information, but to combat the glut of information, so that we may find time to attend to that information which is most relevant to our tasks — something that is possible only if we can find our way expeditiously through the morass of irrelevancies that our information systems contain."[3]

Christine Bullen[4] also associated CSFs with information reporting systems and described the spread of the concept to many industries worldwide. She pointed out that CSFs helped usher in the age of executive information systems by demonstrating that when managers have relevant computer-based information, they can use it to improve their decision making.

A broader description of CSFs explains, "The Critical Success Factor (CSF) concept is a formal process of establishing and maintaining corporate priorities. CSFs are internal or external events or possible events that can affect the firm either positively or negatively and thus require special attention. CSFs provide an early warning system for management and a way to avoid surprises or missed opportunities."[5]

While a business needs to perform many tasks to be successful, the CSFs are the things they "must" do. Each business must decide for itself those CSFs that are critical to their success. A business cannot decide on its CSFs in a vacuum. The environment in which they operate must be considered. In Chapter 3, we will describe the environment more fully when we review the open systems concept. This concept describes how businesses must consider forces external to their business, such as economic conditions, competitors' actions, technology developments, social customs, environmental concerns, and government actions. Managers must also recognize that increased industrialization of a country carries with it an increase in the CSFs that must be met.

The Evolution of CSFs in the United States

In this section, we will describe how global CSFs have evolved in the United States. The evolution described is representative and a way to define further the nature of CSFs. CSFs vary over time, among industries, among companies in the same industry, and among managers within the same company. Ultimately, critical success factors must be specific to the need at hand.

To compete effectively, manufacturing companies should include those CSFs that relate to improved customer service. Likewise, service companies have to include CSFs that relate to improvements in cost efficiencies and quality.

We will describe the effect of the competitive situation in the United States on CSFs; however, similar trends are occurring in other countries, particularly those engaged in global markets — the European Union and the Far East, such as China, India, and Japan. It is especially relevant to see CSFs in light of the economic impact of political revolutions, such as the transition of communist countries to market-driven economies. In the book *The Commanding Heights,* Daniel Yergin and Joseph Stanislaw provide a fascinating description of how individual countries are making this transition.[6]

The United States is an example of a country that has witnessed changes in CSFs throughout its history. Table 2-1 summarizes these changes in CSFs over time in the United States. In the table headings, we use the terms "order qualifiers" and "order winners," as introduced by Terry Hill in his book *Manufacturing Strategy.*[7] Order qualifiers are "those competitive characteristics that a firm must

Table 2-1 Evolution of Critical Success Factors in the United States

Time	Competitive Situation	Order Qualifiers	Order Winners
1840s–1890s: Industrial Revolution	Little: The United States was primarily an importer of finished goods; an exporter of raw materials; developing the "American system" of manufacturing	Function	Function, availability
1890s–1930s: Growth and Recovery	Limited competition; the United States was beginning to become an industrialized nation, next to Europe; little competition from Far Eastern countries	Function, availability	Price (Ford), variety (General Motors)
1940s–1950s: Mass Production	Little competition after WWII (Germany and other European countries rebuilding; Japan not a major factor; the United States could sell all it could make)	Function, availability	Price, quality, delivery
1960s–1970s: Arisings	Stirrings from Germany and Japan in automobiles and basic industries (steel); the United States living off successes of the 1950s and acquisitions	Availability, price	Quality, delivery, service
1980s: Awakening	Inroads from Japan and other Far East countries in consumer electronics and other basic industries; the United States not prepared to compete	Availability, price, delivery (JIT)	Quality, service, product variety
1990s: Globalization I	U.S. firms countered in some industries; Japan and other Far East countries faltered; Europe solid; rapid movement to global market	Availability, price, delivery, quality (TQM)	Product variety, service, flexibility, integrated systems

Table 2-1 Evolution of Critical Success Factors in the United States (continued)

Time	Competitive Situation	Order Qualifiers	Order Winners
2000+: Mass Globalization II	Global market; offshore outsourcing; strategic partners; financial integration; mass customization; integrated systems; lean manufacturing	Availability, price, quality, delivery, flexibility; variety, integrated systems	Mass customization, paperless transactions, integrated supply chains, virtual corporations

exhibit to be a viable competitor in the marketplace" and order winners are "those competitive characteristics that cause a firm's customers to choose that firm's goods and services over those of its competitors."[8]

- Beginning (from First Settlements through 1800). In the first stages of the country's industrial development, there was little internal competition among businesses. Most of the competition for finished goods and services came from other countries; however, without some industry, there was little money available for residents to buy the finished goods. Agriculture was the principal industry. Residents were largely self-sufficient and any goods they produced, such as furniture or fresh vegetables, were for their own use. If they produced more than they needed, they sold the excess or traded it to their neighbors. The order qualifiers were simply the ability to make something; order winners were having some additional goods available. In the United States, settlers cleared the land and exported timber and farmers grew tobacco and cotton for export. In return, they imported tools, clothing, and equipment for use in developing increased industrial capability.
- Industrial Revolution (1840s–1890s). During most of the 19th century, the United States developed its industrial capability to record levels and made the transition from a primarily agricultural economy to one that included a more significant industrial capability. The Industrial Revolution saw the beginnings of major industries, such as telegraph, telephones, railroads, textiles, oil, and steel. The country needed all of these basic industries to feed, clothe, and transport the rapidly expanding population. Although there was still competition from other countries, primarily Europe, American companies could be successful if they were able to produce the goods (e.g., steel) or provide a service (e.g., railroads). The market was expanding so rapidly that it was difficult for an organization's available capacity to meet the demand for its products.

- Growth and Recovery (1890s–1930s). During this period, the United States made the transition to an industrialized country and started to become a major player in the global marketplace. The automobile industry played a major role in this transition. Henry Ford set the stage by emphasizing his company's ability to mass produce a standard automobile and sell it at a low price. General Motors used a different strategy — that of offering an array of choices. The competition between the two companies created the need to meet new CSFs — that of lower prices and greater variety. It also provided for the transition of the United States from an agricultural- to an industrial-based economy. This growth period also set the stage for the origins of the service economy.
- Mass Production (1940s–1950s). During this period, the United States emerged from the Great Depression and became involved in World War II. As a result, the country increased in terms of industrial capability, especially in the areas of automobile and aircraft manufacturing as well as atomic energy. Immediately following the war, the United States was able to sell almost anything it could produce in excess of domestic demand. This feat was possible because other industrialized countries in Europe and Japan had their industrial capability destroyed during the war. Competition was almost nonexistent and U.S. companies did not have much to worry about except making sure they had the capacity to meet demand. As a result, companies did not strive to make improvements in their technology or management methods. Indeed, life was good!
- Arisings (1960s–1970s). As Europe and Japan recovered from World War II, they rebuilt their industrial capacity with the latest technology and looked for new markets to develop, once they satisfied their own domestic markets. Japan transformed itself from a maker of low-quality goods to a producer of high-quality products. Two areas of particular improvement were in the automobile and consumer electronics industries.
- Awakening (1980s). The need for quality products reached a new level during this period and became a CSF that was recognized in other countries. However, because of the lack of competition, producing quality products was not a CSF that many companies in the United States acknowledged as important. For example, the steel industry resisted modernization because it did not need additional capacity and could not justify new technology. Consequently, this was a period of financial wizardry as conglomerates abounded and synergy was the magic word. Manufacturing lost its luster and automobile companies tried to compete on styling and reduced prices. The need for additional CSFs such as quality and response times became a reality.
- Globalization I (1990s). Although some degree of globalization had existed before this period, a movement to expand more aggressively was emerging. This movement progressed, despite the fact that some remnants of "Buy American" campaigns were still active. Wal-Mart became the leader in low-price

retailing, but at the expense of buying many products overseas. Many other companies watched and waited before finally deciding that buying overseas was not a bad thing, especially because Americans wanted low prices.
- Globalization II and Mass Customization (2000 and after). The United States is currently in a transition mode. Traditional manufacturing businesses are diminishing as basic production operations are being outsourced to offshore companies. The major U.S. automobile manufacturers — General Motors and Ford — are losing market share. The apparel industry is also disappearing as companies purchase more clothing from foreign countries that offer dramatically lower labor costs and generally acceptable quality. Understandably, different constituencies have mixed reactions to these trends. Some are pessimistic and want tighter controls on imports; others are optimistic and want industry leaders to help make the transition from traditional industries to futuristic industries. To make this transition requires a change not only in the type of products and services offered, but also in the attitudes and knowledge of the workers.

New CSFs are imperative. In addition to low prices, high quality, and reduced response times, businesses must be agile, flexible, and virtual. The ante is going up; successful businesses have to offer a larger variety of products and services. However, offering a wider variety of products or services is not always the ultimate answer. The ideal situation, at least in the minds of some, is to treat each person as a market of one, and design and deliver products and services unique to each person. Although this scenario may appear overly idealistic, many see it as the direction of the future. What will this transition involve for the United States? It is a transition from one life cycle to a new life cycle. In fact, that transition is already taking place when a country abandons the former life cycle built around traditional manufacturing, with service support, to a new life cycle — one that is built around a service/manufacturing package of goods and services that is systems-oriented and holistic with respect to customers' needs.

Other Changes during a Country's economic Life Cycle

In addition to an increase in competition, the United States and other developed countries face the following changes during their national economic life cycles:

- From agriculture to manufacturing to services. There is a natural progression from an agricultural society to an industrialized one and then on to a service-oriented society. Although agriculture has always been a fruitful component of the U.S. economy, the number of workers in this sector has declined steadily. During the Industrial Revolution, factories attracted workers from the farms, which led to production facilities with large numbers of employees, many of them represented by labor unions. The advances in production machinery replaced many of these workers; their fate resulted in retraining into other fields of work, often service oriented. Today, offshoring has further exasperated this shift so that the U.S. economy is heavily service-based at present.

- From simple to complex. Large company-operated farms replaced the small family farm. Small businesses have grown from local units to large, global entities. Ironically, though, many formerly large companies have downsized as massive layoffs have resulted from international competition. Nonetheless, regardless of the size of the business, relationships within and between entities have increased in complexity. Technology and global business expansion are the key reasons why.
- From tactical to strategic. As complexity increases, the time horizon for decisions also increases. It is no longer possible to take a short-term perspective and focus on tactical decisions that look only a few days to a few months into the future. Businesses have to deal with strategic issues that can involve looking several years down the road. This long-term perspective can have huge impacts on the future survival of businesses.
- From raw materials to integrated systems. Many developing countries begin their globalization efforts by exporting their raw materials or specialty products. In exchange, they hope to achieve the means for more advanced production in the form of equipment and technology knowledge. As they advance in their level of industrialization, they become increasingly adept at making products with more advanced features and capabilities. These advances enable them to develop a greater range of services. Finally, when they can produce advanced products and provide extensive services, they can begin to assemble the components into integrated systems.
- From labor-intensive to equipment-intensive to a blend of the two. Industrialization requires equipment to assist the worker and, at some point, to replace the worker. However, as complexity increases and customization becomes a requirement, it is becoming apparent that an appropriate blend of workers and equipment provides the best combination. Workers in modern factories do less physical work (machines and robots handle this) and more monitoring tasks, such as the regulation of those machines and robots.
- From job specialization to self-directed teams. Job specialization originated from Frederick Taylor's Scientific Management philosophy during the early 1900s. The goal was to increase worker productivity by requiring employees to perform a specific task in the production sequence. Job specialization became the human backbone of production up through the 1970s. However, as the demand for faster response times and increased customer service grew, businesses had to allow employees to make more decisions in the production process. The resulting movement led to entire production teams having the authority to make major decisions on the shop floor. This evolution requires

not only a change in the workers' job functions, but also a change in the structure and job content of managers.
- From "bigger is better" to "small is beautiful." Although it is true in the United States that manufacturing evolved from small business units to large industrial structures, the opposite has been true in the services realm. The move to a service-oriented economy requires managing smaller business units. Service businesses must be near their market and this requires more outlets. Manufacturers that incorporate more services in their offerings will also move closer to their markets to reduce lead times.
- From local optimizing to global satisficing. Decisions are getting too complex to make with optimizing technology alone; there has to be a combination of information, intuition, and initiative. Herbert Simon (1997) coined the term "satisficing" to describe the need to select a course of action that is "good enough." Satisficing is necessary because there is not the time or capability to make an optimal decision. Even with this less demanding criteria, managers will require holistic and systems integrating skills.[9]

Some view the evolution of a country's economic growth as progressing through several phases: agriculture, industrial, service, and integration. Each phase includes some of the past and requires higher levels of achievement. Figure 2-1 illustrates this concept graphically. As the economy moves from one phase to the next, the demand

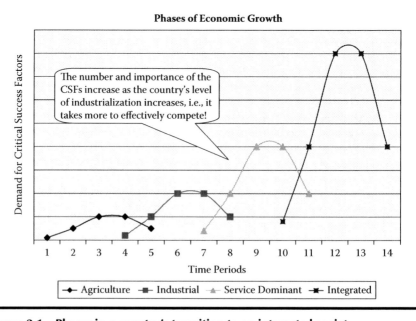

Figure 2-1 Phases in a country's transition to an integrated society.

for higher performance increases. This increased demand means that the number of strategic CSFs must increase also.

The Need to be Effective

CSFs help management to focus on the top priority decisions and actions necessary to keep the business operating as desired. The quote from Herbert Simon in Managerial Comment 2-2 emphasized the need to concentrate most of the effort on sifting through the available information to find the right things to use. Three other renowned management thinkers add to this admonition to identify the important areas for management's attention — Peter Drucker, Peter Senge and Joseph Juran.

Drucker points out that the manager's job is to optimize the yield from the available resources of employees, equipment, facilities, and capital. He makes a careful distinction between effectiveness and efficiency. One of his most famous quotes states, "Effectiveness is the foundation of success — efficiency is a minimum condition for survival after success has been achieved. Efficiency is concerned with doing things right. Effectiveness is doing the right things."[10] Thus, efficiency is what we commonly call a process characteristic, and effectiveness is the manifestation of efficient practices. Effectiveness is often labeled an outcome variable by management scholars.

Senge describes how systems thinking can organize the complexity of today's business environment. The goal is for management to be able to identify the causes of organizational problems and to formulate remedies that will be successful and enduring. One of the problems that managers face in their organizations is the overabundance of information, much of it generated internally. The challenge is to use the information that is most important and set aside the rest. As Senge states, "What we most need are ways to know what is important and what is not important, what variables to focus on and which to pay less attention to — and we need ways to do this which can help groups or teams develop shared understanding."[11]

Juran describes control as maintaining the status quo. In comparing business strategies to process control charts, he points out that businesses should not focus on control alone, but should also be concerned with how to continue to improve. He emphasizes that, although control may be desirable, it can be a cruel hoax, a built-in procedure for avoiding progress. Companies can become so preoccupied with meeting targets that they fail to challenge the target itself. He offers an alternative "breakthrough," which he describes as a dynamic, decisive movement to new, higher levels of performance. He gives a compelling description of the "vital few" versus the "trivial many."[12]

The term "efficiency" was developed first for commercial businesses and was associated with measurable inputs and outputs. It is more difficult to measure efficiency in noncommercial organizations where the outputs are more intangible.

Simon suggests that this is best done by comparing actual progress against the goals or program objectives established by the organization.[13]

How do managers cope with these challenges? They must determine the critical success factors for their business. Once they have identified their CSFs, they can prepare plans at all levels of the business. CSFs help to identify the right things to do, the "vital few," as opposed to attempting to do everything well. With the availability of so much information today and the capability to create even more information quickly, it is easy to see where managers immerse themselves in too many details and have difficulty sorting out what is important from what is unimportant.

A Hierarchy of the Planning Process

CSFs are an important part of the planning process in a company. To show how CSFs fit into the company's planning, we will briefly describe a typical planning process for a company. Figure 2-2 shows a planning hierarchy in a business. The top portion of the diagram illustrates a structured planning process to develop strategic

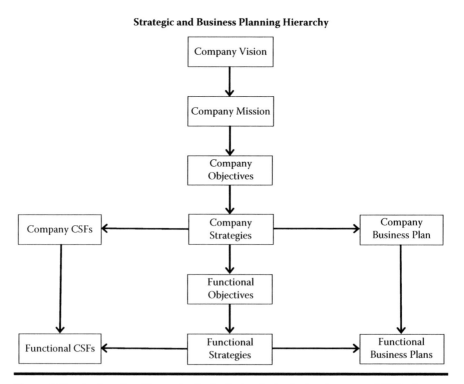

Figure 2-2 A planning hierarchy to include strategic, business, and CSF plans.

and business plans. However, the planning process in most companies is not likely to proceed in such an orderly fashion. More likely, the planning process will have to recognize that plans tend to be static, although the operations of a business are dynamic. As a result, there is always the problem of how best to match actual results with planned. The bottom portion of the figure depicts the need for interaction among the various plans to keep them current and meaningful.

Once the various plans have been prepared, companies should expect that it would be appropriate to develop links among the various plans. The individual plans should receive inputs from each other and provide results and changes to each other. Although it is impractical to change plans every time new inputs are received, the inputs should be available when the next plan is developed. Figure 2-3 shows these interactive relationships.

The Strategic Planning Process

In theory, most businesses develop a vision of why they exist, usually expressed in a few general statements. They expand their vision into a mission statement that adds some specifics to the general direction they want to pursue. Although these are important in providing a reference for subsequent planning, vision and mission statements evolve over time and do not necessarily result from a single meeting.

Developing company objectives is necessary. Although the objectives should relate to the company mission, in practice most of the company objectives are probably developed by comparing past strategic plans with the actual results. The objectives should include all of the goals needed for the strategic planning period; however, they will be somewhat general to allow for further definition later.

In developing company strategies, businesses begin to get more specific. Strategies are the actions necessary to achieve the company's objectives or goals. They are necessary to provide direction and performance expectations for the functional areas of the company. Working with the guidance from the company strategies, each functional area of the business, such as sales and marketing, finance and

Figure 2-3   Interactions among various plans.

accounting, engineering, operations, human resources, information technology, and others, prepares its own objectives and strategies.

With inputs from all areas of the company, top management puts together a strategic plan. It is important that all managers responsible for achieving the plan's objectives be fully involved in preparing the key elements of the plan. Top management defines the planning process and facilitates the assembling and preparation of the plan. Although it sounds straightforward, it is often difficult and requires a great deal of interaction, reconciliation, collaboration, and adjusting among those levels of management that are involved in this part of the process.

A plan should consider a platform for growth and profits as well as take into consideration the following critical success factors:

- Financial factors: Positive cash flow; debt ratios
- Marketing factors: New customers; new products and services to offer
- Production factors: Capacity availability; responsiveness to demand; trained work force
- Logistics factors: New supplier networks; transportation issues — getting the product to the customer and getting supplies and raw materials to the facility
- Quality factors: Satisfied customers; standards of excellence for products and services
- Human resource factors: Increasing intellectual capital; attracting and retaining excellent employees

The Business Planning Process

In general, strategic plans will have more words than numbers. Because most strategic plans are multi-year plans, it is unrealistic to plan at the micro level. Most business plans, sometimes called annual plans, begin with financial expectations and provide the detail, especially in the financial area, in the form of sales forecasts and expense budgets on a monthly time scale. As a result, business plans often have more numbers than words. This difference is both a strength and a weakness of business plans. It is a strength in that it provides specific targets against which to measure progress. It is a weakness because the assumptions on which the plans are built often change throughout the year, and revising the business plan can be a major interruption. As the fiscal year progresses, the business plan becomes increasingly misaligned from the actual activities of the business. As a result, it is important that the company have other ways of measuring performance and a seamless strategy to adjust the business plans if necessary. The authors have both experienced situations where the business plans became inaccurate because of unforeseen circumstances. In one case, a week-long weather crisis closed down a facility unexpectedly, causing a major shortage in revenue. Needless to say, the business plan was inaccurate for the remainder of the fiscal year.

Project and Program Plans

In addition to the strategic and business plans, it is important to identify the specific programs and projects implemented during the planning period. The terms "programs" and "projects" are sometimes used interchangeably. Although it may not be essential to differentiate between them, we will consider programs as broader than projects. For example, a company may decide to implement a Six Sigma program.

Within the program, there may be a number of separate projects, such as training employees in the use of statistical process control (SPC) techniques or holding a series of kaizen blitz sessions. Some companies may not prepare separate program and project plans; instead, they embed these activities within the strategic or business plan. The value of having separate plans is that it focuses more attention on the completion and separates the physical improvement activities from the everyday operations, which are often wrapped within the bundle of financial budgets. It is possible, and even desirable, to prepare project schedules that can be used to execute the project plans and measure performance against the project schedule.

Another justification for formulating a separate project schedule lies in the reality of day-to-day operations. Project implementation often stalls when an operational problem arises. When this occurs, the tendency on the part of many managers is to put the project on the back burner "for today" and run the business because, after all, that is the priority. For example, there may be a need for additional staff to teach employees new statistical process control techniques. However, if a key employee is out sick, a decision may be made to delay the training "to another day." As the days become weeks, momentum for the project declines and before long, the project stalls. Project scheduling should thus include contingency plans for these types of situations.

CSFs Plan

The plans described above should be comprehensive because they need to include all of the components a company hopes to accomplish during the year. Out of that all-inclusiveness, they need to identify those things that they must do — their CSFs. In the following sections, we will describe how CSFs fit into each level of the planning process.

A Hierarchy of Critical Success Factors

CSFs should be planned at all levels of a company:

- Strategic: At the corporate or strategic business unit (SBU) level
- Functional: By function within the company
- Project: By major project or program

To be most effective, the CSFs should be linked together to provide effective coordination among the different plans. Developing this hierarchy makes it possible to link strategic, functional, and project planning at all levels of an organization. This framework makes it feasible to set priorities and to utilize resources in the most effective way.

Table 2-2 shows an example of this kind of hierarchical planning. The strategic plan provides an overall objective to improve quality. At each level below the strategic plan — operation planning, program planning and project planning — activities that are more specific are designated to achieve an increased quality level.

This example offers a straightforward approach to developing a hierarchy of CSFs. However, in practice it is a challenging task because it requires a parallel planning process. Not everything can be a CSF; to attempt this would be equivalent to making every order a "rush order" or viewing all customers as equally important. Indeed, the essence of a CSF is that it must be done successfully. There must be a way of setting priorities in the planning process, and identifying CSFs is one way of doing it.

The Role of CSFs in Strategic Planning

CSFs represent a way to identify the things that a business must do. A good starting point is with the strategic planning process. If a business can select CSFs at the strategic level, the business can then develop a hierarchy of CSFs to carry down through the operational and program planning levels.

Table 2-2   Examples of Interrelated CSFs

Level of Plan	Critical Success Factors
Strategic planning	Increase the overall quality of products
Operation planning	Strategic business unit (SBU): Increase the quality level of Plant X Department: Increase the quality level of molding Individual: Increase the quality level of each operator
Program planning	Introduce a Six Sigma program
Project planning	Provide statistical process control (SPC) training to all operators Schedule six kaizen blitzes during the year

What Is Strategic Planning?

Strategic planning focuses on managing for the future. Typically, companies are looking at a time line down the road that may range from two to five years, although some planning may account for longer periods. The basic questions that strategic planning addresses include

- Where do we want our company to be in the future?
- What should it look like in three to five years?
- How will the company look the same? How will it be different?
- What new markets should the company seek?
- What business ventures should we abandon?

Management writer, John Parnell, sums up the essence of strategic planning in Managerial Comment 2-3, The Essence of Strategic Planning.

Managerial Comment 2-3. The Essence of Strategic Planning

"Strategic management (planning) refers to the continuous process of determining the mission and goals of the organization within the context of its external environment and its internal strengths and weaknesses, formulating and implementing strategies, and exerting strategic control to ensure that the organization's strategies are successful in attaining its goals."[14]

Strategic planning begins by taking a hard look at the environment outside of the company. Specifically, top management needs to ascertain the opportunities and threats that face the business. Opportunities represent new markets to pursue. For example, a company may seek to expand its existing products to new markets outside of the country. On the other hand, it may seek to develop new products it can sell to existing domestic markets. Threats, on the other hand, are forces that can take away market share or impact profitability in a negative manner. Obviously, competitors are one example of a threat. Rising fuel costs, bad weather, the raising of taxes, or the possibility of war also represent possible threats. Determining the opportunities and threats is the catalyst that begins the strategic planning process. Reliable, long-range plans depend on this type of analysis.

Knowledge of the external environment must also be supplemented by an internal assessment, determining the strengths and weakness of the company. Strengths are things the company does well and are often referred to as distinctive competencies. These strengths are often unique and help set the company apart from its competition. Some of these strengths may originate from the personality of the founder. For example, Southwest Airlines is well known for its happy employees and corporate sense of humor. Having fun on the job was one of the values that the airline's founder Herb Kelleher stressed at the onset.

Knowledge of strengths is necessary in formulating strategy because these are qualities the company most likely wants to perpetuate. Weaknesses, on the other hand, may or may not be surmountable. Some weaknesses of a company should be overcome, such as poor training of employees or negative relations with certain customer groups. Including strategic plans to overcome these weaknesses would be advisable. However, companies may have to live with some of their weaknesses. For example, a company may not have the best location for attracting top graduates. Moving the company may not be feasible either, so the weakness may just have to remain a weakness. A company can develop a strategy to deal with that weakness, recognizing that it will be a problem. Consequently, strategic planning would focus not on relocating the company, but finding innovative ways to sell the appeal of the company and its locale to various types of college graduates.

Background of CSFs in Strategic Planning

Determining the CSFs for a company is part of the strategic planning process. CSFs fit with strategic planning at the macro level; they can be used at any level of functional planning, such as information systems (IS); and at the program planning level, such as the implementation of a continuous improvement program.

CSFs become a subset of the strategic plan's goals and strategies. They compose the activities of the company to be performed successfully. There might be other goals and strategies that enhance the welfare of the company, but these are not necessarily CSFs. For example, a goal of the company may be to reduce absenteeism among its employees, but this goal is not a CSF in most cases. Once CSFs have been identified at the strategic level for the entire company, it is then possible to use them as guidelines so that the functional areas of the company can develop CSFs.

The term "critical success factors" originated in connection with executive information systems. Rockart[15] identified procedures for providing executive management information needs:

- The by-product technique: Regardless of the executive needs, all reports are essentially by-products of a particular system, such as sales or payroll, designed primarily to perform routine paperwork processing. This is probably the predominant method and often provides the executives with more information than what they need.

- The null approach: Advocates of this approach conclude that all computer-based reports — no matter how they are developed — are useless, and the executive must dynamically acquire information through personal contact and informal means.
- The key indicator system: This approach involves (1) selecting key indicators to measure, (2) exception reporting of results, and (3) expanding availability of better, cheaper, and more flexible visual display techniques (for example, providing graphs in a board room).
- The total study process: This approach looks at all the existing information systems, compares the results with the information needs, and designs changes to fill in the gaps. The amount of data and opinions are staggering and the approach is expensive.

Rockart described the CSF approach as occurring in two steps: (1) Interviewing key executives to identify the goals and the underlying CSFs, and (2) deciding how to measure progress. With this approach, he believed the information would be of greater value to top management. He credits Daniel[16] and Anthony, Dearden, and Vancill[17] with providing background to the CSF approach. Rockart offered examples of CSFs for different industries. For the automobile industry, he suggested excellent automobile styling, quality dealer systems, effective cost controls, and meeting energy standards. In a service industry, such as a government hospital, he proposed successful integration with other hospitals in the area, efficient use of scarce medical resources, and improved cost accounting procedures.[18]

Although Rockart's work was early, the concepts have transcended in use over the decades. In another example, Berry et al.[19] showed how the use of CSFs could aid in the linking of marketing and service operations strategies. In a study of retail banks, they explored industry CSFs along two dimensions, one is market-oriented and the other is competitor-oriented. The choice of operations strategy is dependent on the marketing strategy.

Characteristics of Strategic CSFs

How do companies determine their strategic CSFs? Some top managers begin by looking at their outside threats and opportunities — a competitor has introduced a new product or the federal government has passed a law that requires compliance with financial reporting for public companies. Some CSFs result from an examination of a business's internal strengths or weaknesses — the development of closer supply chain relationships or the need to extend the employee empowerment program. Some strategic CSFs remain the same over time; they become part of the vision and mission of the company. Other strategic CSFs change over time; they become part of the dynamic plans of the business.

CSFs should be developed through the interaction of key employees at all levels — individual, department, division, and corporate. Once top-level managers in an organization have identified their CSFs, these can be compared to other managers' CSFs in the company to determine where there are overlapping concerns, related issues, and obvious conflicts. Put another way, individual managers can think through their personal CSFs, groups of managers can think through their organization's CSFs, and groups of organizations can think through their industry CSFs. Eventually, it should be possible to come up with a hierarchy of CSFs for all levels of a company. Sometimes, the CSF development process can help in aligning the thinking of all levels of management, as described in Managerial Comment 2-4.

Managerial Comment 2-4. Churning out the CSFs

Christine Bullen describes an example of how the CSF process can help to resolve conflicts. In one high-technology engineering company, the CEO called a meeting of his top managers to discuss their progress in developing their CSFs. In the interest of moving the program along, the CEO offered his list of CSFs. Although initially upset at being presented with what they interpreted as a mandate from the CEO, the top mangers soon saw that the CEO's critical success factors were very similar to their own, and operational in nature. During the subsequent discussion, it became obvious that the CEO, who was also the entrepreneur who founded the firm, was reluctant to give up hands-on control of the day-to-day activities. The CSF process, by focusing on what each person viewed as critical, revealed the conflict and made it possible for this management team and their CEO to move ahead in a productive manner.[20]

The initial success of CSFs led to their application in other ways. The CSF method helps individual managers think through the vast number of activities and issues they must perform, reduce that number to a manageable few, and set priorities on the activities so they can focus on the most critical ones.

To be most useful, CSFs must be:

- Important to achieving overall corporate goals and objectives
- Measurable and controllable by the organization
- Relatively few in number — remember, not everything can be critical
- Expressed as things that must be done — not the endpoint of the process
- Applicable to all companies in the industry with similar objectives and strategies
- Hierarchical in nature — some CSFs will pertain to the overall company, and others narrowly focus in one functional area.[21]

CSFs play an important role in identifying the areas of strategic importance to businesses.

How do you distinguish CSFs from strategies? A CSF identifies an area of emphasis; a strategy defines an action that relates to the CSF emphasis. CSFs are not the same as strategies — strategies are ways to achieve CSF goals.

Critical success factor analysis is most effective when done from the top down:

- Analyze the corporate mission, objectives, and strategies to pinpoint the success factors of the overall business.
- Determine the CSFs for each business unit's functional areas.
- Develop strategies to leverage competitive strengths and overcome weaknesses in each area.
- Develop measurement tools that will enable managers to monitor performance against the plans.
- Finally, establish processes and procedures to report performance information in a timely fashion.[22]

CSFs are the characteristics, conditions, or variables that can have a significant impact on a company's success. Their analysis can aid in the strategy development process. Possible sources of CSFs include environmental analysis, analysis of industry structure, industry/business experts, analysis of competition, analysis of the dominant firm in the industry, company assessment, temporal/intuitive factors, and profit impact.[23]

The Temporal Nature of CSFs

CSFs change over time. Among other factors, they change over a product's life cycle. Figure 2-4 shows a typical product life cycle for both a new product and a new service. In general, the emphasis shifts during each phase of the life cycle and

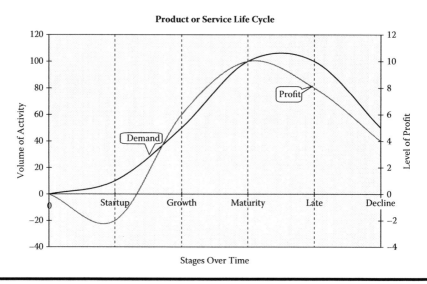

Figure 2-4 Product life cycle.

the functional areas of the organization need to adapt their own CSFs to keep pace with the changes.

Start-Up

In the early, or start-up, stages of a product's life cycle, the primary CSFs will be function and appropriate presentation. A company may be test marketing the product in a small market; it is essential that the product function as designed and be acceptable in appearance. It is also important that the product be presented in an attractive, or marketable, form.

Growth

If the product gains acceptance and begins to move into the growth part of the product life cycle, the CSFs shift to availability and quality. The manufacturer must have the product available in sufficient quantities to meet the demand. This is often a treacherous time because it is difficult to forecast demand for a new product; yet, if there is unsatisfied demand, the product may never reach its full potential. In addition to having the capacity to meet demand, the manufacturer must be diligent in providing a high-quality product over longer production runs. Just as lack of availability will kill a new product, so will a record of defective products, especially if they are hazardous to the user, such as toys for children or medications for adults.

In addition, the manufacturer must have the flexibility to adjust to the refinements needed in the product as customer reactions and feedback become available.

Maturity

At some point, the demand for the new product or service will begin to level off, often because other competitors enter the marketplace and gain a share of the demand. When this happens, the CSFs shift again, this time to cost and reliability. Although there is a continuing need for availability, most of the companies have developed sufficient capacity to handle the demand. With increased competition, there is increasing pressure on prices and, as a result, on product costs. Because of global competition and the move to outsourcing, the need for competitive costs is becoming a vital CSF. In addition, product quality enters a new phase of product reliability. The product must not only work as expected, but it also has to operate over extended periods without an undue number of failures.

Decline

Many products eventually reach the decline stage of their life cycle. At this point, the CSFs that are emphasized may vary from one company to another. Some may continue to focus on reducing the price of the product, such as in clothing styles or older appliance models. Others may maintain or even increase the price, but emphasize availability, such as with replacement parts for older-model automobiles. Other companies may hasten the decline of the existing product by promoting a replacement product that is "bigger and better," such as for PCs or new software.

The product life cycle is an inevitable sequence for all products and services. Some life cycles are short, others are long. Even though some products may appear to have an indefinite life cycle, such as aspirin or chewing gum, a closer examination would reveal that these products experience a series of life cycles, where the existing product is replaced with a newer version of the old product and, as a result, enters a new life cycle.

While the product life cycle is changing, the process that is used to make the product or provide the service must also change. We will describe this transition later in the chapter.

The Role of CSFs in Operational Planning

Once companies develop CSFs at the strategic level, they can extend them into operational plans for each functional area, such as marketing, accounting, finance, information technology, and operations. As shown in Figure 2-2, the CSFs become more specific as they move lower into the organizational structure.

We will look at some of the CSFs recommended for specific functional or industry areas, including purchasing, human resources, accounting, information systems, and hospitals.

- Purchasing. Most purchasing programs concentrate on reducing costs and use purchase price variance as a measure of how well this is being done. However, it may be equally important to try to reduce the purchase price variability of items. This requires a cross-functional effort among purchasing, marketing, and finance to form an agreement on the specifications of the product or service. In addition to agreeing on the common requirements, each function can consider other factors. Purchasing can use such techniques as risk sharing between customer and supplier or smoothing formulas to average the fluctuations in prices, such as in some commodity items. Marketing can set up an indexing scheme in which the price to the customer varies with the price of the critical supply item, such as in retail gasoline sales. Finance can use hedging techniques, such as future options, for commodity items like energy, metals, and agricultural items.[24]
- Human resources (HR) is evolving into something called organizational effectiveness (OE), in which HR professionals are full partners in their companies' decision-making processes. Many companies see the relevance of linking OE to business strategy, and they have begun to use a relatively simple model to reshape their organizations. The model includes such components as a corporate mission, a strategy, and the organization's CSFs. The overall performance of a company may depend on how well it fits its CSFs to its strategy and how well it blends the HR implications to the technology, structure, systems, policies, and skills of the company. As HR evolves into OE and becomes more strategic and less reactive, its practitioners must become more business-focused to help their organizations deal with the business side of employee issues. Successful OE professionals must possess the talents and mind sets to be strategists, consultants, and facilitators. In their emerging role, OE professionals must upgrade their skills and capabilities to meet the demands of their new jobs.[25]
- Accounting. In the past, managers have often tried to use financial accounting information to manage their organizations. Financial accounting is more externally oriented to meet regulatory, financial analysts, and shareholder needs, and often is not satisfactory for internal, or management, purposes. Although financial accounting information is important and somewhat helpful in reporting on the financial status of a firm, it does not provide the right kind of information to help managers identify problems and to take corrective action. In the past, accounting-based information was the only form of management report available in organizations. Critical success factors changed that. The method encouraged line management, accounting, and information technology to settle on the information that would be most

beneficial in the management process. The use of CSFs to determine what should be reported and how it should be measured revolutionized computer-based information reporting systems. As a result, the phrase has become widely used and the concept has been applied in many industries around the globe. To a certain extent, CSFs helped in the development of executive information systems by demonstrating that when managers are provided with relevant computer-based information, they can use it to improve their decision making.[26]

- Management information systems. The CSF method works well at the policy, operational, and strategic planning levels of information resource planning. Although CSF methods applied to requirements analysis help build conceptual models of roles in organizations, some managers have trouble using conceptual thought processes. Case studies show that the CSF technique helps identify information infrastructures and gives insight into information services that impact a firm's competitive position.[27] Another study of planning participative information systems, extends CSF methodology to facilitate participation by many people within and around the organization. The resulting new methodology, called critical success chains (CSCs), extends CSF explicitly to model the relationships between IS attributes, CSFs, and organizational goals. A practical procedure is defined for data gathering and analysis to uncover and model CSC in the firm.[28]

- Hospital industry. The health care industry, especially hospitals, is under increasing pressure to reduce costs. To be responsive, management needs to address changes in hospital payment mechanisms to make their operations more efficient and profitable. Accounting information is vital, and the CSF approach can be a valuable tool in accomplishing cost control and effective management. CSFs can be used in planning and evaluating flexible budgets, cost per diagnostic-related group, and cost per unit of service.[29]

- Operations. Production or service operations management is a functional area that is greatly influenced by the CSFs selected by other functional areas. In selecting its own CSFs, the operations function affects other functional areas. As the product moves through its life-cycle stages, CSFs change and the operations function must adapt to keep pace with these changes. Accordingly, changes in the manufacturing processes may be necessary. They may change their positioning strategy, their manufacturing flows, or their manufacturing strategies. Figure 2-5 describes how a company chooses among alternative processing strategies.

Positioning Strategy

The positioning strategy reflects the position of the product on the life-cycle curve. In the early stages, the manufacturer uses a process-focused strategy that groups

Figure 2-5 The effect of CSFs on operations deployments.

like functions together to enable the production of a wide variety of products. This configuration is necessary when trying to determine the acceptance of different product variations in the marketplace. As the product demand increases for some varieties, the company can set up additional capacity in what is known as a product focus, where all of the equipment and support services needed to make a single product are grouped together. At this stage, they have a hybrid, or combination, of processes, some geared to small volume and variety, the remainder making high volume, low-variety output. If the progression continues, they may eventually move completely to a product process strategy, making high volumes of a small variety of products. The dilemma comes when the demand begins to decline. At this stage, there may also be a need to begin customizing products for selected customers. This process may require an adaptation of the product focus to make it more flexible in making a wider variety of customized products.

Manufacturing Flows

A process focus is usually associated with a job-shop environment. Job shops make a wide variety of products at relatively low volumes. As the volume increases, there is a need to move to a batch flow process, where batches of like products move through the production process. The equipment can still be grouped by function and there is an intermittent flow of goods from one function, or department, to the next. The batch flow process has been very successful and durable in most production environments. However, it is not as efficient as a flow shop, where the goods are processed along a production line or in some form of organized flow that is the lowest-cost form, but limited in flexibility and variety.

Manufacturing Strategies

The manufacturing strategies selected by companies may not be tightly related to the product life cycle; however, there is some relationship. Most companies that make low-volume, high-variety products and use a process positioning strategy, use the make-to-order (MTO) strategy. As they move along the life cycle toward high-volume, low-variety production, a make-to-stock (MTS) strategy becomes more practical. Today, the assemble-to-order (ATO) strategy is associated with a maturing of the MTS strategy, where the manufacturer can maintain its high-volume capability, but begin to provide greater variety by using modular subassemblies and adding the customizing features after an order is received. This strategy will be described more fully in later chapters under the mass customization label.

Role of CSFs in Selecting Management Programs

As a result of identifying their CSFs, companies may decide to implement incremental or radical improvement programs. We will discuss these programs more fully in Chapter 4. Each program requires that CSFs be developed to guide the program. For example, a Six Sigma program would have as a CSF the need to improve quality. It may also include a CSF to increase the empowerment of employees.

The following studies on CSFs illustrate their relationship with various management programs that have been implemented in recent years:

- Lean Manufacturing. Achanga et al. recently looked at the CSFs needed in implementing lean manufacturing. They found six CSFs that were essential: leadership, management, finance, organizational culture, skills, and expertise, among other factors.[30]
- ERP. Stephen King and Thomas Burgess also developed CSFs for ERP implementations. They identify the top 10 ERP critical success factors:
  1. Top management support
  2. Project team competence
  3. Interdepartmental cooperation
  4. Clear goals and objectives
  5. Project management

6. Interdepartmental communication
7. Management of expectations
8. A project champion
9. Vendor support
10. Careful package selection.[31]

- Web-based supply chain management systems (WSCMS). A survey revealed five major dimensions of CSFs for WSCMS implementation, namely, communication, top management commitment, data security, training and education, and hardware and software reliability.[32]
- Information center (IC). A study of an information center identified potential CSFs that ended up as composite CSFs: commitment to the IC concept, quality of IC support services, ease of end-user computing, role clarity, and coordination of end-user computing. Tests indicate that the importance of these composite CSFs tend to vary among themselves, but they are comparatively constant individually across IC stages.[33]
- Interorganizational information system (IOS). A case study examines the benefits of the IOS for the companies studied. The research reveals seven CSFs for the IOS: intensive stimulation, shared vision, cross-organizational implementation team, high integration with internal information systems, interorganizational business process re-engineering, advanced legacy information systems, and infrastructure and shared industry standard.[34]
- Technology absorptive capacity. Lin, Tan, and Chang found a significant relationship between technology absorptive capacity and the critical factors: technology diffusion channels, interaction mechanisms, and R&D resources. Different organizations will experience different technology transfer performance.[35]
- IBM is another company that employs the use of CSFs. Hardaker and Ward describe the use of CSFs at IBM. They describe them as follows: "Like the mission, CSFs are not the how-to of an enterprise, and they are not directly manageable. Often they are statements of hope or fear." They go on to stress that the CSFs would be necessary for the success of the company and sufficient for the needs of the company.[36]

Almost every project manager can list the main factors — the CSFs — that distinguish between project failure and project success. However, despite the fact that CSFs are well known, the rate of failed projects remains high. This may result when CSFs are too general and do not contain knowledge specific enough to better support project managers' decision making. A field study that involved 282 project managers found that the most critical planning processes, which have the greatest impact on project success, are "definition of activities to be performed in the project," "schedule development," "organizational planning," "staff acquisition," "communications planning," and "developing a project plan." The study also found that project managers usually do not divide their time effectively among the different processes, following their influence on project success.[37]

Organizations today are constantly searching for even the slightest competitive advantage and cost savings. The performance of an organization is important in achieving either outcome. Unfortunately, most organizations take a tunnel-vision approach to improving performance. Employees often resist the implementation of costly enterprise resource planning implementations. Despite training and management's efforts to explain the benefits of the program to the company, such programs are viewed as "this year's" attempt to frustrate the employees. Accordingly, process improvements are rarely rigorous enough to provide sustained results. Yet, when improvement efforts to these key factors — culture, process, and technology — are managed in an integrated manner, success can be achieved and sustained over the long term. Only by understanding the inter-relationship between these business-critical factors can lasting performance be sustainable.[38]

Performance Measurement and CSFs

When selecting CSFs, it is important to measure the results, or performance, against the CSF goals. This means that the CSF must have measurable goals. Quantitative goals are easier to measure; however, managers should attempt to measure the progress against qualitative goals.

CSFs often take the form of a project or a program. Although there may be some financial goals that will tie into the company's financial reports, it is more likely that the goals will be expressed in terms of expected outcomes, or events, such as the completion of a program to adopt an activity-based costing system. In this case, the performance will be based on completion times or quality of the program outcomes, as opposed to financial results. Most accounting systems are not designed to measure the effectiveness of programs. Consequently, supplemental measurement programs are needed, especially in measuring less quantitative performance indicators, such as those represented by CSFs.

A way to handle this type of performance is to use key performance indicators (KPI). KPIs are not the same as CSFs; they are a way to measure the accomplishments of a CSF. KPIs are flexible and can be used in any kind of environment. Often they are the way that operating managers measure the performance of their own responsibilities. Although they cannot always be made to relate directly to the firm's financial results, they are a good way to monitor the progress of CSF projects.

Summary

In this chapter, we described one of the basic elements of a business — strategic planning — and showed how this extends into operational planning and program planning. At each level of planning, companies need to identify those strategies, or actions, that are essential for the company's success. These essential goals or strategies are the critical success factors and they are unique to each business. We will build on this concept in Chapter 3 when we describe the evolution of supply chains, because a supply chain is a CSF for practically every business today. We will further develop the idea of CSFs in Chapter 4 as we describe the evolution of continuous improvement programs, also a necessity for successful businesses.

References

 1. Bullen, C.V., Productivity CSFs for knowledge workers, *Information Strategy*, 12, 1, 14, 1995.
 2. Rockart, J.F., Chief executives define their own data needs, *Harvard Business Review*, 57, 2, 81, 1979.
 3. Simon, H.A., *Administrative Behavior: A Study of Decision Making Processes in Administrative Organizations* (Fourth Edition), The Free Press, New York, 1997.
 4. Bullen, C.V., Productivity CSFs for knowledge workers, *Information Strategy*, 12, 1, 14, 1995.
 5. Dickinson, R., Ferguson, C., and Sircar, S., Setting priorities with CSFs, *Business*, 35, 2, 44, 1985.
 6. Yergin, D. and Stanislaw, J., *The Commanding Heights: The Battle for the World Economy*, Touchstone, New York, 2002.
 7. Hill, T., *Manufacturing Strategy*, Irwin, Homewood, IL, 1989.
 8. Blackstone, J.H. and Cox, J.F. III, *APICS Dictionary* (Eleventh Edition), APICS — The Association for Operations Management, Falls Church, VA, 2005.
 9. Simon, H.A., *Administrative Behavior: A Study of Decision Making Processes in Administrative Organizations* (Fourth Edition), The Free Press, New York, 1997.
10. Drucker, P.F., *Management: Tasks, Responsibilities,* Practices, Harper & Row, New York, 1974.
11. Senge, P.M., *The Fifth Discipline, The Art and Practice of the Learning Organization*, Doubleday, New York, 1990.
12. Juran, J.M., *Managerial Breakthrough, A New Concept of the Manager's Job*, McGraw-Hill, Inc., New York, 1964.
13. Simon, H.A., *Administrative Behavior: A Study of Decision Making Processes in Administrative Organizations* (Fourth Edition), The Free Press, New York, 1997.
14. Parnell, J., *Strategic Management: Theory and Practice* (Second Edition), Atomic Dog, Cincinnati, 2006.
15. Rockart, J.F., Chief executives define their own data needs, *Harvard Business Review*, 57, 2, 81, 1979.

16. Daniel, D.R., Management information crisis, *Harvard Business Review*, 39, 9, 111, 1961.
17. Anthony, R.N., Dearden, J., and Vancil, R.F., *Management Control Systems*, Irwin, Homewood, IL, 1972.
18. Rockart, J.F., Chief executives define their own data needs, *Harvard Business Review*, 57, 2, 81, 1979.
19. Berry, W.L., Hill, T., Klompmaker, J.E., and McLaughlin, C.P., Linking strategy formulation in marketing and operations: Empirical research, *Journal of Operations Management*, 10, 3, 294, 1991.
20. Bullen, C.V., Productivity CSFs for knowledge workers, *Information Strategy*, 12, 1, 14, 1995.
21. Freund, Y.P., Planner's guide: Critical success factors, *Planning Review*, 16, 4, 20, 1988.
22. Freund, Y.P., Planner's guide: Critical success factors, *Planning Review*, 16, 4, 20, 1988.
23. Leidecker, J.K. and Bruno, A.V., Identifying and using critical success factors, *Long Range Planning*, 17, 1, 23, 1984.
24. Bruning, B. and Lustig, M., Weathering the storm, *Inside Supply Management*, 17, 7, 26, 2006.
25. Edwards, M.H. and Hyer, R.M., The organizational effectiveness evolution: Wizardry or humbug? *The Human Resource Professional*, 5, 1, 28, 1992.
26. Bullen, C.V., Productivity CSFs for knowledge workers, *Information Strategy*, 12, 1, 14, 1995.
27. Boynton, A.C. and Zmud, R.W., An assessment of critical success factors, *Sloan Management Review*, 25, 4, 17, 1984.
28. Peffers, K., Gengler, C.E. and Tuunanen, T., Extending critical success factors methodology to facilitate broadly participative information systems planning, *Journal of Management Information Systems*, 20, 1, 51, 2003.
29. Hanks, G.F., Rx for better management: Critical success factors, *Management Accounting*, 70, 4, 45, 1988.
30. Achanga, P., Shehab, E., Roy, R., and Nelder, G., Critical success factors for lean implementation within SMEs, *Journal of Manufacturing Technology Management*, 17, 4, 460, 2006.
31. King, S.F and Burgess, T.F., Beyond critical success factors: A dynamic model of enterprise system innovation, *International Journal of Information Management*, 26, 1, 59, 2006.
32. Ngai, E.W.T., Cheng, T.C.E., and Ho, S.S.M., Critical success factors of Web-based supply-chain management systems: An exploratory study, *Production Planning & Control*, 15, 6, 622, 2004.
33. Magal, S.R., Carr, H.H., and Watson, H.J., Critical success factors for Information Center managers, *MIS Quarterly*, 12, 3, 413, 1988.
34. Lu, X.H, Huang, L.H, and Heng, M.S.H., Critical success factors of inter-organizational information systems – A case study of Cisco and Xiao Tong in China, *Information & Management*, 43, 3, 395, 2006.
35. Lin, C., Tan, B., and Chang, S., The critical factors for technology absorptive capacity, *Industrial Management + Data Systems*, 102, 5/6, 300, 2002.

36. Hardaker, M. and Ward, B.K., How to make a team work, *Harvard Business Review,* 65, 6, 112, 1987.
37. Zwikael, O. and Globerson, S., From critical success factors to critical success processes, *International Journal of Production Research,* 44, 17, 3433, 2006.
38. Marquardt, M., Smith, K., and Brooks, J.L., Integrated performance improvement: Managing change across process, technology, and culture, *Performance Improvement,* 43, 10, 23, 2004.

Chapter 3
The ITO Model

John dreamed on. In the next window, he looked out at a flurry of activity involving women and men of all descriptions. They were working on a beach and, in the distance, he could see a body of water. Although they were all working together, he noticed that some of them were dressed differently than others. He looked more closely. Some had green shirts and caps with "Accountant" lettered across them. Some had "Sales," others had "Marketing," still others "Engineer," and a large number with "Operations" on them. He looked harder. Now he could see some with "Finance" and a few with "Executive" on their clothes. There were even some with "Consultant" on them. Although some were working harder than others, all of them seem to be vitally interested in the bustle of activity. But what were they doing? It looked as if some were just making small modules that had some kind of connector on each end and a space in the center that looked as if it held something, but John couldn't make out what it contained. It was as if the thing in the middle was some kind of black box. As he looked further out on the shore, he could see that they were connecting the modules together to form some sort of a bridge. When he looked even further out, he thought he could see some people on the far side of the water building a bridge toward the side that John was on.

His curiosity got the best of him, so he asked one of the Consultants, who looked as if she knew what was going on, "What are they doing?" She replied, "They are building a supply chain. This side of the bay has goods and services that the other side of the bay wants to buy, but there hasn't been a good way to get them across. We tried going around the bay, but that took so much time and cost so much that we were asked to recommend a better way. We recommended that they build a supply chain and that is what they are doing. We will use pontoons to provide a way to get from this shore to the first island, where we can transfer the goods to another pontoon bridge to get it to the next island, and so on, until the goods get to the far shore. The network should be ready to use in a short time. We think it will cost less, be much faster, and allow us to ship smaller quantities each time."

"What are the modules?" John asked. "I see different things"—John noticed that they were labeled 'processes'— "being loaded into each one and yet they all seem to end up looking about the same and they fit together very nicely." They were the little black boxes that John had noticed earlier. The Consultant replied, "You are very observant. We found that a lot of different activities, or processes, have to be performed on each good to transform it from the raw materials stage to a finished goods stage. We also have to have processes to add the services necessary to provide a complete package for the customers on the other side. If we have a way to connect the processes together, we can make a smooth flow of the goods and services along our supply chain. It also works very well in providing a way for the customers to send their payments back along the supply chain and to provide information about what they like or don't like about the goods and services they receive." John was not sure he understood all of this, but the part about paying back along the supply chain was intriguing to him. He remembered that some of his customers were not always paying on time, a problem he still had not figured out how to solve completely.

John persisted. "But why do you have all of those different kinds of people working together on this program? Wouldn't it be better to separate them into their own special areas and let them work away from each other instead of mixing them up with other types of workers?" "Ah," replied the Consultant. "That is one of the new concepts that we introduced. We found that putting the workers together makes it easier to prevent product problems from coming up and to gain the benefit of the team's interactive thinking. It was a little hard to get started, but it has been a wonderful innovation." The Consultant called this arrangement a cross-functional team. John thought it reminded him of that old adage, "two heads are better than one." In this case, though, the more heads involved, the better, or at least that is what it looked like was happening.

John still wasn't satisfied. "Why do customers care about the supply chain? Weren't they just as happy when the people on this side traveled around the bay to bring them their goods and services? Doesn't it cost more by building *this supply chain thing, and going directly over the water with the goods? Surely the customers wouldn't like that." He paused, sure that he had finally stumped the Consultant. But the Consultant only smiled. "The customers love the new arrangement. One of the big advantages to them is that they don't have to carry as much inventory waiting for the next trip around the bay, which used to take weeks. Now they can get replenishment orders in days, sometimes in hours. The customers found they did not have to build as big a storage area as in the past, and this saved them money. In addition, we have found that costs actually decrease because of the increased availability of information that can be used in demand forecasting and the elimination of waste as we streamline the processes." John understood the comment about bigger storage areas. He had recently rented a small warehouse to keep some of his inventory, a move he thought was clever because he did not seem to run out of those goods as much as before. Still, the warehouse was costing him extra money, and he was not sure if he was coming out ahead financially. The other comments about increased availability of information and the elimination of waste were still not registering in his mind. It was, after all, a strange dream and he was tired.*

Now John was perplexed. He remembered he had heard about this idea of supply chains at one of the last meetings of his professional society and he had even decided that it might make some sense, but getting started on it just seemed always to be something he could put off until a later date. Maybe he had better reconsider its importance in his thinking. Could it be one of the Critical Success Factors he had heard about earlier? Because he was in the retail business, he probably should try to find out what was happening in the supply chain that related to his business.

Just as it was at the first stage, there was a basket with several pamphlets, entitled "The ITO Model and Supply Chains." John took one and put it alongside the first one.

The ITO Model

The Input–Transformation–Output (ITO) Model is the DNA of the supply chain. The supply chain is composed of a series of linked elements, whether in a manufacturing or a service business. This chapter describes the ITO Model and how these elements can be linked together to become a functioning supply chain. In this chapter, we will discuss the following topics:

- Types of models
- Uses and benefits of these models

- The basic ITO Model
- Extensions to the basic model
- The concepts of open and closed systems
- Supply chain formation
- The ITO Model and its application to management functions

A single ITO module is equivalent to a transaction, such as the receipt of an order from a customer. A linkage of several ITO modules is a process, such as the complete handling of an order from receipt to delivery. A supply chain involves the linking of a process from one company to an input or output process in another company. To help us understand the ITO Model, it would be beneficial to examine why models are used in the first place. Managerial Comment 3-1 by Martin Starr begins our discussion on why models are useful.

Managerial Comment 3-1.
The Nature of Models

"Operations management is a many-problem field. In other words, the manager concerned with operations encounters one problem after another. These many problems arise because many operations must be accomplished to assure that the result of the operations is of agreed-upon quantities, on schedule, with specific qualities and costs. Clearly, these operations create many problems, and it is not obvious how to treat such great variety. Seldom is the manager faced with one big problem for which one major strategy decision will suffice.

By simplifying reality in a systematic and organized fashion, it is possible to study systems that are too complex to be understood by intuition. Then, for such systems, we can diagnose errors to bring about useful remedial changes. The idea of a model is crucial to this reasoning, because a model is a simplified representation of reality. It is constructed in such a way as to explain the behavior of some but not all aspects of that reality. The reason that a model is employed is that it is always *less complex* than the actual situation in the real world. It must, however, be a good representation of those factors or dimensions that are strongly related to the systems objectives; otherwise, it will not be a useful model and, therefore, it should not be used.

Models have always existed, but the recognition of model properties is just yesterday's achievement. This recognition of *properties* has made an enormous difference to operations management as a field because the properties of models were found to be very general, so general, that all products and any services could be studied.[1]

Introduction to Models

This chapter shows how we can use ITO models to describe the basic elements of service operations management. It will also show how managers use a variety of models to help them in their decision making and management of their companies.

Types of Models

Three types of models exist — physical, mathematical, and conceptual. Physical models are smaller objects that represent larger objects. Wind tunnels test model airplanes to predict how the full-size airplane will react in various flight conditions. A model building and its surrounding parking area can help planners decide the size and layout of the full-scale facility. A model of a sales counter for a retail store can be set up for use as a training facility for new sales personnel. Physical models are widely used in almost every kind of business to help us better understand the composition and relationship of the elements that compose a given situation.

Mathematical models represent relationships among variables that we can quantify. Some of these relationships include:

total costs equals fixed costs plus variable costs, or TC = FC + VC;

or

the percentage of process defects equals the total number of defects divided by the total number of units tested, or % p = d/n;

also

the number of inventory turns equals the annual cost of sales divided by the average inventory, or Turns = COS/AI.

As with physical models, we use mathematical models to help us understand and improve our business.

Conceptual models represent ideas. They are usually more difficult to design and understand than physical and mathematical models. However, conceptual models are often simple when compared with the real world they represent and are, therefore, subject to more liberal interpretation. Do you believe that increased knowledge leads to improved performance? If you do, how would you show this relationship in a model? You could simply state the idea in equation form as

$$\text{Increased Knowledge} \rightarrow \text{Improved Performance.}$$

You might also display the idea as a graph, with Knowledge as the independent variable represented on the horizontal axis with increases to the right. The dependent variable, Performance, is shown on the vertical axis with increases moving from bottom to top. When the two variables are graphed together, an upward sloping line would result, starting in the lower left part of the graph and moving toward the upper right. How rapidly would it increase? Would it be linear or curvilinear? Would it continue to increase indefinitely or would it reach a peak and start down? You can see that the curve could vary under different conditions. Yet, all conditions support the basic idea that improved performance follows increased knowledge.

In this chapter, we introduce a conceptual model that describes the basic operation of a business. We will use this model throughout the book; in fact, the remaining chapters will explain the model in more detail and show how to use it in managing a business. The model will also be useful in explaining the similarities and differences among different kinds of businesses.

Why Are Models Used?

Operations managers face many problems. They must deal with some problems rather quickly; others, involving strategic decisions and larger investments, require more critical analysis before making a decision. The use of models can help in this latter kind of decision. An effective model can help answer several questions:

- What are the important parts of the problem? A model identifies important components of a problem. Recognizing these allows decision makers to focus on the components that should be considered from those that can be ignored. For example, the manager who is experiencing high employee turnover would want to identify the causes of the turnover. These causes are the important components of the model. One such cause may be job dissatisfaction.
- How do these parts relate to each other and the problem at hand?. A model also helps to specify relationships among components of a problem, especially the relationship between causes and effects. Models often designate

independent variables that represent causes of particular symptoms. Dependent variables represent events influenced by the independent variables. For example, job dissatisfaction, an independent variable, can influence employee turnover, a dependent variable, in an unfavorable way. In other words, high job dissatisfaction can lead to high employee turnover.

- Why are these parts important? Some models explain reasons for the relationships between causes and effects. In the example of employee turnover, many models have been offered by management researchers that indicate which independent variables are the most important influencers of turnover. The manager who wishes to decrease employee turnover will seek to identify the variables to change to reduce turnover.
- Under what conditions will this problem exist? Models specify conditions such as when events occur, where they occur, and who makes things happen. For example, there are employee turnover models that vary according to the industry under study as well as the particular job.

Answers to these questions make it possible for us to use models to study other business problems such as identifying the key variables in the application and training of inexperienced employees. We can also examine the model itself for opportunities to make improvements in its structure and operation.

Benefits of Using Models

Managers find several benefits from using models. Models can help them to control and manipulate the large number of details that surround business problems. A model clarifies the distinction between critical elements that require close attention and extraneous factors that often distort or confuse real relationships between inputs and outputs. Models allow managers to experiment with changes to see what effects they would have in real management situations. In other words, managers can benefit from asking "what if" questions. Without models, managers would have to make changes in real systems, a time-consuming and expensive process with a high cost of failure. Models can help problem solvers to evaluate the effectiveness and efficiency of various alternatives, separating poor solutions from better ones. Models can improve intra-organizational communications, as managers often discuss operating systems using models. Models help to focus on information-gathering activities. By identifying the key factors of a real situation, a model directs managers' attention to information they need to make decisions.[2] As with any management tool, the value of the model depends on the competence of the user.

Figure 3-1   The Basic ITO Model for both manufacturing and services.

The Basic ITO Model — Inputs, Transformation, and Outputs

Figure 3-1 is the most basic depiction of the ITO Model. It shows the transformation of things, such as wood or steel, into goods, such as furniture or automobiles. It also shows the transformation in service industries, where the inputs are often people.

The transformation process can be as simple as building a table out of pieces of wood or cashing a check at a bank. It can be as complex as building an airplane or changing a customer's anger over poor service to delight when an employee resolves the complaint with courtesy and promptness. The models described in this chapter examine how the transformation process applies to a wide variety of business situations.

Table 3-1 shows a number of examples of the ITO Model in service applications. Inputs can be both things and people. However, measuring the outputs is more difficult in services. For example, the output of a hospital can be counted as the number of patients processed, but there is greater variation in the patients than the output of a manufacturer, making wood desks. This variation makes it more difficult to measure both the effectiveness and efficiency of service transformation processes.

The General Systems Model

We can extend the simple linking relationships shown above as a building block for a more general integrated system. Outputs are the goods or services produced by a business. If the output is a tangible product, the business usually fits into one of the extractive or transformative industries. In any of the other industries, outputs are less tangible; consequently, the business usually fits into a services category. Outputs, as shown in the general systems model, often become inputs to another ITO module. A general system consists of several ITO modules linked together.

Because inputs are often outputs from another part of the system, inputs are also both tangible and intangible. Iron ore, for example, is both an output from a mining operation and an input for the refining process that produces iron as its output. The iron then becomes input for the next process that adds other materials to produce outputs of steel. Steel becomes an input for other processes and the linking continues until we have a final product for the consumer.

Table 3-1 Examples of ITO Modules

Type of Business	Inputs	Transformation Process	Outputs	Customers
Retail supermarket	Unserved customers	Provide facility, inventory, system, and employees to make food and other items available	Served customers	Individual consumers
Electric utility	High voltage, undistributed power	Transport power from generating plant to distribution points; reduce voltage for use	Low voltage, distributed power	Individuals and industrial customers
Hotel	Arriving customers	Provide food, accommodations, and related items to a variety of customers	Departing and satisfied customers	Individuals and groups
Hospital	Sick or injured patients	Provide diagnosis, tests, and treatment for a variety of illnesses	Treated patients (hopefully, well)	Individuals
Public accounting firm	Unaudited companies	Provide a series of professional services, as needed	Audited company	Public companies

On the services side, high school graduates (output) become an input to a university, where we transform them into first-year students (output). Following a series of intermediate transformations, they become a college graduate (output). This graduate becomes an input to a company where additional transformations take place in an organizational setting. This individual will also experience transformations in the social and political environments of which they are a part.

The transformation process requires the use of resources. The resources include the employees who do the work, the equipment they use, the facilities in which they work, plus the systems and information they use. Figure 3-2 shows this addition to the ITO Model.

To expand the model a bit, we can add customers as the recipients of the goods and suppliers as the providers of the input materials. These additions show the beginning of a supply chain, or the linking of different entities into an integrated unit (Figure 3-3).

For a long time, many have viewed manufacturing and services as unrelated processes. However, in practice, they are almost inseparable. Most manufacturers find that they must provide some degree of services to accompany their goods. For example, an automobile manufacturer sells a product, but also provides services to the consumer through the dealer network, or a computer company provides an online help center for customers having problems with their computers.

Let us look at another example in a service business. Consider a distribution company that receives an order from a retail store. The transformation process must deal with the order as an input. However, the customer is also an input, in the sense of having an unmet need and expecting to become a satisfied customer (output). What does the distribution company do to transform the customer? The distribution company may send an order acknowledgment confirming the order when the customer wants it, and then ensure that it is shipped on time. Suppose that the dis-

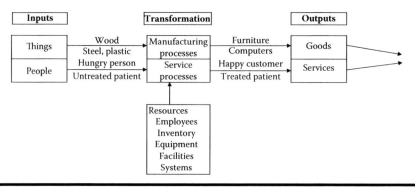

Figure 3-2 The basic ITO Model with resources used in the transformation process.

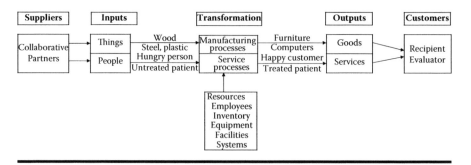

Figure 3-3 The extended ITO Model with customers and suppliers added.

tribution company dissatisfies the customer because the order shipped late with two of the ten items ordered missing. The company may send a message to the customer to explain that the backordered items would be shipped on a certain date at no extra charge to the customer, and then follow up to assure that the items were shipped and received by the customer. This example illustrates how parallel processes can work on different inputs and produce different, but related, outputs at the same time. Consequently, most businesses are transforming people at the same time they are transforming things.

The Basic Model Extended

In Figure 3-4, we extend the model to show that the output from one transformation becomes an input to another ITO module. Suppose the first input is an order from the customer of a distribution company. The distribution company transforms the customer's order into an approved order ready to be shipped (output), accomplishing this task through a series of actions that include approving the customer's credit, verifying that the ordered items are in stock and available for shipping, and scheduling the order's shipping date. The approved order then becomes the input for the warehouse, where the items are retrieved from stocking locations, loaded onto the delivery truck, and delivered to the customer. The output from this transformation process is the completed order. This linking of ITO modules illustrates a common series of activities, or a process, in any kind of business.

Linking two ITO modules is just a beginning. If we can link two ITO modules, we can link a third module, a fourth, a fifth, and so on, until we complete the desired configuration that we need.

Now, let us add another dimension to the model. We have shown the customer as the recipient and evaluator of the goods and services provided. Customers can also be inputs when they are waiting to be treated at a hospital or waiting for a haircut at the hair salon. However, in recent years, customers have also become partici-

Figure 3-4 Linking ITO Models to form a process.

pants in the transformation process, such as when they select their own groceries at a supermarket and use the self-checkout terminal. Customers can also help design their educational process. They can select the format in which they will receive instruction — by traditional attendance in a classroom or online in a distance education environment. Figure 3-5 shows this extension of the model.

Components of the Model

The models described above show how we can link simple ITO modules together to make a more complex model. This integration process is an important idea — one that we will discuss more fully later in the chapter. At this point, however, we will disaggregate the basic ITO Model into its components and look at each component separately.

Customers

The objects of all service activities are satisfied customers. In a sense, services require that a business transform its customers, or potential customers, into satisfied customers. The transformation process may affect customers minimally, such as in the delivery of electric power, or substantially, such as when a surgeon performs heart-bypass surgery. Because customers vary, most companies may need to develop processes for the best handling of each type of customer.

An important consideration is whether the customer is adaptable and willing to participate in the service process. Customer participation can make the process more productive, but more importantly, the customer helps determine the quality

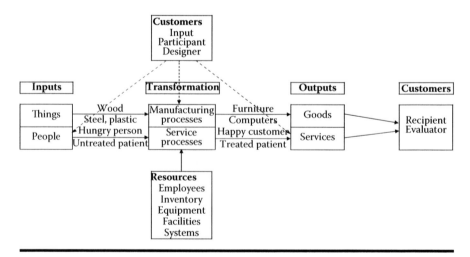

Figure 3-5   A closed ITO Model with multiple customer roles.

and customization of the final service. Examples of customer participation include the use of salad bars at restaurants or planning a two-week vacation online by selecting various travel and entertainment options.

Customers will possess different levels of knowledge about the services they receive. Some customers will have little knowledge, like the new member who joins a fitness club and needs assistance in learning to operate the various exercise machines. Other customers will possess a great deal of knowledge about the services, but will still need some minimal amounts of assistance. An example is the fitness club member who knows how to operate all of the machines, but needs advice on how to set up a specific exercise program. This example illustrates how the service plan must satisfy both extremes, even in the same business.

Customers may receive services as individuals or as part of a group. Usually, the level of customer contact and customization of service is greater for individuals than for groups. The effect of the service will vary for each person within the group. This variation makes it more difficult to satisfy all of the members of the group. In a tour of the Gettysburg battleground, those tour group members who are not at all familiar with the battle may be very interested in the tour because the information being presented is new. Other members may be bored by the presentation if it is information with which they are familiar.

Outputs

Outputs can take the form of goods, or things, which we can specify and measure to see if they conform to specifications. Thin plastic paper is an example of an output that is used to wrap after-dinner mints. It must be attractive and meet other physical specifications; simultaneously, it provides part of the service package at a restaurant. The other form of output is the customer, each of whom is unique. Measuring the output for a customer may be a complex undertaking. The measure of a completed output may be as indefinite as the perception of the customer. For example, attending a training course on the use of a word processing software package transforms a person with little, or no knowledge, to a person with more knowledge. However, the knowledge gained varies with each person. Clearly, defining the output for a pure service is difficult and the achievement of a satisfactory output requires variations in the process used. An understanding of the expectations of the customer before providing the service may preclude subsequent customer dissatisfaction or rework of the service.

The amount of customization required of a service varies among outputs, especially when the output is a person instead of a thing. Another variation is the rate of demand by the customer. Demand is rarely at a constant rate; therefore, the transformation process has to be able to accommodate both customizing the service for the customer and adjusting the demand rate.

The success of any business depends on how well the outputs reflect the critical success factors of the business. This requirement includes the cost of service, the timeliness of service, and in recent years the quality of services. Careful planning is required in providing the service package to maximize the benefits of critical success factors.

Transformation Process

Much of the research in management has focused on the transformation process. Some variables in the transformation of services are tangible, such as determining the number of store sites in a retail business. Other variables are intangible, such as the level of judgment required by a server in a custom flower-arrangement shop.

The transformation process can be viewed as a technical function, capable of standardizing or customizing the service. Here, we attempt to adapt the transformation to any of the variations that may be encountered from the operating environment. For example, the service transformation can be rigid, such as in the use of ATMs, or flexible, such as with the providing of human bank tellers.

Automation is an example of the active use of technology in the providing of services. Automation usually increases over time as the service becomes more standardized, such as in the case of travel insurance policies sold through vending machines. The amount of automation also increases as the technology capability increases, such as in airline reservation systems now available to every travel agent and online to the Internet user.

The transformation process must be adaptable to variations in tasks and rates of service. The service can be continuous, such as in an electric utility, or discrete, such as a lawyer preparing a house purchase agreement. Companies can service customers at their place of business, such as in a bakery; in the customer's house, as by an interior decorator; and over a communications system, such as in a home shopping channel. The presence of many variables in the transformation process precludes adopting a "take two and call me in the morning" type of approach for most businesses. Each business has a uniqueness that results from a combination of variables that works for that particular service and owner.

Inputs

Inputs can be things or people. With things as inputs, businesses attempt to standardize their inputs to reduce undesirable variations in the transformation process. With people as inputs, usually as trained employees, businesses attempt to build adaptability into the transformation process to accommodate the increased variations that often arise. Such variations may result when we introduce additional changes in tasks and work flows. Consequently, the ideal transformation process must have sufficient flexibility to adapt to input variations.

We can combine transformation variables in hundreds of ways, each of which may require a different approach in the design and management of the resulting service business. Although inputs, processes, outputs, and customers are presented as separate elements, they operate within the framework of an integrated system. As a result, a change in one element will impact other elements as well. For example, an adjustment in input, often the result of a changed customer need, requires a subsequent change in the transformation process. Furthermore, these changes will also result in changes in one or more of the output variables.

Although describing a separate case for each possible combination of variables is impossible, Table 3-2 shows three possible combinations. The top section of the table describes a simple service, the processing of customer checks, where things (the checks) are the primary input. Even in some operations where there is no direct customer encounter, some people are involved, mainly the employees who process the checks. Many manufacturing plants also fall in this category, particularly those that are isolated geographically from their customers. In terms of technology, the transformation process is standardized and geared towards economies of scale.

The middle section of the table describes a combination of a product and service output, such as the delivery and installation of a refrigerator. In this transformation, a large technical core (the factory) manufactures a small technical unit (the refrigerator). The service function is illustrated through the delivery, installation, and demonstration of the refrigerator. This combination of output, both things and persons (or alternately, product and service), is common in most business transactions; what varies is the amount of service to add to the sale of the product.

The bottom section of the table depicts an example of a pure service; the design and performance of an accounting audit at a small business. The technology core includes a vast amount of individual and group knowledge and interaction with the customer is extensive.

The general approach that we can glean from Table 3-2 is that businesses need to (1) develop a total service concept, (2) design the service package, and (3) manage the transformation processes. Later chapters will present each step in more depth.

Extending the Basic ITO Model into Supply Chain Configurations

The preceding discussion shows how the basic ITO Model works. It also provides a glimpse of the ideal supply chain by showing the addition of customers on the receiving end and suppliers on the providing end. This section describes how basic ITO models can be linked together to form a supply chain. The clothing industry is a good example. It includes a combination of both manufacturing and service businesses. It also illustrates the complexity of many supply chains in that a number of independent businesses must work together to make the supply chain effective.

Table 3-2 Examples of the ITO Model

Inputs	Transformation	Outputs	Customer
Things (checks)			
Consistent task content Consistent work flow	Processing mortgage payments in the back room of a commercial bank Large technical core with low customer contact and dedicated resources to do standard services	Product - notation in customer's file that payment has been made; no formal report to customer	Knowledgeable (at least for this task) individual who has adapted to the bank's system of payment
Things (refrigerators)			
People (delivery and install crews)	Delivery and installation of a refrigerator into a household Large technical core to make; small crew to install The thing is transformed and the customer is also transformed, hopefully to a delighted new refrigerator owner	Delivered, installed, and demonstrated product; Apprehensive prospective owner becomes a delighted new customer	Individual with moderate knowledge, at least as user of the product Discriminating buyer of service Some flexibility in timing of delivery
People (accounting audit at a small business)			
Uncertain arrival rate Uncertain work flow	A CPA must design and conduct an audit of a small business that has never been audited Small technical core with high customer contact by servers who must perform highly customized services	A completed audit with a variety of accompanying work papers and financial statements	Relatively lacking in knowledge and with limited resources to participate in or understand the audit process and the results

A Look at the Supply Chain

Figure 3-6 illustrates how a number of ITO modules can be related to each other in that one ITO output becomes the input for another, and so on. In this example, the process begins with the growing of cotton, and ends with the recycling of worn clothing.

The Closed and Open Systems Model

Up to this point, we have been discussing the ITO Model and its linking with other ITO models in what is referred to as a closed system model. That is, essentially all transformation activities are under the control of the parties designing the system. In the real world, such a scenario does not exist because businesses operate in an

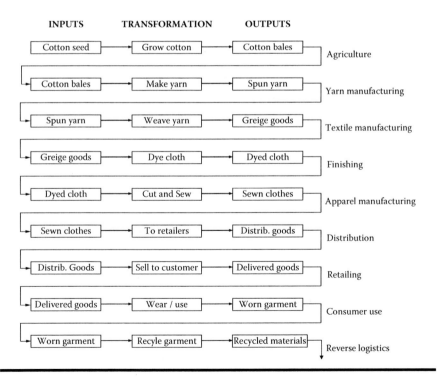

Figure 3-6   Linked ITO Modules to form a supply chain.

open system environment. The open system concept acknowledges that external factors, such as competitors, economy, technology, society, environment, and government, may have a significant effect on the operations of the supply chain.

In addition to the differences between closed and open systems, general systems theory includes several other attributes of systems, including holism (view the whole system, not just the parts), goal seeking (the need for tangible objectives), entropy (the tendency of a system to fall into disorder over time), regulation (the need for management), hierarchy (components within subsystems within systems), differentiation (the uniqueness of a system to fit its application), and equifinality (alternate ways to reach the same objectives). "Of all the distinctions noted, that of open and closed systems appears to be most fundamental and most useful. This classification rests upon the basis of resource availability."[3]

Closed Systems

The closed system model is a concept that is useful in describing a system because it makes it possible to screen out a number of variables that are difficult to consider. It may be thought of as saying "all other things being equal" to confine the analysis and planning to those activities within the control of management. One assumption is that all of the system's resources are present within the system under study and no additional resources will enter the system from the outside environment. The resources include the employees, equipment, facilities, technology, processes, and information used in impacting the transformation process. "Problem solving by business executives is often done in a quasi-closed system with the aim of simplifying the situation enough so that a rough estimate can be arrived at."[4]

Entropy is a useful concept that relates to closed systems. Entropy suggests that a system will eventually fall into disorder and become less effective over time unless it is re-energized with additional resources. This means that the resources must come from outside the system, so although the closed system concept is useful, it is not adequate in the business environment.

Open Systems

The general systems model shows that every system has a boundary that limits its sphere of application. However, the environment outside the system, no matter how closely constrained, affects the transformation processes of all systems. For example, a university may have considerable control over curricula, course content, classroom procedures, and acceptable student academic performance. However, external factors outside of the university, such as the student's family, economic conditions, funding sources, social pressures, the music industry, books, and many other factors, also influence the university and its relationship with the student.

Figure 3-7 shows the relationship of the ITO Model to external factors that may influence the effectiveness of a system. Elements in the open system include competitors, the economy, technology, the government, environmental concerns, and societal forces. It also illustrates the idea that the organization has little control over these factors, but does have some ability to anticipate their effects in the planning process. As a result, these factors should be considered in the design and operation of a business.

Here are some ways these external forces could affect a business. Competitors may introduce new products, raise or reduce prices, offer sales promotions, or introduce a new kind of store. The economy may reflect increased interest rates, a recession in housing starts, or a reduction in fuel oil because of increased demand during cold weather. Technology advances could provide new software for interorganizational communication, a new material that is stronger and lighter than steel, or a breakthrough in the use of RFID tags. The government could raise the minimum wage, enact legislation that would limit sales of critical items, or reduce spending on weapons systems. Environmental considerations could enhance sustainable development efforts, reduce the use of toxic materials, or encourage the use of recycling or remanufacturing processes. Society could lose its interest in high-fat foods, increase its acceptance of new technologies, or begin to save money instead of spending it. Any business, regardless of size, faces at least some of the challenges presented by an open-system environment.

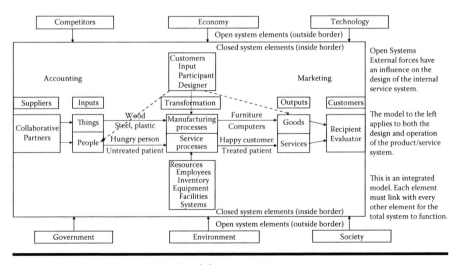

Figure 3-7   Open system ITO Model.

Closed-System Strategy

James Thompson[5] studied the effect of closed and open systems on the organization. The system within the organizational boundary represents the closed system and the opening of that system to the environmental factors represents the open system. The rational model results from a closed-system strategy and the natural-system model, discussed below, results from an open-system strategy.

The closed system is comparable to a determinate system in which we know all relevant variables; therefore, we can accurately predict the future performance of the system. "Scientific management," "administrative management," and "bureaucracy" are examples of rational models. We will discuss them in more detail in Chapter 4. They suggest that the ingredients of the organization are deliberately chosen for their necessary contribution to a goal and the internal structures established are those deliberately intended to attain highest efficiency.

Managerial Comment 3-2.
Isolation of the Technical Core: Classical Management Thinkers

James Thompson developed a series of propositions to explain how the technical core of the organization must be isolated as much as possible from the environment to assure maximum efficiency.

Proposition 1. Organizations seek to seal their core technologies from environmental influences.

Proposition 2. Under norms of rationality, organizations seek to buffer environmental influences by surrounding their technical cores with input and output components. If an organization can buffer its technical core from the environment, then it presumably can enjoy the conditions for maximum efficiencies. However, if buffering the technical core is not possible, then the organization has to seek other devices for protecting its technical core.

Proposition 3. Organizations seek to smooth out input and output transactions. This means that if the company cannot use inventory to buffer its technical core, it must attempt to smooth the demand in a way that reduces the fluctuations in the environment. This would involve changing the demand patterns that exist in a company's business, and because this is not always possible, the technical core must settle for a lower level of efficiency.

Proposition 4. Organizations seek to anticipate and adapt to environment changes that they cannot buffer or level. This means that organizations must forecast the demand and attempt to set their technical capability (capacities) to gain much of the efficiencies of a closed system, though with fluctuations in the environment.

Proposition 5. When buffering, leveling, and forecasting do not protect their technical cores from environmental fluctuations, organizations resort to rationing. Rationing represents the attempt by the organization to schedule demand or to allocate available output to the demand. Rationing is an unhappy solution, for its use signifies that the technology is not operating at its maximum, yet some system of priorities for the allocation of capacity under adverse conditions is essential if a technology is to be instrumentally effective.[6]

Open-System Strategy

The rational model, or a closed-system strategy, works when we know and can control the variables and relationships at hand. This is seldom the case in practice; therefore, most businesses must resort to a natural-system model and employ an open-system strategy. This assumption occurs when we expect uncertainty, when the number of variables exceeds comprehension, or when we cannot control or predict some variables because of outside influences.

Rational models produced important insights, but did not provide an adequate understanding of complex organizations. Herbert Simon proposed that in dealing with complexity, the organization must develop processes for searching, learning, and deciding. They called this process making decisions in bounded rationality. This requirement involved replacing the maximum efficiency criterion with one of satisfactory accomplishment. Put another way, the perfect outcome did not have to be attained; instead, a decision that was sufficient would be satisfactory. Thus, decision making now involved "satisficing" rather than maximizing.[7]

How do we use these ideas in business operations?

1. The notion of closed systems and open systems suggests that most businesses conform to the open-system model because their interactions involve external customers.
2. Although the total business requires an open-system model, isolating the technical core may be possible and viewed as operating under a closed-system model. For a fuller explanation of the technical core, see Managerial Comment 3-2.
3. Service organizations can employ strategies to isolate the technical core when the environment is uncertain or fluctuating. These strategies can include buffering, smoothing, anticipating, and rationing.[8]
4. The activities of the organization can be viewed as gathering inputs, transforming the inputs by some technology or process, then distributing the outputs of the transformation process in an effective manner.

Both Perrow[9] and Thompson[10] incorporated the ITO Model in their work. They pointed out that minimizing variations in inputs allows greater control in the transformation process and results in standardized outputs. Allowing large variations in inputs requires greater flexibility in the process and results in more customization of the output. Table 3-3 shows a summary of these relationships.

Both Perrow and Thompson placed heavy emphasis on the inputs used and the transformation process; they did not focus as directly on outputs and the direct contribution of the customers. They portrayed the customer as part of the outside environment, not as a direct participant in the company's organizational structure.

Managerial Comment 3-3. Core Competency

"What is a core competency? The answer I hear from many executives is, 'It is what we do well.' My response is, 'What if you are really good at running the company cafeteria?' This may be well done, but it does not help the company make money or achieve advantage over the competition. A more accurate answer to this question is, core competencies are the thing we have to do well to achieve competitive advantage. This may include superior R&D, superior production, superior marketing or, you guessed it, superior supply chain management. In the last case, this is a process of identifying what your company has to do well and, at the same time, identifying what your supply chain partners have to do well to see the overall supply chain is successful."[11]

Table 3-3 The Effect of Variation on Process

Condition	Inputs	Transformation	Outputs
Closed system with an isolated technical core	Small variation, standardized	Controlled, routine	Standard, high volume
Open system with an external focus	Large variation, nonstandardized	Flexible, less controlled	Custom, low volume

Feedback

In the interest of keeping our diagrams as simple as possible, we have not mentioned an essential element of any system — feedback. Feedback completes the loop in what is called a closed-loop system. All transformation processes produce an output, and managers of the transformation process are interested in the quality of that output. The first feedback should come from the managers within the company as they evaluate their own performance, based on the results of their performance measurement systems. Additional feedback should be garnered from the customers in both formal and informal communications. Although feedback can provide useful information, it is often difficult to gather and analyze.

The Fine Art of Building Relationships

We have shown how individual ITO modules can be linked together to form complex supply chains. There is a challenge in the actual process of linking these ITO modules together. What transformation process is required to accomplish this monumental task? It is not manufacturing or services in the traditional sense. Figure 3-8 illustrates this new kind of transformation, a process we call "relationships building."

We have transformation processes for manufacturing and for providing services. But do we have transformation processes for building relationships? Probably not at the present time; however, we will address this topic later in this book. But

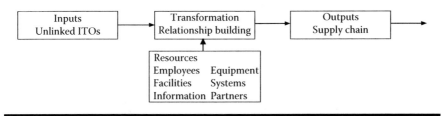

Figure 3-8 The role of the ITO Model in business relationships building.

first, we will describe a hypothetical example of why it is both necessary and difficult to link individual ITOs into a complex supply chain.

Our example includes three separate entities — a manufacturer with one supplier and one customer. Rather simple, but if we can create it for this configuration, we can extend it to multiple configurations. Within each of the entities, we have three separate processes, and between the entities, we have two interfaces. For illustration purposes, we will depict them as shown in Figure 3-9.

The arrows represent the process steps and generally point to the right. However, they are not exactly aligned, which illustrates that a smooth flow of goods or services from supplier to producer to customer is not always possible.

Figure 3-10 shows a progression of alignment steps. First, there is a need to align the internal functions within a single business entity. Lest this appear too

Original Conditions (lack of focused direction, both internally and externally)

| Supplier | Interface | Transformer | Interface | Customer |

Figure 3-9 The beginning of relationship building

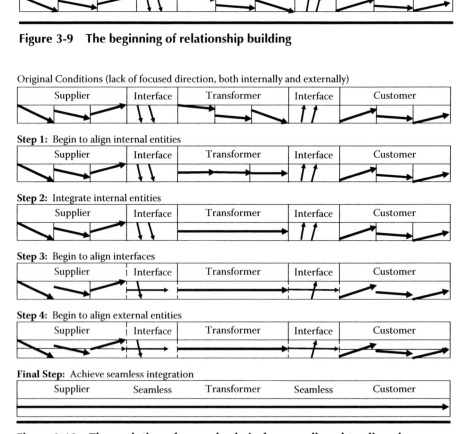

Figure 3-10 The evolution of a supply chain from unaligned to aligned.

obvious and simple, one of the authors recalls a discussion with a VP of Operations for a tool company. When asked what his major problem was when he first joined the company, he replied, "Getting the production department to make what marketing was selling." He went on to explain this seeming incongruity. It seems that the production department employees, including the supervision, were paid on an incentive system based on output. Consequently, they wanted to produce large volumes, regardless of demand, to maximize their piece rate incentives and production bonuses. Usually the production department favored manufacturing the older products, while marketing was pushing the newer products to keep the company competitive. This example illustrates how conflicts of interest caused by compensation or performance measurement systems are commonplace; they must be resolved to smooth the flow of materials along the internal supply chain.

Once a reasonably good alignment is achieved within a company, it can begin to work on aligning the interfaces between itself and their customers and suppliers. Often, this re-alignment includes not only the flow of information, such as orders, but also the flow of physical goods. To attain an adequate flow of orders, there must be compatibility within the information technology functions and this process often does not occur easily. Ideally, all businesses are moving toward electronic information flows to achieve speed, accuracy, and lower costs; however, not all businesses are moving at the same pace or with the same kind of hardware and software. In addition to the information flow alignment, there is a need to decide on how best to transport the goods. These decisions require agreements on timing, cost, and selection of intermediaries.

In this re-alignment process, a customer may ask for changes within a supplier's operation. Factors to be resolved include the quality level of the product, the amount of capacity to be allocated, the response time for orders, the performance level of the product, the type of packaging to be used, and the level of readiness for use. If these issues are not reconciled, inventory levels will build up and service expectations will not improve.

We have shown the difficulties in smoothing the flow of goods and services through a relatively simple supply chain. Think of the difficulties in extending this example through all of the possible supply chain configurations that are likely even in small companies. (An analogy can be drawn to human DNA, where each person is different because of the way their DNA strands are linked together). If that were not enough, we have only been talking about the forward flow in one supply chain. Most companies will have multiple supply chains to smooth. Developing well-defined supply chains for all products or services is probably impractical, if not impossible. Consequently, a company must choose which supply chains will have priority. In a sense, they will decide which supply chains will be critical to their success.

Up to this point, we have only described supply chains moving toward the customer. In the next section, we will describe the added complexity of the reverse logistics concept.

The Concept of Reverse Logistics

A reverse logistics supply chain is defined in the *APICS Dictionary* as "a complete supply chain dedicated to the reverse flow of products and materials for the purpose of returns, repair, remanufacture, and/or recycling."[12] Like forward supply chains, few companies are very far along in the implementation of their reverse logistics programs. If building the reverse supply chain is as difficult as the forward supply chain, and most writers suggest this is true, it represents another challenge for most businesses.

A distinction should be made between "reverse logistics" and "green supply chains." Reverse logistics refers to the process for handling returns and how best to reintroduce the returns, as whole units or component parts, back into the forward supply chain, through repackaging, remanufacturing, or reprocessing. The intent is to avoid costs or, in more forward-looking companies, generate added revenue. A green supply chain implies a concern with the eventual disposition of the materials in the finished products and the effect on the environment. Although both elements may be part of the same system, they have different drivers.

Why the Interest in Reverse Logistics?

One of the major trends in warehousing in the 21st century will be reverse logistics. Brockman identified key drivers of this movement including increasing regulatory pressure to recycle and reduce landfill waste, the growing focus on customer relations and the potential for lowering costs by reusing packaging and reclaiming unsold or damaged product drives this trend.[13] Another trend is the increase in returns, especially at the retail level. More liberal returns policies and the rapid rise of e-commerce contribute to these increases. In light of these increases, retail businesses will need to develop more formal processes to handle them. In addition, they, in turn, will seek to return some of their product to their suppliers, the manufacturers. What results is that the product moves backward along the supply chain, in other words, in reverse. The interest in sustainable development practices is a strong trend that encourages reverse logistics.

In his fact-filled book, Blumberg[14] lists scenarios where products will need to be recovered and pushed back into the supply chain:

- Products in the field that have failed and need to be repaired or properly disposed of
- Parts and subassemblies of products that can be reused, either because they are perfectly good (no trouble found), or that can be repaired or reworked
- Products that are perfectly good, but nevertheless have been returned by the purchaser as well as products sitting on retailers' shelves that have not been sold
- Products and materials that have been recalled or are obsolete, but still have a useful life

- Products, materials, and goods that have been thrown away, but can be recycled and reused
- Products at the end of a lease, but not at the end of life.

Blumberg points out that reverse logistics is essentially the responsibility of the original manufacturer of the product. In this situation, a closed loop supply chain (CLSC) exists because the manufacturer both makes and recovers the produced product.

In addition to the economic implications, public awareness of environmental concerns is motivating governments to enact legislation mandating companies to meet stringent requirements in the handling of hazardous materials presently used in products, such as automobiles and electronic products. The European Union (EU) has enacted two major pieces of legislation — the Waste Electrical and Electronic Equipment (WEEE) Directive and the Restriction on Hazardous Substances (RoHS) Directive. The WEEE Directive applies to a huge spectrum of electronic and electrical products. It sets criteria for the collection, treatment, recycling, and recovery of waste equipment and makes producers responsible for financing most of these activities.

The RoHS Directive bans placing on the EU market new electrical and electronic equipment containing more than agreed-upon levels of lead, cadmium, mercury, hexavalent chromium, polybrominated biphenyl (PBB), and polybrominated diphenyl ether (PBDE) flame retardants, effective July 1, 2006. It is the responsibility of the manufacturers to ensure that their products and their components comply.

One of the significant requirements of both of these laws is the supplier of the product (usually the manufacturer) has responsibility for the eventual disposition of the product. Although U.S. manufacturers may avoid this requirement over the short run, it appears likely they will have to meet this requirement at some point in the future. Efforts launched collaboratively by the Environmental Protection Agency, the U.S. Commerce Department, and at the industry level, seek to lessen the environmental impact of small and midsize manufacturing suppliers while boosting efficiency, productivity, and profitability. Called the Green Supply Network, this program offers assessment reviews and advice in implementing "Lean & Clean" methodologies at all levels of the manufacturing supply chain.[15]

Benefits of Reverse Logistics

The most direct benefit from reverse logistics is cost avoidance. Ironically, upper management is often the main obstruction to achieving that benefit because they do not understand the significant costs that returns represent to the company. Even the recycling of buildings is considered in the reverse logistics process. The Pillowtex complex of buildings in Kannapolis, North Carolina, is being demolished to make way for a research campus. The wrecking company will recycle 10 million bricks (enough for 625 brick houses), 10,000 tons of steel (enough for 6,667 sedans), and 5,000 pine beams (enough to floor 500 living rooms) during the recovery phase of this project.[16]

There are also environmental benefits to reverse logistics. The private sector will move to recycling, both for profit and for public relations purposes. Dave Rosenfield of Roman Industries, Inc., comments, "I have no doubt that every industrial product will be subject to recycling and remanufacturing. At some level, state or federal, guidelines will be established to influence people to do it. The environmental political capital achieved by pressing on recycling issues can't be ignored. Politically it's hot to support recycling."[17]

Barriers to Reverse Logistics Implementation

As with all programs, there are barriers to their successful implementation. The Reverse Logistics Executive Council, found the following major barriers:

- Perceptions that reverse logistics is relatively unimportant
- Lack of company policies that support reverse logistics
- Lack of systems that can implement reverse logistics
- Competitive issues that make reverse logistics cost prohibitive
- Management inattention
- Lack of personnel resources to carry out reverse logistics
- Lack of financial resources
- Legal issues[18]

In a more recent study, Andel[19] found similar barriers:

- No executive overseer or champion who takes responsibility for reverse logistics and drives it
- The attitude that reverse logistics is "just a cost"
- The potential of missing the customer satisfaction and service differentiation opportunities
- An inability to quantify cost of returns
- Lack of systems/information visibility
- A silo versus cross-functional mentality
- Poor manual processes

Perhaps the greatest barrier is in thinking that returns can be handled on an ad hoc basis instead of using a systems approach.

System Design and Implementation

Figure 3-11 depicts a typical system of both forward and reverse supply chains. The top row depicts the forward stream of raw materials to the final product for the consumer. The second row illustrates typical steps involved in the backward stream from consumer to final disposal.

To get the most out of a reverse logistics channel, excellent information management resources are necessary. The emerging importance of reverse logistics requires the execution and management of new inter- and intra-organizational processes. Given the complexity and uncertainty of these processes, information and communications technology (ICT) support is necessary, perhaps even more so than in traditional (forward) supply chain management. Moreover, the handling of reusable materials requires the involvement of new chain partners, such as material recovery organizations and e-marketplace intermediaries.

ICT vendors have not yet given high priority to incorporate reverse logistics in their systems. The importance of managing reverse flows in the supply chain, and Internet-based technologies has created new opportunities. Besides the development of new design, planning, and control components in existing ERP and APS systems, electronic marketplaces can be developed to manage return flows in the supply chain effectively.[20]

A recent study looked at the timing of reverse logistics programs (early or late adoption) and the commitment of resources. The research team found that early entrants have an advantage over late entrants because it is difficult for companies to catch up, although late entrants can somewhat offset their delay by committing additional resources to the effort. They recommend that:

- Organizations that have not yet started a formal reverse logistics program should do so.

Figure 3-11 Forward and reverse logistics supply chains.

- Organizations that have implemented reverse logistics need to be sure resource commitments are adequate to gain the potential benefits.
- Organizations that are unable to commit sufficient resources to reverse logistics should consider outsourcing.[21]

Michigan State University professors, Diane Mollenkopf and David Closs provide their insights on the importance of reverse logistics in Managerial Comment 3-4.

> ### Managerial Comment 3-4.
> ### Reverse Logistics Update
>
> "Reverse logistics has often been viewed as the unwanted stepchild of supply chain management. It has been seen as a necessary cost of business, a regulatory compliance issue, or a 'green' initiative. But more companies are now seeing reverse logistics as a strategic activity — one that can enhance supply chain competitiveness over the long term. To understand how reverse logistics can create value, it is necessary to understand both the marketing and logistics components of this process. Companies can use the returns process to enhance marketing efforts by analyzing reasons for returns and conducting ongoing defect analysis. In many cases, the returns management process had been ad hoc, meaning that the strategies, operations, and guidelines were not well defined or thought out. However, all the organizations surveyed have quickly seen compelling benefits, to the point where they have all opted to build reverse logistics capabilities into their overall logistics and supply chain strategies and processes."[22]

Summary

This chapter introduced the basic Input–Transformation–Output (ITO) Model. We showed how to expand the basic ITO Model into more complex models through linking individual ITO models, both in a series and in parallel. We described open and closed systems, recognizing that most businesses operate in the open system mode. Finally, we showed how the basic ITO module could be linked together into an integrated supply chain, although with considerable difficulty. We also showed that supply chains are like automobiles; companies need both forward and reverse capabilities.

In Chapter 4, we will show how a business performs a variety of transformation processes. In addition to the normal transformation processes of running the day-to-day business, they also have to solve problems and make continuous improvements. All of these processes require using the ITO process.

References

1. Starr, M.K., *Systems Management of Operations,* Prentice-Hall, Inc., Englewood Cliffs, NJ, 1971.
2. Melnyk, S.A. and Denzler, D.R., *Operations Management, A Value-Driven Approach,* Irwin, Chicago, 1996.
3. Schoderbek, P.P., Schoderbek, C.G., and Kefalas, A.G., *Management Systems Conceptual Considerations* (Fourth Edition), BPI Irwin, Homewood, IL, 1990.
4. Schoderbek, P.P., Schoderbek, C.G., and Kefalas, A.G., *Management Systems Conceptual Considerations* (Fourth Edition), BPI Irwin, Homewood, IL, 1990.
5. Thompson, J.D., *Organizations in Action,* McGraw-Hill Book Company, New York, 1967.
6. Thompson, J.D., *Organizations in Action,* McGraw-Hill Book Company, New York, 1967, pp. 19–23.
7. Simon, H.A., *Administrative Behavior, A Study of Decision-Making Processes in Administrative Organization* (Fifth Edition), The Free Press, New York, 1997.
8. Thompson, J. D., *Organizations in Action,* McGraw-Hill Book Company, New York, 1967.
9. Perrow, C., A framework for the comparative analysis of organizations, *American Sociological Review,* 32, 194, 1967.
10. Thompson, J. D., *Organizations in Action,* McGraw-Hill Book Company, New York, 1967.
11. Mentzer, J.T., Achieving competitive advantage through supply chain management, www.industryweek.com/PrintArticle.aspx?ArticleID=13355, January 10, 2007.
12. Blackstone, J.F. III and Cox, J.F., *APICS Dictionary* (Eleventh Edition), 2005, APICS – The Educational Society for Resource Management.
13. Brockmann, T., *21 Warehousing Trends in the 21st Century,* IIE Solutions, 31, 7, 36, 1999.
14. Blumberg, D.F., *Introduction to Management of Reverse Logistics and Closed Loop Supply Chain Processes,* CRC Press, Boca Raton, FL, 2005.
15. Anonymous, Lean and clean auto industry pilot expands to other verticals, *Manufacturing Business Technology,* 23, 2, 11, 2005.
16. Bell, A., Rubble full of riches, *The Charlotte Observer,* Sunday, February 12, 2006, p. B1.
17. Andel, T., Rethinking recycling, *Material Handling Management,* 59, 5, 35, 2004.
18. Caldwell, B., Reverse logistics, *Information Week,* Apr 12, 1999, issue 729, pp. 48–52.
19. Andel, T., Rethinking recycling, *Material Handling Management,* 59, 5, 35, 2004.

20. Von Hillegersberg, J., Zuidwijk, R., and van Nunen, J., Supporting return flows in the supply chain, *Communications of the ACM,* 44, 6, 74, 2001.
21. Richey, R.G., Chen, H., Genchev, S.E. and Daugherty, P.J., Developing effective reverse logistics programs, *Industrial Marketing Management,* 34, 8, 830, 2005.
22. Mollenkopf, D.A. and Closs, D.J., The hidden value in reverse logistics, *Supply Chain Management Review,* 9, 5, 34, 2005.

Chapter 4

The Role of Management Programs in Continuous Improvement

John moved along to the next stage in the long corridor. First he saw a complex of houses. In front of one house, he saw a truck with a sign that said, "Home improvement programs — built to fit." He remembered that he needed to put new windows in his house. Most likely, this program was about windows, he reasoned.

John approached one of the workers and asked, "What are you doing?" The worker replied, "We are installing improvement programs in these houses. I am working on one that will cover the floors; we call it a cost reduction program. Tom is working on one that will make the walls an attractive color; he calls his a quality improvement program. Susan has a really good program for window dressings; she calls hers response time reduction. Alex has one of our newest programs that can be used in just about any room in the house. It is called process flexibility. Our company has a number of other programs, but that should give you some idea of what we do. Here is a pamphlet called 'Improvement Programs' that explains more about each type of program. You can build your own program, but you need to be sure that you cut and fit it to work in your house. Too many people don't understand that different programs have different uses; they just put them into their house without adapting them to their needs." "Hmmmm," John thought, "this is more than just about putting windows in a house."

John thanked the worker and sat down on the nearest bench to study the pamphlet. That had been a surreal experience. Those were certainly strange names to call improvements to a house. They almost seemed to be a better fit in a business setting. Why, these names sounded almost the same as some he had heard at some of his professional society meetings. The workers seemed to be busy; there must be a lot of demand for these types of programs.

John looked at the home improvement contractor's truck. On the side was the name of the contractor, the address, phone number, and even a Website address. That was a good idea, thought John. Putting the Website address on the truck meant the contractor was sort of up to date on his technology. John was proud of the fact that he had created his own Website using his home PC and some software he had purchased. Still, his Website was not as impressive as some of the others he had seen on the Internet.

As he looked at the side of the truck again, he saw a list of items underneath the address information. These were the services the contractor provided, but they looked different from other contractors' trucks he had seen. The list did not say things like, "new construction, additions, remodeling, roof repair, or custom woodworking." Instead, it looked something like this:

- *Cost reduction programs*
- *Quality improvement programs*
- *Response time reduction programs*
- *Process flexibility programs*

This was unlike any home improvement contractor vehicle he had ever seen. John reasoned that he would have to study the pamphlet carefully.

In Chapter 2, we described how critical success factors (CSFs) help a business in its strategic planning process and, as a result, enable the company to become more effective by doing the right things. In this chapter, we will show how the selection of continuous improvement programs will help a business become more efficient by doing things right.

We will examine several questions:

- What are management programs? How do they differ from normal, everyday operations?

- What is the life cycle of a management program? What happens to management programs at the end of the life cycle?
- Why are management programs important? What are their benefits? What are the obstacles to their implementation?
- Where do management programs come from? Who invents them? For what purpose? Are they original or adaptations from some other management concepts?
- Why are some programs successful and some not? Although there have been a number of well-publicized successes, many companies have achieved limited success.
- What does the future hold? Will management programs continue to be an active and worthwhile source of ideas to stimulate improvement in businesses?

By addressing these questions, we expect to provide managers with a better understanding of how to use management programs to meet their companies' needs.

What Are Management Programs?

In most entities, management is concerned with several different kinds of activities. On one hand, there is the normal, everyday operation of the business. However, in this book, we will not be trying to tell you how to run the normal activities of your business. Instead, we will describe how the use of management programs — those activities outside the course of your normal business — can play a vital role in keeping your business prosperous.

Management programs are those coordinated activities in a business that are usually assigned a name to distinguish them from the normal operations of a business. Often they are known by an acronym, such as Just-in-Time (JIT) or Total Quality Management (TQM) programs. In most cases, they are concentrated efforts to improve some part of the business operation, such as to reduce costs, improve quality, shorten customer response time, or to provide greater customization of product or service. They may involve a part or all of an organization. Usually, they are of a project nature, with a beginning, a life cycle, and an end. Table 4-1 lists examples of management programs.

Management programs usually originate as an attempt to introduce improvement into a business. They may be original for a particular company or they may be an adaptation of an existing program that has become popular, or at least reasonably successful, in another company. For example, Toyota started a program to reduce inventory and improve cash flow by revamping their production system. This program was first known as the Toyota Production System (TPS)

Table 4-1 Examples of Management Programs

The following list is representative of the management programs that have been developed over the past two decades. They are grouped by their primary focus, or objective.

Process Improvement
- Just-in-Time (JIT)
- Lean production or manufacturing
- Business process re-engineering (BPR)
- Agile manufacturing
- Computer-integrated manufacturing (CIM)
- Theory of constraints (TOC)
- Mass customization

Production Planning
- Materials requirements planning (MRP)
- Manufacturing resources planning (MRP II)
- Enterprise resources planning (ERP)
- Warehouse management system (WMS)
- Manufacturing execution system (MES)
- Advanced planning and scheduling (APS)
- Sales and operations planning (S&OP)

Quality Improvement
- Statistical process control (SPC)
- Total quality control (TQC)
- Total quality management (TQM)
- Quality function deployment (QFD)
- Six Sigma

Supply Chain Management
- Quick response system (QRS)
- Efficient consumer response (ECR)
- Vendor managed inventory (VMI)
- Collaborative planning forecasting and replenishment (CPFR)
- Supply chain management (SCM)

Performance Measurement
- Activity based costing (ABC)
- Activity based management (ABM)
- Balanced scorecard (BSC)

and then by a variety of names, such as stockless production or zero inventories. Eventually, this program achieved widespread acceptance and became known as the JIT system.

Most programs begin small to address a specific need. If successful, they usually expand into a much broader program to the point that it becomes a management philosophy. For example, JIT started as a kanban scheduling system to reduce in-process inventory. Later, it broadened to a waste reduction program. From that, it expanded to a continuous improvement system and, in recent years, has blossomed into a management philosophy whose primary mission is waste elimination. Even with all of this growth, lean production replaced JIT in recent years as the lean concept has grown in scope and magnitude since its inception in the early 1990s.

Where do management programs fit in with running a business? Let us look first at the "other" kinds of activities in a business. Figure 4-1 shows three major types of transformation activities for a business — normal (or sustaining) day-to-day operations, problem-solving activities, and improvement programs. The middle section of the diagram illustrates the normal, or sustaining, operations in which the business transforms inputs into goods or services for sale. This depiction is the concept introduced in Chapter 3 as the ITO Model. The bottom portion shows that a business must correct problems that occur or, in the ITO vernacular, transform problems into solved problems. The top portion illustrates the concept of transforming ideas into improvement actions.

Normal or Sustaining Day-to-Day Operations

This is not a book about the normal operation of a business. It is an essential part of a business; however, each industry is different, and each company within an industry is different. If you are a member of a successful business, you already know a

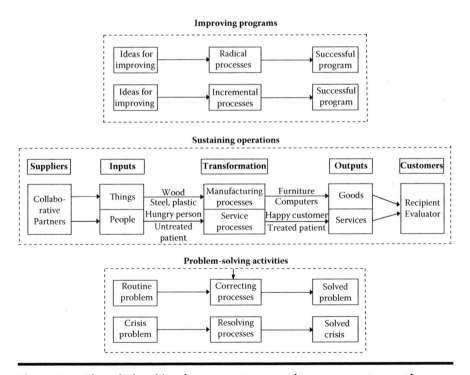

Figure 4-1 The relationship of programs to normal management operations.

great deal about how to run the normal operations. Therefore, we will be spending the bulk of this book in describing the other two types of business activities — solving problems and making improvements. They are equally important to the normal operations, but require a somewhat different approach (see Figure 4-1).

Problem-Solving Activities

Let us look at the problem-solving part of the diagram first. There are two problem-solving activities. Ideally, all of a company's problems could fall into a category of everyday types of problems — machine breakdowns, late orders, unsatisfied customers, and product defects. Although some of these can be serious, their solutions are within the capability of the existing employees, facilities, and infrastructure of the organization. Companies typically use their own resources, such as employees and facilities, to solve these types of problems. Management programs, such as those listed in Table 4-1, are not normally used to address this category of normal operating problems.

However, there is another type of problem that can destroy a company. A crisis is a situation that develops unexpectedly; most companies do not have the experience or in-house capability to handle these types of problems with ease. Crises do not occur often, but they are high impact events that can seriously hamper the smooth running of the organization. It can be an oil spill from an ocean-going tanker, food poisoning of multiple customers, tampered medications, or accounting scandals. When handled promptly and properly, the company may actually prosper, such as in the case of Johnson & Johnson's Tylenol cyanide incident in the 1980s. In this example, Johnson & Johnson received positive publicity for the effective management of this crisis. Consequently, sales of the product eventually rebounded. If neglected or denied, the company may be destroyed, such as in the case of Arthur Andersen and Enron. This is a special type of management activity and is not considered a program in the context of this chapter. There are a number of practitioner books on this subject, such as Steven Fink's *Crisis Management: Planning for the Inevitable*,[1] and Timothy Coomb's work, *Code Red in the Boardroom: Crisis Management as Organizational DNA*.[2]

Improvement Programs

This book is primarily about management programs that lie outside the realm of normal operations and crisis-related activities. We will focus on those management programs that address improving some aspect of the business. As Figure 4-1 illustrates, such programs can be classified as incremental or radical. Although both types of programs are in popular use, most of the programs probably would fit best in the incremental or continuous improvement category. The use of TQM is an example of an incremental improvement program. These programs enable a company to improve by implementing small-step interventions alongside their normal operations.

However, some programs are major, or radical, and are capable of transforming a company in substantial ways. They are often disruptive and usually require changes in the normal way of doing business. Such examples include Business Process Re-engineering (BPR) or lean production. When they are effective, they can save a failing company or make a winner out of a mediocre company. These types of programs also fall within our classification of improvement programs. Childe, Maull, and Bennett make the following distinction between continuous and radical change programs: "TQM-led programmes are usually associated with continuous improvement efforts and not with the radical change brought about by re-engineering. TQM may thus represent a necessary but insufficient condition for a successful BPR programme."[3] Another comparison of incremental versus radical change is described in the following excerpt from a book by Richard Luekue[4]:

Managerial Comment 4-1. Discontinuous versus Continuous Incremental Change

The company, having done its thing for many years, suddenly throws the cards in the air, everyone gets involved in reform and improvement, and then it's over. We call this "discontinuous change" — a single, abrupt shift from the past. The momentum of the organization is shifted, hopefully to a higher level of performance or in a more promising direction.

But the benefits of a successful single fix don't last forever. Change initiatives that accomplish stated goals often lead to complacency in senior management. Units that developed market-beating products and services during the change gradually shift their attention from innovation to defending their turf. Employees settle into routines and become more inward-looking. On top of all this, with each passing day the environment of competition and technology is altered. This combination of complacency, defensive behavior, routines, inward focus, and ever-evolving competition is the enemy of progress; ultimately it creates a situation in which major reform is needed once again.

If change programs are eventually followed by periods of organizational complacency and stasis — as described above — the alternative situation is one in which the organization and its people continually sense and respond to the external environment. Their radar is attuned to signals of change from customers, markets, competitors, and technologists. And they respond in appropriate ways. Simultaneously, they monitor internal activities to assure continuous improvement in key processes. Open communication assures that new ideas have a forum in which they are heard and objectively evaluated. Change is ongoing and takes place through many small steps — that is, through continuous incremental change.

On the surface, the advantages of continuous incremental change are many: small changes are easier to manage, and small changes enjoy greater probability of success than big ones.

Disruption is short-term and confined to small units at any given time the organization and its people are kept in a constant state of competitiveness and change-readiness.[4]

We can also show how to restructure the normal transformation process of a business through problem solving and the improving process. Figure 4-2 illustrates how a solved problem or an incremental improvement can lead to a restored or improved transformation process. A further extension to a restructured transformation process could result from resolving a crisis or implementing radical improvements. These cycles could repeat as a company continues to adjust to its changing environment.

In Figure 4-2, we start with the normal transformation process, converting inputs into outputs, shown in the diagram as Condition I. The company begins an incremental improvement program, shown as Activity B. The program is successful and, when completed, the company assimilates the improvement into the normal operation of the business, shown as Activity D. On the other hand, suppose a problem develops, as depicted by Activity A. It could be a customer's order is late being shipped or the food is served at the wrong temperature in a restaurant. The problem-solving process is used to correct the problem. As a result of solving the problem, the company discovers a way to improve the process. This link is shown as Activity E in the figure. In the case of the late order, management determines it can implement an early warning system to prevent late orders. By working the idea through the Incremental Improvement process, the problem can actually lead to an improvement. Even without Activity E, if the problem-solving process works as expected, the situation returns to a restored normal operation, shown as Condition

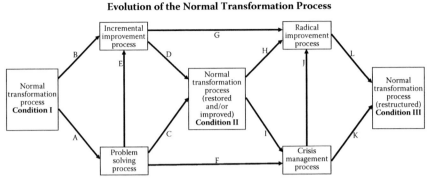

A	An everyday problem develops
B	A program of incremental improvement is implemented
C	The problem is solved and normal operations are restored
D	The improvement program improves the normal transformation process
E	The problem solving process feeds ideas to the incremental improvement process
F	The known problem is not solved and a crisis develops
G	The incremental process triggers a radical improvement program
H	A program of radical improvement is implemented
I	A crisis develops from an unexpected source
J	The crisis stimulates the need for a radical improvement program
K	The crisis is resolved and normal operations are restored
L	The radical improvement program restructures the normal transformation process

Figure 4-2   Transitions in the normal transformation process of a company.

II. If the incremental improvement process works as designed, the improvement is implemented into the improved transformation process through Activity D.

The restored normal transformation process — Condition II — may find that competitive conditions require change. As a result, management may decide on the need for a radical improvement, such as the decision to completely restructure their production processes from a batch flow to a line flow. Activity H provides this link. The idea for a radical improvement could also have originated during an incremental improvement process, as shown by Activity G. As a result of the radical improvement process, the normal transformation process is restored, through Activity L, into a restructured process, shown as Condition III.

However, suppose that a number of customers at a restaurant eat improperly prepared food and incur food poisoning. The small problem becomes a crisis, as shown as Activity F. Now the crisis management process takes over to restore the normal transformation process through Activity K to Condition III. The crisis could have arisen from the normal transformation process, shown as Activity I, necessitating the activation of the crisis management process. Crisis management seeks to address problems that disrupt the normal operational flow of the business. Bringing the business back to normal and seeking to prevent future occurrences of the event may result in a transformation of some aspect of operations.

Figure 4-2 should not suggest that there is a rigid, sequential process for correcting problems or introducing change. Most businesses do not operate in such a controlled environment. The diagram should reinforce the idea proposed in Figure 4-1 that there are three streams of transformation activities in most businesses: (1) normal transformation processes, (2) problem-solving processes, and (3) improving processes. Our emphasis in this book is to describe how improving processes are a necessary part of business operations.

Management Program Life Cycles

Perhaps the most common point that researchers agree on is that management programs follow a life cycle. Scholars use bibliometric data — number of articles written about a particular program — to study the shape of these program life cycles. Most research indicates a bell shape as the most common cycle form.[5,6] However, Ponzi and Koenig have indicated an S-shape is also a possible pattern.[7] Carson and associates acknowledge that shapes will vary in terms of slope rates because of the existence of other management fashions on the market that may impact the particular fashion under study.[8] However, a shape of some kind is plausible, most likely one that resembles a bell curve.

Life-Cycle Stages

In addition to the shape, the stages of the life cycle are also of interest. Abrahamson wrote extensively about management fads and fashions.[9] The widely cited model by Ettorre shows fashions progressing through a five-stage life cycle:

1. Discovery: "A buzzword is born."
2. Wild acceptance: "The idea catches fire."
3. Digestion: "The concept is subject to criticism."
4. Disillusionment: "The idea does not solve all problems."
5. Hard core: "Only true believers remain."[10]

The Gartner Research Group developed a "hype cycle" with the following phases: Technology Trigger (beginning), Peak of Inflated Expectations (growth), Trough of Disillusionment (decline), Slope of Enlightenment (revival), and Plateau of Productivity (sustained level).[11]

In Table 4-2, we compare the stages of program life cycles for publication and for implementation in businesses. The length of the life cycle varies. In the management research literature, programs with short life cycles are deemed fads and the more durable programs are considered fashions.

What, or who, are the drivers of a program's life cycle? Many of the popular management programs originate as a focused effort within a company, such as JIT

Table 4-2 Publication and Management Program Life Cycles

Publication Life Cycle	Management Program Life-Cycle Phase
• Start-up: The initial articles published on a program will reflect the birth or start-up of that program. Usually, trade publications will dominate the articles first published.	• Recognize a need for the particular program. • Select and develop the program. • Gain acceptance of the program with employees and other key stakeholders. Implement pilot projects.
• Growth: The number of articles on that management program increases. On a graph, this would be represented by a positively sloping curve. While trade publications may still dominate, scholarly journal articles are beginning to appear.	• Implement the program on a larger scale. • Analyze and adapt the program to other industries. For example, JIT moves into the service industries.
• Maturity: The management program is written about in a number of publications. Although there is no longer a growth phase of articles, practitioner and scholarly interest remains high.	• Capitalize on the program and bring in subsequent versions or technologies related to that program.
• Decline: The number of articles related to that program decreases substantially. Often, there is a more significant decrease in trade publication articles than in scholarly journal articles.	• Rationalize or refine the program. The program may be abandoned and replaced by a different program. Or, the program may be so entrenched in the management philosophy and operations that it is no longer necessary to keep reading new articles on that program.
• Reorientation: The program spins off related programs and applications.	• Extend scope of the management program into new applications.

at Toyota or Six Sigma at Motorola. The program may be designed internally or with the aid of a consultant. Often, consultants package the program as an addition to their product line and promote the program to other potential clients. In the early stages of a program, consultants and trade publications are often the primary sources of information about the program. A typical way for other practitioners to find out about the program is to attend conferences or workshops offered by consultants or trade associations. In the early stages, the reports about the program are usually positive and describe the benefits of implementing such a program. As time goes on, business researchers begin to study the program and view it with greater objectivity. They begin to analyze the program elements and identify the major causes of success or failure. They often compile survey information that summarizes the actual results achieved, often reflecting a range of results, from high success to low success or even failures of the program. At this stage, many companies that are considering the program may become reluctant to even try the program, even though it still has many enthusiastic supporters. The keys to success may be in selecting the correct program and then implementing it correctly. We will discuss both of these decision areas in later sections of this chapter.

At the End of the Life Cycle

What happens to programs at the end of their life cycle? Those considered as fads fade away quickly, although some remnants of the program may remain as traces of something worthwhile. Other programs may also appear to fade away because the number of articles written about them diminishes. This may mean that the program was not of value and companies discontinued its use. However, it is also possible that the program did have value, but had lost its premier status because of the appearance of new programs.

Many programs morph into a new program, such as JIT being succeeded by lean or TQM by Six Sigma. This morphing shows up in the changes in the number of publications about the programs. Figure 4-3 shows this transition for JIT and lean, and Figure 4-4 illustrates the same transition for TQM and Six Sigma.[12]

Some companies assimilate management programs into their normal practices. Although they may not have a definite identity as originating in a specific program, basic elements of the program remain as standard practice. For example, some companies may introduce self-directed work teams as part of either a TQM or a JIT program and continue the use of those teams even after discontinuing the formal program.

Implications of Program Life Cycles for Management

What are the implications for managers? Why should they care about program life cycles? Several reasons seem most evident:

Figure 4-3   Number of articles for JIT and lean production. (*Source:* **Crandall, R.E. and Crandall, W.R. 2007.[12]**)

Figure 4-4   Number of articles for TQM and Six Sigma. (*Source:* **Crandall, R.E. and Crandall, W.R. 2007.[12]**)

- It is beneficial to know where a program is in its life cycle. Is it just getting started? If so, many of the articles will address the positive attributes of a program with little about the potential problems or hidden surprises. At this stage, companies should view the program optimistically, but with caution. As the program develops, more insightful articles will appear and help make an evaluation of the program more rational. As the end of the life cycle nears, management may replace the program with an even more appropriate program.
- As programs develop along their life cycle and mature, successful implementation and further research make them more beneficial. A company can successfully implement a program early in its life cycle; however, it should continue to look for ways to improve that program. Often, improvement ideas are addressed in the more thoughtful articles that emerge later in the publication life cycle of a program.
- New programs will replace old programs. Just as lean production may succeed JIT, or Six Sigma follows TQM, most programs spawn new programs that are designed to correct problems or capture opportunities not addressed in the original program design. Although it is important to be progressive, it is also important to recognize that true differences between the old program and the new one will exist. What are the incremental differences? What are the incremental benefits? What are the incremental costs or inherent problems?

Implementing a program is a major project. It requires a company to commit employees and other resources to the effort. There is a great deal of information available about the most popular management programs, from the very beginning of the program's life cycle. Companies need not enter into a program uninformed.

Companies need management programs to help them improve. It is difficult to operate in a normal fashion and make improvements at the same time; therefore, many companies find it useful to mount special efforts — programs — to focus attention on their improvement efforts.

Why are Management Programs Important?

Why do companies like to use programs to implement improvement efforts? Is it because they are "driven" to use them by developers and promoters, such as consultants or major customers, or do managers "pull" programs because of their needs?

As described earlier, most businesses deal with three types of activities: (1) normal day-to-day operations, (2) correcting problems (and even crises) that arise, and (3) making improvements in their operations. Joseph Juran, a noted author and scholar in the quality improvement movement, presents an interesting insight into the functions of a business. He uses the terms "control" and "breakthrough." Control represents the normal operation of a business when things are running smoothly, or in control. If the process goes out of control, it must be fixed, or a problem solved. However, Juran warns that control implies maintaining the status quo, and that is never enough to satisfy the competitive pressures on a business. Instead of being satisfied with control, a business should be looking for breakthroughs, or improvements, in the way it does business. His book, *Managerial Breakthrough*, contains ideas about how to achieve breakthroughs in a business.[13]

The day-to-day operations represent the control part of the business in which managers perform the required functions of the business in a relatively stable environment using accepted practices without major variances in either the demand or supply side of the business. An example of this could be a retail store that serves customers with only normal variation among the customers and the services they require.

If there is a problem such that normal practices are insufficient to handle it, and the process goes out of control, there is a need to exercise the problem-solving elements in the business. Using a retail store again as an example, this scenario could be illustrated by running out of a heavily advertised sales item, a leak in the roof that is damaging merchandise, or a customer that slips and falls while shopping. Resolution of these problems is essential to the continued prosperity of the store. The actions required to solve a problem — restore the process to a controlled situation — are somewhat or even dramatically different from the normal operation itself.

However, there is another type of management problem that does not lend itself to normal, every-day problem-solving methods. This type of problem occurs when a company encounters a crisis. The crisis may be operational in nature, such as when an industrial accident occurs and an employee is severely injured or killed. It could also be strategic, as when a key executive, such as the CEO, dies unexpectedly. The crisis could be public relations oriented, such as when a product malfunctions and a subsequent recall is issued. In some crises, it may be necessary to call in a consultant with crisis management expertise.

Finally, there is the type of problem where a business needs to improve. Most businesses are fighting a constant battle to remain competitive, and to accomplish that, they have to keep improving. This is where management programs enter the picture. These programs can help a business to improve.

Where Do Management Programs Come From?

Are management programs original or do they have earlier origins in management practice? McMahon and Carr offer a growing concern that students today are reading less from the early writers and more about what today's scholars are attributing to the early writers. They suggest this may lead to a distortion of the original intentions, especially if the current writings have been assigned a different label.

They recommend a return to reading the original writings as a part of the business educational process.[14]

In line with the feelings expressed by McMahon and Carr, we will show that some of today's most popular management programs had their origin in earlier management theories. There are a few basic management theories that have evolved over the past century as separate streams of research and application. Businesses have learned to apply these theories into management programs. This section describes the process of packaging management theories into effective improvement programs. Let us look first at some of the foundation management theories.

Overview of Employee Management Theories

Prior to the end of the 19th century, job design and the management of workers was somewhat informal and left to the individual employee and the "boss" to figure out. Little attention was paid to how the work was accomplished; only that it was done. At the end of the 19th century and the beginning of the 20th century, a body of management knowledge began to emerge that sought ways to improve productivity. Henry Towne, of Yale and Towne, is usually credited with moving the study of management to the forefront in his presentation "The Engineer and the Economist," given to the American Society of Mechanical Engineers (ASME) in 1886.[15]

Scientific Management

Frederick W. Taylor is often referred to as the father of scientific management because of his research in work methods studies. His approach was based on the theory that any job can be improved by breaking it down to its basic elements and finding ways to improve work methods. However, Taylor was not alone in his quest for productivity improvements. Henry Gantt developed management principles and procedures that began to take a humanistic approach to management. Frank and Lillian Gilbreth studied methods and motion techniques. Harrington Emerson championed efficient operations as well as advocating a bonus pay plan for employees.

Taylor saw their work and attempts to develop basic principles as a scientific approach to management. This approach later evolved into a philosophy of work productivity known as "scientific management." Eventually, these principles became the foundation for the field of industrial engineering.

Taylor's philosophy soon led to job specialization, a job with a narrow range of tasks that have high repeatability, resulting in greater efficiency. Jobs designed in this manner are simple, require little training, and are easy to measure and manage. Indeed, the scientific management approach made possible high-speed, low-cost production that plays a great part in the standard of living we enjoy today. Conversely, job specialization carried to the extreme can have significant adverse effects on employees, such as absenteeism, lack of motivation, and employee turnover. These in turn may cause serious problems in other parts of the production system.

Deciding on the degree of specialization is one of management's toughest decisions. Even though high specialization has resulted in tremendous productivity gains for many companies, there are still widespread opinions regarding the optimum degree of specialization. In service operations, work content seems to be less specialized using broader job designs.

Human Relations Management

Because of the adverse effects of job specialization, some behavioralists and managers began to look for alternative ways to design and manage work. In the 1950s Eric Trist's studies in Great Britain's coal fields led to his theory that work has both technical elements and behavioral elements. This dichotomy implies that management should seek a balance between the technical and behavioral aspects when designing jobs. Analysts should study the entire work system, not just individual tasks. Most writers believe that the Hawthorne studies at Western Electric provided the framework for the human relations movement. Although the results of the Hawthorne studies have been debated for decades, one area that seems to represent a consensus is that job design is an important element of a business operation.

Over time, more humanistic job design techniques evolved — job enlargement, job rotation, and job enrichment. Job enlargement gives an employee more tasks to perform, thereby increasing the need for more skills and presumably reducing boredom and increasing job satisfaction. For example, instead of just assembling a product, the employee may also perform preventive maintenance on the work equipment.

Job rotation allows employees to exchange jobs with fellow employees from time to time, usually on a predetermined schedule. This practice provides some diversity for the employee, enlarges job skills, and makes the employee more valuable to the company. The benefit for management is a more highly skilled, flexible workforce. An example of job rotation involves rotating bank employees between working as tellers and as loan processors. Restaurants have long recognized the need for job rotation as a training practice for aspiring managers. Typically, managers in training will learn to cook, wait tables, wash dishes, and perform other functions relevant to the restaurant.

Whereas job enlargement is the horizontal expansion of a job, job enrichment expands an employee's tasks vertically into aspects of managerial functions. Job enrichment not only expands tasks upward, but also expands responsibility. It is the most comprehensive of the humanistic approaches to job design, and embodies the three factors that Frederick Herzberg's research indicates enhances job satisfaction: achievement, recognition, and responsibility.[16]

In modern management, aspects of the human relations approach are well entrenched in the manufacturing of products and the providing of services. Regardless of the job design approach used, operations managers retain their responsibility to increase their company's competitive advantages. Business publications and literature suggest that top-performing companies organize to meet the needs of their people; consequently, they attract better applicants, and their employees are self-motivated to do an excellent job. Chapter 9 will describe how improvement programs depend on the successful involvement of employees and consideration for the organization's culture.

Administrative Management

Scientific management started with a shop-floor emphasis and gradually expanded over the years into a more general management theory. The administrative management stream of research was more of a top-down approach to management. One of the early thinkers in this area was Henri Fayol, a French engineer who progressed through the management ranks in the coal and iron industry during the later part of the 19th century and the early part of the 20th century. He considered the following groups of activities as applicable to all general managers:

- Technical (production/manufacturing, adaptation)
- Commercial (buying, selling, and exchange)
- Financial (finding and using capital)
- Security (protection of property and persons)
- Accounting (stocktaking, balance sheet, costing, statistics)
- Managerial (planning, organizing, commanding — later modified to directing — coordinating, controlling)

He believed that the managerial functions needed further study and expanded his view of this area by defining 14 principles of management that seem so obvious to us today (see Managerial Comment 4-2).

Managerial Comment 4-2.
Fayol's List of Management Principles

1. Division of Labor. Work should be divided to permit specialization.
2. Authority. Authority and responsibility should be equal.

3. Discipline. Discipline is necessary to develop obedience, diligence, energy, and respect.
4. Unity of Command. No subordinate should report to more than one supervisor.
5. Unity of Direction. All operations with the same objective should have one manager and one plan.
6. Subordination of Individual Interest to General Interest. The interest of one individual or group should not take precedence over the interest of an enterprise as a whole.
7. Remuneration. Rewards for work should be fair.
8. Centralization. The proper degree of centralization/decentralization for each undertaking is a matter of proportion.
9. Scalar Chain. A clear line of authority should extend from the highest to the lowest level of an organization.
10. Order. A place for everything and everything in its place.
11. Equity. Employees should be treated with kindness and justice.
12. Stability of Tenure of Personnel. Turnover should be minimized to assure successful goal accomplishment.
13. Initiative. Subordinates should be allowed the freedom to conceive and execute plans to develop their capacity to the fullest.
14. Esprit de Corps. Harmony and union build enterprise strength.

"The familiar ring of Fayol's ideas suggest how thoroughly they have penetrated current managerial thinking. While many of them may seem relatively self-evident today, they were revolutionary when first advanced. They remain important not only because of his enormous influence on succeeding generations of managers, but also because of the continuing validity of his work. His concepts continue to have a significant impact on managerial thinking. For this reason, Fayol is known as the Father of Modern Management."[17]

Another pioneer in building the administrative management theory was Max Weber, a German sociologist who published his work at the end of the 19th century, but was largely unknown in English-speaking circles until the 1920s. He outlined the characteristics of what he called "bureaucratic management." He used the term "bureaucratic" as an ideal modern and efficient method of organizing, not in the modern sense of a tangled, inefficient governmental organization. He described the following characteristics for an ideal bureaucracy:

- Division of Labor. Divide labor to define authority and responsibility clearly.
- Authority Hierarchy. Offices or positions are organized in a hierarchy of authority.
- Formal Selection. All employees are selected on the basis of technical qualifications demonstrated by formal examination, education, or training.
- Career Orientation. Managers are professionals rather than owners of the units they administer. They work for fixed salaries and pursue "careers" within their respective fields.
- Formal Rules and Controls. All employees are subject to formal rules and controls regarding the performance of their duties.
- Impersonality. Rules and controls are impersonal and uniformly applied in all cases.[18]

Although Weber considered his model to be efficient, recent experiences may weaken its effectiveness because of the rigidity of its structure. However, as with other pioneering work, some of its basics are still valid. Chapter 8 will describe more fully how improvement programs require the appropriate selection and use of organizational structure concepts.

Systems Theory

Systems theory provided a way to blend elements of the major management theories into packages, or programs. Systems theory was formalized in 1954 when the Society for General Systems Theory, later renamed the Society for General Systems Research, was founded under the leadership of biologist Ludwig von Bertalanffy, economist Kenneth Boulding, biomathematician Anatol Rapoport, and physiologist Ralph Gerard.[19] Prior to that time, most researchers and practitioners used a reductionist approach in which they broke a large problem into small parts and attempted to solve the small problems and then reassemble the components into a more workable process. Systems theory encouraged viewing not just the components, but also the relationships among those components. It has had wide application in medicine, as in the development of vaccines, gene splitting, DNA analysis, and organ transplants. Science and technology provided applications in space travel, weather forecasting, and digital data transmission. The computer has facilitated the design and implementation of systems not only in the sciences, but also in business applications. Systems theory has evolved over the latter part of the 20th century into an ever broader and more complex topic.

We will consider some of the more recent developments in systems theory later in Chapter 12, such as catastrophe, chaos, and complexity theory. These ideas are still in the formative stages, but will no doubt become important components of business management in the coming years.

In the early part of the 20th century, scientific management, administrative management, and human relations management were viewed as complete in themselves and independent of each other. Proponents tended to subscribe to one of these philosophies as a primary managerial approach to running their businesses. Applying systems thinking made it easier to select applicable elements from the different management theories to form a complete systems approach. It made it possible to select the best ideas from each of the basic management theories.

Contingency Theory

Contingency theory makes it possible to apply a concept, technique, or program in a modified format to a particular company to fit their specific needs. Contingency theory originated in the information systems area of management and has been widely extended to other management areas. For example, it supports the position that no single organizational structure — bureaucratic, decentralized, free form — is best. Instead, the structure should be adapted to the situation. We will show that the most effective applications of management programs are to design and implement them to fit the specific needs of a business.

Figure 4-5 outlines the evolution of management thought described above. It illustrates how the popular management programs of today have received much of their content from earlier management theories. This approach suggests that a business could package an assortment of basic ideas into a program to fit its own specific needs. Western Electric developed statistical process control in the 1920s to improve quality processes. Toyota developed what became JIT in the 1960s to reduce waste. Motorola developed its Six Sigma program in the 1980s to set new targets in quality. These are just a few examples of businesses that introduced programs that would achieve widespread acceptance and application.

In Table 4-3 and Table 4-4, we show a number of examples of the specific concepts or techniques that originate from each of the main management theories and how they fit together in a program such as JIT and TQM. Table 4-3 lists the evolution and goals for each program. Although they evolved as separate programs, they have a number of similar objectives. Table 4-4 shows how program elements of JIT and TQM could be derived from the basic management theories. These examples are illustrative and not meant to be all-inclusive.

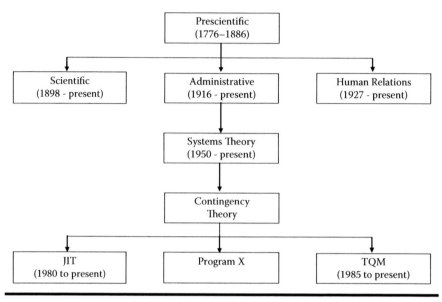

Figure 4-5 Evolution of management thought and programs.

Table 4-3 Comparison of JIT and TQM: Evolution and Objectives

	JIT	TQM
Evolution of the program over time	Scientific management: 1890 Ford's River Rouge plant: 1905 Industrial engineering: 1940 Toyota production system: 1970 Just-in-Time (system): 1980	Statistical process control: 1925 TQC (statistical): 1930 TQC (meet specs): 1950 TQC (satisfy customer): 1980 TQM (integrated system): 1990
Major objectives of these programs	Reduce inventories Reduce lead times Eliminate waste Pursue continuous improvement Recognize customer needs	Reduce cost of defects Offer competitive advantage Eliminate waste Pursue continuous improvement Recognize customer needs

What can we conclude? Many of the basic components of "new" management programs are elements from aspects of earlier management theories. Does this make these new programs less valuable? No, because they have been adapted (an application of contingency theory) to represent the best answer to an organization's specific needs.

Table 4-4   Program Concepts Derived from Management Theories

Management Theory Source	JIT	TQM
Scientific management	Pull method of material flow Standardized work methods Uniform workstation loads	Continuous improvement Cost-of-quality Problem-solving process
Administrative management	Product focus Close supplier ties Group technology	Quality as a competitive weapon Benchmarking Quality as customer's perception
Human relations management	Flexible work force Horizontal organization Teams/employee empowerment	Self-managing teams Quality at the source Cultural change

Why Are Some Programs Successful and Some Not?

Some management programs enjoy great success as they are widely heralded both in trade publications and scholarly journals as leading-edge evidence of how companies can become more competitive. The same program, in another company, may achieve only limited success or even be considered a failure. Why is there so much discrepancy in success among companies that allegedly implement the same program?

The movement of a management program from one company to another involves a great deal of knowledge management (KM). Knowledge management systems (KMS) require liberal amounts of technology; however, their eventual success depends on supplementing technology with systems and people skills. In Managerial Comment 4-3, Robin Wakefield articulates this point further.

Managerial Comment 4-3.
Knowledge Management Strategies

"The development of KM strategies for knowledge transfer is a dynamic and complex undertaking. A principal belief within organizations is that the ability to compete based on knowledge depends primarily on people, rather than processes or technology. Strategies that guide the sharing of internal knowledge represent great challenges."[20]

The originators of individual programs considered them a success; otherwise, they would not have promoted them. Have followers, or later adopters, of these programs been as successful? Some have, but many organizations have had limited success or considered the programs as failures. Why is this so? Is it the program or the situation? Figure 4-6 suggests that the answer must be in the fit between the program, as a proposed solution, and the problem or need of the business. Lack of success must arise from the incorrect matching of the program to the need, or the failure to implement the program correctly. Let's look at each of these possibilities.

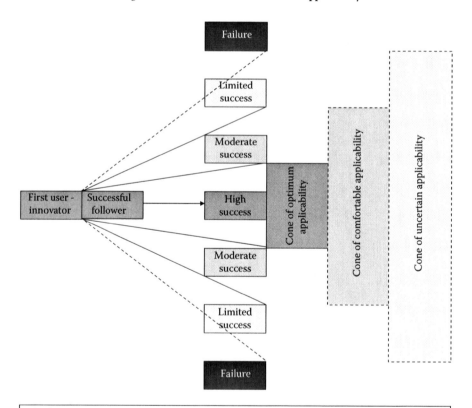

The first users or innovators of a management program do it successfully because they apply it to a specific set of conditions. As the program is applied in more remote conditions, the likelihood of success dimishes.

Figure 4-6 Program extensions and their chances for success.

Failure to Match Program with Need

Figure 4-6 shows conceptually what happens when programs are applied. The originator (or first company) is successful; a close follower using that program is also successful, most likely because it has an operation that is very similar to the originator. Additional followers similar to the first company can implement the programs and also achieve success. As the program grows in popularity and the success stories abound, other companies implement the program. As the program is extended into businesses that are different from the originator, however, the level of success varies and, in some cases, the program may actually be considered a failure. Conclusion: managers should match the program carefully with the needs of the organization.

As always, there is an alternative, and that is to adapt the basics of the program to the different conditions of the business. Figure 4-7 shows a conceptual model of this approach. As the conditions change, the adopter modifies the program elements or selects those elements that fit the new need. In general, the greater the difference in conditions in the new environment, the greater the need for adaptation.

Conditions vary for a number of reasons. We describe several that are among the most likely differences that could affect the eventual success of a management program:

- Strategic objectives. Strategies vary among companies. One company may want to focus on cost reduction, and another company may focus on quality improvement. Each has different needs; therefore, each company may utilize a different program.
- Types of products or services. Product volumes and variety vary. A business that thrives on high-variety, low-volume products should not expect that a program developed for high-volume, low-variety products would fit their needs, at least not without some modification. The same is true for a line of service offerings. An inventory management program developed for widely fluctuating demand patterns, such as in a seasonal retail business, may have more variations than needed for a stable-demand business, such as the bread and milk departments of a retail grocery store.
- Types of processes. Manufacturing processes can be generally classified as job shop, batch, repetitive, and process. Service processes have been classified as a function of the degree of customer contact, such as high contact (hospitals) versus low contact (computerized banking). The wide range of requirements make it difficult to apply the same program equally well to all types of processes. As with product variations, some modification to the program would be necessary to address different conditions.
- Centralization versus decentralization. Decision making is generally classified as centralized or decentralized. There are somewhat opposing trends evident today. The influence of the human relations movement is moving companies toward empowered employees, which suggests decentralized deci-

Figure 4-7 **Adaptation of programs to new conditions.**

sion making. However, the widespread availability of information at all levels of detail is making centralized decision making feasible as well. It is important that management programs recognize the possible differences that can exist between these two approaches.

- Cultures. Cultures among businesses range widely. This topic will be covered in more depth in Chapter 9. Internal factors may cause some of the differences, such as level of employee empowerment, level of job scope, types of wage payment systems, and management styles. External factors that influence the culture of a business include the demographics of the region in which the business is located and social trends. However, it is essential that the management program be adapted to the existing culture, or the existing culture modified to be compatible with the management program. It should be noted that modifying an organizational culture may be more challenging than modifying the management program.

- Top management support. Almost unanimously, authorities stress the need for top management support in the design and implementation of management programs. However, what top management support means can vary widely from business to business. This type of support is dependent on management styles, the relative importance of the program, the background and experience of top managers, and the relationship of top management to the external stakeholders in the program, such as consultants, customers, and suppliers.
- Industry traditions. Some industries have unique origins, practices, language, and peculiarities. Many improvement programs arise and are successfully installed within a particular industry. For example, JIT and lean manufacturing are associated with the automobile industry; Quick Response systems with the retail industry; and MRP/ERP systems with repetitive manufacturing of all types. To implement a management program in one industry closely related to another may be a small step, such as from manufacturing washing machines to refrigerators. However, to move the same management program from automobile manufacturing to railroads or the health care industry may be a major transition because of the embedded practices within each of these industries. In some cases, a program will not gain any degree of acceptance until the terminology is changed to fit the industry where the program is being introduced.

Because of the foregoing reasons, a management program needs to be adapted to its proposed application area if it is to have a chance of success. Even if it fitted correctly, it then must be correctly implemented, as we will see in the next section.

Implementing the Program Correctly

Limited success or failure can also result from an inability to implement the program correctly. In the following sections, we outline some basic steps that are important in successfully implementing a management program.

Planning and Preparation

Implementing a program consists of two phases: (1) planning and preparation, and (2) execution and evaluation. Many businesses are eager to see results from a management program and, as a result, jump into the execution phase too quickly. Management programs are long-term projects and they warrant a careful and systematic approach. The planning and preparation phase is essential.

- Identify the need for improvement. The first step is to identify the need correctly. If the need is to improve quality, then a program designed to reduce cost through reduced inventory may not be appropriate. Certainly, a program should not be attempted just because another company has reported great success with it.
- Select, or design, the right program. We indicated earlier that many companies fail to achieve the desired results because they choose a program that was designed for different conditions. If that is the case, take the time to select those elements of the program that will fit your company's needs, even if you have to omit some of the more attractive parts of the program. For example, the concept of problem-solving teams has widespread application. On the other hand, a kanban scheduling system designed for repetitive production may not be appropriate in a process industry. If you can see a way to adapt other program elements or develop new approaches, do not hesitate to do that. Just avoid trying to fit a square peg in a round hole.
- Create a receptive environment. The first two steps of planning and preparation do not necessarily involve the part of the organization that will be responsible for the actual implementation. Now is the time to get that part of the company involved. Most people resist change, either actively or passively, unless they have been involved in its design.[21] Many well-designed programs have failed because the employees affected by it resisted its implementation. It is better to prevent a potential resistance problem early than to try to solve it later when the program is supposed to be up and running.
- Organize for successful implementation. You may have the right program and have conducted a good job of involving the employees; however, do not forget to organize the implementation effort to make it work. This phase could involve the organization of teams, deciding whether project team members will be full or part time, deciding whether to use consultants, determining the level of training required, or selecting the area in which to test and pilot the program.

Execution and Evaluation

If the planning and preparation steps have been managed well, the execution phase has a better chance of succeeding. This is not to say that it will not be time consuming, costly, and fraught with its share of problems and frustrations. However, these problems are normal in the execution of new programs, and are surmountable. The following guidelines should be considered in the executive and evaluation phase:

- Outline the steps to be accomplished in the implementation process. This phase involves proper scheduling, communication, training, and all of the other factors involved in getting the program up and running. Any program is a complex project; use project management techniques in getting it right. If there is a lack of experience within the business, supplement it with outside help. Try to be consistent; too much wavering conveys a feeling of uncertainty and unease among the participants in the program. Expect that the implementation will be a multi-period (weeks or months) effort and adjust the pace accordingly. Keep in mind that the longer the program, the more challenging it will be to retain interest and enthusiasm. Above all, be patient.

- Evaluate and refine. The evaluation of the program should assess both tangible and intangible criteria. Tangible items include the costs, savings, and benefits of the program. A number of intangible factors can also be considered, depending on the program and the needs of your organization. These can include such items as changes in the company image in the eyes of customers, competitors, and stakeholders; the morale of the work force and their continued acceptance; and the compatibility of the program with other processes in the company. If actual results do not match expectations, take the necessary actions to correct or improve the situation.
- Assimilate the program into normal operations. At some point after a successful implementation, it will be appropriate to plan for the end of the program as a separate and discrete project in the company. The evaluation phase should make it possible to select those program elements that should be assimilated into the normal operations of a business. Assimilation implies that the program elements that should be retained are continued on a long-term basis.
- Prepare for the next program. After assimilating the selected features of the program, it is time to plan for the next program. New opportunities or threats may have emerged that will signal new areas for improvement. There is always a need to improve processes and that is what management programs can accomplish. If the previous program was successful, identify those factors that you think contributed to its success. Likewise, mistakes should be noted so that the implementation of the next management program will be more seamless.

Future of Management Programs

What is the future for management programs? Ideally, all of the workable components of the various management programs will be assimilated into the normal operations of a business. Figure 4-8 shows this conceptually. All of the programs, regardless of the initial stimulus, become part of a master program containing all of the desired objectives. Once integrated into a single program, the elements can then be absorbed as part of the ongoing operations of the business.

	Time Line (Approximate Origin of Program)							
	1975	1980	1985	1990	1995	2000	2005	Beyond
Planning	MRP		MRP II		ERP		ERP Exp.	FLAQRACE SCM (Flexible, Lean, Agile, Quality, Responsive, Accurate, Collaborative and Efficient)
Execution		CIM		MES	WMS	APS		
Measurement		ABC	ABM		BSC			
Costs		JIT			Lean			
Quality		SPC	TQM			Six Sigma		
Response			QRS	ECR	VMI	CPFR		
Flexibility				Mass Cust	Agile			
Communications		EDI			I-EDI	B2B	SOA	
Integration				S&OP	SCM	CRM,SRM	SCM Exp.	
	1975	1980	1985	1990	1995	2000	2005	Beyond

Figure 4-8 Time line for management programs.

Management for the 21st Century

As changes take place in society, changes also occur in the various key roles which individuals in society play. One rapidly changing role is that of the contemporary manager, especially the operations manager who must deal with employees on a day-to-day basis. Many of the routine activities, which were formally managerial tasks, have been computerized. Examples of these tasks include the scheduling of employees and purchasing supplies and raw materials. Other tasks have been delegated to empowered employees and self-directed work teams. The time needed to gather data to compile in reports has been shortened as well as the task of interpreting these reports, which are now often accompanied by notices of key variances that are being violated. (An example of this type of report is when your doctor hands you the printed results of your lab results. Cholesterol levels and other key items are "flagged" so you can easily make sense of the results.) Relieved of the task of accumulating and interpreting this information, managers must effectively communicate the findings to their employees. This change means that managers must be more highly skilled in interpersonal communications and human relations to perform their roles effectively.

Just as business and management have undergone change, so too have the employees managers lead. Many of these workers have come to expect and demand much more out of their jobs and life in general. For instance, there is more emphasis on nonmonetary rewards, and employees are looking increasingly toward their work as a source of fulfilling their needs for self-esteem, accomplishment, challenge, and involvement. That is why top-performing companies organize to meet the needs of their people, and not just serve as a place to "work."

Organizing around the needs of people is not as easy as it sounds. It means going against the traditions of what managers do. Instead of giving orders and gathering and massaging information (managing), managers must set direction and provide resources and encouragement to their employees (leading). It also means aligning the systems, structure, and culture so that the needs of employees are linked with the needs of the business. The business environment is simultaneously faced with the challenges posed by customer service, total quality, continuous improvement, re-engineering, and assorted other hot topics of the month. As a result, many top-performing companies are turning to work teams as the way to link their employees' needs with company needs.

Why teams? Work teams provide the opportunity for employees to get involved in the production processes on a more comprehensive level. In addition, teams can draw on a broader mix of skills, experience, and know-how than any one person could offer. You do not have to look far to find proof of the stunning success of teams within corporations. Motorola used teams to produce the world's smallest and highest-quality cellular phones. Ford relied on teams to create its popular Taurus model. 3M credits work teams as the source of its incredible stream of innovations.[22] Although the team approach to achieving results started primarily in the manufacturing sector, it is now becoming more common in service organizations as well.

Figure 4-8 shows conceptually how the management programs of today could eventually end up as part of a single master program that incorporates all of the features of the individual programs.

Would the master program shown in Figure 4-8 end the need for management programs? Probably not. There always seems to be a need to consider some new technique or concept. Therefore, we need to expect that we will always have new management programs to consider. Table 4-5 lists a few of the possibilities for the future.

Table 4-5   Future Program Emphases

Aspect of the Program	Possible Future Emphases
Paradigm	Mass customization Virtual corporation The Learning Organization
Area of focus	Manufacturing core competency (outsourcing) Strategic and collaborative alliances (trust) Design for adaptability (equipment/employee interface)
Global issues	Ethics/values equilibrium Environmental equilibrium Social equilibrium

Summary

In this chapter, we described the concept of management programs, why companies use management programs to effect improvements, how management programs involve packaging basic management concepts and practices, how systems and contingency theories help to focus the program on the need of a company, and how failure to adapt the program to the company's needs can lead to limited success. We also posed some of the possibilities for future programs. In Chapter 5, we will describe management programs that originated in the manufacturing sector and have been successfully transferred to service businesses. In Chapter 6, we will describe programs that have moved from services to manufacturing.

References

1. Fink, S., *Crisis Management: Planning for the Inevitable*, Backinprint.com, Rev Ed Edition, 2000.
2. Coombs, T., *Code Red in the Boardroom: Crisis Management as Organizational DNA*, Praeger Publishers, 2006.
3. Childe, S.J., Maull, R.S., and Bennett, J., Frameworks for understanding business process re-engineering, *International Journal of Operations & Production Management,* 14, 12, 22, 1994.
4. Luekue, R., *Managing Change and Transition*, Harvard Business School Press, Boston, 2003.
5. Abrahamson, E., Management fashion, *Academy of Management Review*, 21, 1, 254, 1996.
6. Spell, C.S., Management fashions: Where do they come from, and are they old wine in new bottles? *Journal of Management Inquiry,* 10, 4, 358, 2001.
7. Ponzi, L. and Koenig, M., Knowledge management: Another management fad? *Information Research,* 8(1), paper no. 145, 2002. [Available at http://InformationR.net/ir/8-1/paper145.html]
8. Carson, P.P., Lanier, P.A., Carson, K.D., and Guidry, B.N., Clearing a path through the management fashion jungle: Some preliminary trailblazing, *Academy of Management Journal*, 43, 6, 1143, 2000.
9. Abrahamson, E., Management fashion, *Academy of Management Review,* 21, 1, 254, 1996.
10. Ettorre, B., What's the next business buzzword? *Management Review,* 86, 8, 33, 1997.
11. Fenn, J. and Linder, A., Gartner's Hype Cycle Report for 2005, August 5, 2005, www.gartner.com.
12. Crandall, R.E. and Crandall, W.R., An analysis of management programs: Origins, life cycles, and disappearances, *Proceedings of the 2007 Southeast Decision Sciences Institute Conference*, Savannah, GA.
13. Juran, J.M., *Managerial Breakthrough, A New Concept of the Manager's Job*, McGraw-Hill, Inc., New York, 1964.

14. McMahon, D. and Carr, J.C., The contributions of Chester Barnard to strategic management theory, *Journal of Management History,* 5, 5, 228, 1999.
15. Bedeian, A.G., *Management,* The Dryden Press, Chicago, 1986.
16. Herzberg, F., One more time: How do you motivate employees? *Harvard Business Review,* 65, 5, 109, 1987.
17. Bedeian, A.G., *Management,* The Dryden Press, Chicago, 1986.
18. Bedeian, A.G., *Management,* The Dryden Press, Chicago, 1986.
19. Schoderbek, P.P., Schoderbek, C.G. and Kefalas, A.G., *Management Systems Conceptual Considerations* (Fourth Edition), BPI Irwin, Boston, 1990.
20. Wakefield, R.L., Identifying knowledge agents in a KM strategy: The use of the structural influence index, *Information & Management,* 42, 7, 935, 2005.
21. Beer, M. and Nohria, N., Cracking the code of change, *Harvard Business Review,* 78, 3, 133, 2000.
22. Katzenbach, J.R. and Smith, D.K., *The Wisdom of Teams: Creating the High-Performance Organization,* Harper Collins, New York, 1993.

Chapter 5

How Manufacturing Techniques Are Being Adapted to Service Operations

John turned over in his sleep. While his body was resting, his brain was whirling. He was beginning to think about some of the things he could do to improve his business. He thought it strange that he could get good ideas from some of the bizarre sights he had seen on the stages he had visited. Maybe he should keep going; so this time, he tried to go back to sleep and revisit the dream he had already started.

Sure enough, John fell back asleep and the dream continued. At the next stage, John saw people loading trucks with a variety of goods. There were food packages, hay for animals, clothing of all sizes, and medical supplies. He asked the nearest person he could find, "What are you doing?" The woman he addressed was stern, but pleasant. She turned and said, "My name is Kate. We are sending supplies from our town of 'Manufacturing' to a town in the next state, called 'Services.' They have grown so fast that they can't keep up with their needs. We have been settled longer and have learned to be rather effective and efficient in what we do. We thought that some of the things we have would be useful to them. We are especially good

118 ■ New Methods of Competing in the Global Marketplace

at such basic goods as cost reduction and product quality improvement. We used to call the cost reduction program 'Just-in-Time,' but now we call it 'lean production.' We used to call the quality improvement program 'Total Quality Management,' but now we call it 'Six Sigma.' As we make improvements in our products, we continue to change their names to fit their purpose. I know of a few families that have become more self-sufficient with just a little help from us."

As Kate turned to continue supervising the loading, John noticed that, as before, there was a pamphlet on the edge of the stage. This one was entitled "From Manufacturing to Services." He reflected on his latest experience. This was getting even stranger. He had heard about some of the crops that the town of Manufacturing was raising, but because that town had always been considered somewhat different from his community, he had no idea that their products would be something that other towns might be interested in. He thanked Kate for her answer and took one of the pamphlets on the stage.

In Chapter 4, we looked at how management improvement programs are developed. In this chapter, we will look at how these programs are being applied. The manufacturing sector has developed an impressive body of knowledge to help businesses become more efficient and to improve product quality. As these techniques become more refined, they can be extended from manufacturing to service businesses.

Introduction

Manufacturing companies have long been known for their emphasis on reducing costs and improving quality. As a result, they have developed programs such as JIT and lean production to reduce costs, and TQM and Six Sigma to improve quality. Although there are a number of other examples, we will concentrate on these to illustrate how programs developed in manufacturing can be extended to service companies.

First, let us look at continuous improvement programs such as JIT, or its successor, lean production. In their research, David Bowen and William Youngdahl described how service businesses, such as Taco Bell, Southwest Airlines, and Shouldice Hospital, have applied lean service — the application of lean manufacturing techniques.[1] Similar studies have found applications of JIT in insurance firms, retailers, mail-order firms, hospitals, and finance companies.[2] Although the terminology may vary from manufacturing to services, the concept of JIT as a process-oriented waste eliminator applies to both areas.

Perhaps quality programs have been even more popular, such as Total Quality Management (TQM) or its successor, Six Sigma. Although service quality is more difficult to measure because of the intangibility of the service, some of the improvement concepts and tools used in manufacturing can apply equally well to services. Ron Snee and Roger Hoerl provide a number of examples of how Six Sigma has been extended beyond the factory floor to "the real economy," as they describe nonmanufacturing, or services.[3] Their examples cover financial services, e-commerce, health care, nonprofit organizations, and services within manufacturing organizations, such as delivery, finance, and human resources. They stress that it is necessary to think of processes, not functions or transactions. Once past that hurdle, it is as easy to apply Six Sigma concepts and techniques to service processes as it is to manufacturing processes.

We will use Table 5.1 as the starting point for our discussion. The top row shows the most common classifications of manufacturing processes — continuous, repetitive, batch, and job shop. The left column lists the major characteristics of interest to managers and researchers. Table 5-1 also shows that there may be as much differ-

Table 5-1 Four Types of Manufacturing Processes

Characteristics	Continuous	Repetitive	Batch	Job Shop
Labor content	Low	Low	Moderate	High
Employee skill: mental	High	Low	Specialized	High, versatile
Employee skill: physical	Intermittent	Low	Low	High
Employee flexibility	Limited	Low	Low	High
Equipment intensity	High automation	Specialized automation	Individual machines	General equipment
Facilities type	Special purpose	Special purpose	General purpose	General purpose
Product standardization	Standardized	Standardized	Standardized families	Customized
Process flow	Continuous	Assembly	Sequential intermittent	Variable intermittent
Batch size	Large	Moderate	Moderate	Small
Volume of products	High	Moderate	Moderate	Low
Product orientation	MTS	MTS to ATO	MTS	MTO to ETO
Manufacturing examples	Chemicals, liquids	Automobiles, appliances	Clothing	Replacement windows
Service examples	E-commerce, online banking	Fast food, drive-through banking	Airlines, theme parks, bank check processing	Hospitals, investment banking

ence between types of manufacturing processes as there is between manufacturing and services. For example, online banking may be more like continuous manufacturing processes than job shops are like continuous manufacturing processes.

Description of Manufacturing Process Types

In manufacturing, processes are usually classified as:

- Continuous, such as in making chemicals, steel, beer, bread, and bottling soft drinks
- Repetitive, such as in the assembly of automobiles or appliances
- Batch, such as in cutting and sewing clothing, or in making pies
- Job shop, such as in making tools for special-purpose equipment, or making wedding cakes

It is not unusual for a company to have a combination of processes at the same location. For example, an automobile assembly plant may use a repetitive process to assemble the cars, a batch process to make components for the car, and a job shop to make replacement tooling for the assembly line.

Their classification depends on a number of factors, such as:

- Amount of labor content. This factor is usually expressed as a percentage of product cost. Is the labor content low, as in a continuous process, moderate as in repetitive or batch processes, or high, as in a job shop?
- Employee skill requirements, mental. Continuous processes require a high level of mental skill to operate the highly specialized equipment. Repetitive batch processes have relatively low mental skill requirements; job shops rely on a high level of mental skills among employees.
- Employee skill requirements, physical. Job shops require the highest level of physical skills because of the variety of tasks performed. Repetitive and batch processes have low physical skill requirements as regards the variety of tasks, but do require dexterity and endurance. Continuous processes require only intermittent physical skills, as the employees are primarily equipment attendants and maintainers.
- Level of employee flexibility through cross-training. Job shops have the greatest flexibility in scheduling work because of the need to have cross-trained employees. Cross-training is more limited in batch and line operations, although more of it is being done as part of JIT and lean implementations. Continuous processes tend to have reduced flexibility because of their specialized operations.
- Level of equipment intensity. Continuous processes require the highest level of equipment intensity, in the form of chemical processing systems or baking

ovens for bread. Repetitive processes may require substantial investment in assembly lines or robotics. Batch processes usually have multiples of individual pieces of equipment, such as sewing machines or lathes. Job shops usually have general-purpose equipment capable of handling a variety of products.

- Type of facilities required. Continuous processes require the most specialized facilities, i.e., a chemical plant with connections between buildings. Repetitive processes usually require some specialized facilities, such as a painting department for automobiles. Batch processes do not require special facilities and can be set up in smaller units. Job shops are the least demanding insofar as facility requirements.
- Level of product standardization. Continuous processes require standard products and are not easily changed from one product to another. Repetitive processes can be changed between products in the same family, but also require relatively standard products. Batch processes can handle a wide variety of products within the same general product category, such as clothing. Job shops generally make a wide variety of products with low levels of standardization.
- Regularity of process flow. Continuous processes require an almost continuous flow of product and this is the origin of its name. Continuous processes, such as in steel mills, often operate continuously to avoid disruptions to their furnaces or rolling mills. Repetitive processes can be interrupted without dire consequences; however, they tend to operate with a high level of capacity utilization for economic purposes. Batch flow is intermittent and the flow requirements are more for efficiency than process requirements. Job shops are intermittent with a variety of jobs moving in the shop at the same time.
- Batch sizes. Batch sizes range from very large in continuous processes, where the changeover times are high, to very small in job shops, where changeovers and setups are a way of life. Repetitive processes can be changed from one product to another if they are similar, such as from six-cylinder to four-cylinder Honda Accords. Batch processes can vary batch sizes between different sizes or colors in women's dresses. However, the pressure to reduce batch sizes has been growing to reduce the amount of work-in-process inventories.
- Volume of products. Continuous processes are designed to make a high volume of products with a low unit cost. Job shops make a low volume of products with a high unit cost. Repetitive and batch processes are in between continuous and job shops insofar as the volume of products produced.
- Product orientation. Continuous processes have a make-to-stock (MTS) orientation. They must anticipate (forecast) demand and make both the variety and amount of products required. Repetitive processes have traditionally been MTS producers, although in recent years some have moved more to the assemble-to-order (ATO) mode to become more flexible to demand. Batch

processes have the flexibility to be ATO and even make-to-order (MTO) as it becomes more difficult to forecast demand, especially the product mix. Job shops have typically been MTO or even engineer-to-order (ETO) producers.

Product–Process Relationship

Robert Hayes and Steven Wheelwright were among the first to emphasize the need to match the production process with the product characteristics. They pointed out that, as the product evolves over its life cycle, it tends to move from a high-variety, low-volume state to high-volume, low-variety requirements. To be responsive to the demands of the marketplace, the process has to change.[4]

Although a generalization, this concept illustrates the need to match the process with the product. Not correctly matching the process with the product is expensive. Using a job-shop process to make high-volume products results in unnecessarily high product costs. On the other extreme, using a specialized (continuous) process to make low volume, high-variety products results in higher than needed investment costs per unit of product.

In the early stages of the life cycle, a job-shop approach is appropriate because the demand may be for a wide variety of low-volume orders. As demand grows, the most popular products and processes can be adapted to batch or repetitive flow. Should the product reach very high volumes and low variety, it may be economically feasible to move to a continuous process. This reasoning works when moving along the life-cycle curve until the product reaches the mature stage. Then, a dilemma occurs. What type of process is best in the decline stage of the product as volume decreases? This question does not have a single answer; rather, it has many answers, depending on the situation at each business.

Service Industry Classifications

The manufacturing processes described previously have been used for a number of years and appear to fit the different types of manufacturing processes well. Unfortunately, service classifications are not as well established. We will look at two prominent service classification models and examine some of the characteristics of services that make it possible to apply techniques developed in manufacturing industries.

The Chase Model

One of the best-known service typologies is that developed by Chase, Jacobs, and Aquilano.[5] They proposed that services could be classified by the level of customer contact they had. With this model, the level of customer contact progresses from no direct contact, with standardized processing, to face-to-face contact, with customized processing. Sales opportunities move from low in the low-contact segment of the matrix to high in the high-contact segment. Concurrently, production efficiency moves from high in the low-contact segment of the matrix to low in the high-contact segment.

Schmenner Model

Another well-known model is that presented by Roger Schmenner.[6] Like Chase, he shows the level of customer contact on the horizontal axis, but only as discrete low and high categories. He shows degree of labor intensity versus relative throughput times, with two discrete categories of low and high. The result is a two-by-two matrix in which he classifies type of services. In the upper-left quadrant, he shows a service factory (low contact, low degree of labor intensity, and low relative throughput time). The upper right is designated as a service shop (high customer contact, low degree of labor intensity, and low relative throughput time). The lower-left section represents the mass service group (low customer contact, high degree of labor intensity, and high relative throughput time). The lower-right quadrant covers the professional service group (high customer contact, high degree of labor intensity, and high relative throughput time).

Comparison of Manufacturing and Services

How can the service models be compared with the manufacturing model? If service processes are similar to manufacturing processes, then it makes sense to assume that techniques that worked in manufacturing can also work in services, albeit with a little modification. This approach has worked in many instances although, as we will show later, some manufacturing techniques work better than others in services. We provide another perspective in Figure 5-1.

Figure 5-1 shows product and service characteristics along a horizontal axis. It shows process characteristics along the vertical axis. At the top of the figure, when moving from left to right, product and service volumes increase, variety decreases, and the orientation usually moves from MTO to MTS. The bottom of the figure shows that the length of the product or service life cycle increases when moving from left to right. In addition, the time to develop new products or services also increases.

The left side of the figure shows variations in the process. Moving from top to bottom shows an increase in the importance of a structured flow, the level of capital intensity increases, the emphasis on low unit costs, and an emphasis on efficiency in both labor and equipment.

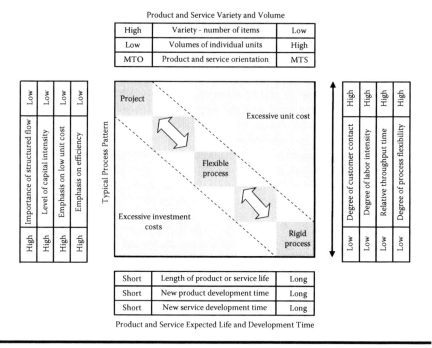

Figure 5-1 A product and service process matrix.

The right side of the figure also shows variations in the process. Here, moving from top to bottom shows a decrease in degree of customer contact, degree of labor intensity, relative throughput times, and degree of process flexibility.

In the figure proper, the diagonal shows a movement from a project focus in the upper-left corner, where the project requirement dictates a special kind of process. As the product or service moves to the right, the process moves along the diagonal toward a flexible process. As product and service characteristics move even further to the right, the process must change to a more rigid process to achieve the high-volume, low-cost objectives. Arrows along the diagonal indicate that the process can change in either direction. Unfortunately, change is easier in moving down the diagonal toward the rigid process that entails heavier investment than it is in moving up the diagonal with emphasis on flexibility and the need for higher labor content.

Figure 5-1 shows that there is a way to relate service processes with manufacturing processes. However, if service processes are similar to manufacturing processes, then it makes sense to assume that techniques that worked in manufacturing also can work in services, albeit with a little modification. This approach has worked in many instances although, as we will show later, some manufacturing techniques work better than others in services.

Manufacturing Objectives

In this section, we will describe the kinds of programs that manufacturers have used to make improvements. They include efforts to reduce product costs, reduce inventories, increase resource utilization, improve quality, reduce response time, and reduce product development time.

Reduce Product Costs

One of the earliest efforts to reduce product costs was the specialization of labor. Instead of having one person do the complete job, such as the making of a musket, the job was divided into smaller tasks that could be done by different workers. One of the earliest recorded examples of this approach was in the making of pins,[7] where the productivity of each operator increased dramatically. It also led to the need for interchangeable parts, although at the early stages of job specialization created a need for filers and fitters to make the parts fit together.

Another major movement in reducing product costs was the use of equipment. David Hounshell offers a fascinating description of the early stages of this effort in his book *From the American System to Mass Production, 1800-1932*.[8] First, hand tools, such as hammers, saws, and augers, were used to assist the operators. Then, power-assisted tools, such as lathes and drill presses, were provided. Finally, in an attempt to achieve the ultimate in product cost reduction, manufacturers moved to automation to replace the operator completely. This procedure worked well in some cases, but not so well in others. The present thinking is that there is a need to design a machine–worker combination that capitalizes on the best from both approaches.

Another major effort was in improving the methods used in making products. Beginning with the work of Taylor, the Gilbreths, and others in the scientific management movement, there was an emphasis on eliminating unnecessary tasks, simplifying those tasks that remained by redesigning workplaces to reduce distances and weights handled, and combining tasks that were performed better together than separate. These efforts were one of the earlier efforts to reduce waste, particularly waste of motion and physical effort. Daniel Wren describes this period in detail in his book *The History of Management Thought*.[9]

Reduce Inventories

Accounting classifies inventory as an asset. At one time, an asset conveyed the impression of something good and a liability as something bad. However, over the years, inventory increasingly is being viewed as necessary, but not always desirable. If you think of inventory as stored expenses that will eventually show up in the income statement as a cost of goods sold, it is easier to understand that inventory should be subject to careful review. As products change more quickly today, it is important that inventory not contain obsolete products.

One of the major efforts to reduce inventory in the latter quarter of the 20th century was the development of materials requirements planning (MRP). The concept of independent and dependent inventory originated with MRP. Dependent inventory has to do with parts of an assembled product, the parent. If the final product is an automobile, then all of the parts that go into an automobile, such as a fender or an engine, are required. Before MRP, most components, the dependent demand items, were managed as individual items without regard to the demand for the parent, or independent, demand item. As a result, it could be possible to have an inventory of dependent demand items even if there was no demand for the parent item. MRP attempted to correct this by calculating the demand for dependent demand items once the demand for the independent item was known. For example, if there was a demand for 50 automobiles, there was a corresponding demand for 50 engines, 200 wheels (four per automobile), 50 steering wheels, and so on. In theory, inventory for dependent demand items resulted only when there was demand for the independent demand item.

A more recent attempt to reduce inventory came from the movement to smaller lot, or batch, sizes. With smaller batches, it was possible to reduce inventories at all stages of completion — finished goods, work-in-process, and raw materials. However, to achieve smaller batch sizes, it was necessary to reduce setup and changeover costs by reducing the time required to make or purchase them. One of the concepts growing out of the JIT movement was SMED (single minute exchange of dies) where the emphasis was on making dramatic reductions in the time to effect a changeover from one product to the next. Many companies found that they could reduce the setup times by 80 to 90 percent.

A third approach to reducing inventories was the development of modular components. This is an extension of the interchangeable parts idea. Automobiles can again provide a good example. Traditionally, each model of automobile had uniquely designed parts, even down to the smallest part. There was little attempt to reduce the total number of parts by standardizing on some parts throughout the product lines, such as with bolts, or even larger items, such as braking systems. By developing modular components and subassemblies, it was possible both to reduce inventories and reduce assembly times.

Although reducing inventories became a popular initiative in many businesses, it was somewhat in conflict with another major improvement effort — increasing resource utilization.

Increase Resource Utilization

As the level of capital investment in the form of equipment and facilities increased, it became more imperative that this investment was fully utilized. A measure of this was called "equipment utilization," which was a combination of both the efficiency in operating the equipment and the percentage of time that the equipment was operating. To achieve high utilization, several approaches were developed. These have been most applicable to the continuous process industries with secondary application to the repetitive industries.

One of the first was the standardization of the product. This often took the form of a reduced variety of final products. When coupled with specialization of labor, this movement produced a high volume of limited variety products. When price was the primary concern, this was an ideal approach. Today, as customers become more selective in their choices, the need for product variety is becoming a greater concern.

Closely coupled with product standardization is long production runs. To achieve long production runs, it is necessary to forecast demand and operate in an MTS mode. When demand is predictable and relatively stable, this approach works well. For continuous process industries, such as chemicals, it is necessary because of the high cost of starting, stopping, and changeover of products, such as from automobile fuels to heating fuels. For repetitive products, long production runs are becoming a thing of the past and companies in this area are learning to adapt.

Capacity management is another way in which companies work to increase resource utilization. This may include careful analysis of the type of equipment best suited to the process needs, timing of the purchase with the capacity needs, careful monitoring of the equipment to assure in-control operation, use of preventive maintenance to avoid unplanned shutdowns, and planning for smooth transitions from one generation of equipment to the next. Table 5-2 offers a summary of the degree of applicability improvement programs can have on reducing product costs, reducing inventories, and increasing resource utilization.

Improve Quality

Although high quality of products has always been desirable, it became an imperative in the latter quarter of the 20th century. This was especially notable in the automobile industry, where Japanese manufacturers made major inroads into the U. S. market, largely because they made higher-quality automobiles. While U.S. manufacturers continued to rely on styling and price, the buying public was moving to more comfortable, reliable, and fuel-efficient automobiles. Repetitive industry businesses especially felt the need for this type of program

Process-control techniques were first developed at Western Electric in the 1920s, with the work by Walter Shewhart and others; disciples of Shewhart include Juran and Deming. Process control means that a consistent product is produced by a consistent process. Statistical techniques, especially the use of sampling techniques,

Table 5-2 Areas of Potential Application of Improvement Programs: Group 1

Type of Improvement Program	Types of Processes and Level of Applicability in Each Type of Improvement Program			
	Continuous	Repetitive	Batch	Job Shop
Reduce product costs				
Worker specialization	Limited	High	High	Low
Automation	High	High	Limited	Limited
Methods analysis	Limited	High	High	Low
Reduce inventories				
Dependent demand	Limited	High	Limited	Low
Reduce batch size	Limited	High	High	Low
Modular components ATO	Limited	High	High	Low
Increase resource utilization				
Product standardization	High	High	Limited	Low
Long runs MTS	High	High	Limited	Low
Capacity management	High	Limited	Limited	Low

were developed to monitor processes and to determine if a process was out of control, or moving toward an out-of-control situation. Over time, automated sensing systems have been developed to monitor time, temperature, pressure, and other variables of sophisticated continuous processes, such as the operation of nuclear reactors.

In addition to process-control techniques, a major movement has included empowering employees to be more responsible for product quality. The reasoning is that the equipment operator may be in the best position to influence product quality, through more awareness of the quality requirements and the assignment of responsibility to produce a quality product. This requires a movement away from job specialization to job enlargement, and even further to employee empowerment. Although it increases the labor costs, many companies have found that it reduces the cost of poor quality, a cost that is largely hidden in most accounting systems.

Another approach to quality improvement is in product design. Companies have realized that no matter how carefully they control the process or how diligent the empowered employees, the product can never have a quality higher than in its inherent design. Consequently, they are now trying to design quality into the product. This effort has to do with better materials, closer fits, improved fabricating methods, and the like.

Reduce Response Time

As the demands of the marketplace have increased, companies have had to address an increasing array of requirements (see Chapter 2 on Critical Success Factors) to be competitive. After reduced price and increased quality, reduced response time has been increasing in importance. Reduced response times not only reduce the level of inventory, but also match production to demand more closely.

For years, both academics and practitioners developed scheduling algorithms that would make it possible to move jobs through the production process in an efficient (low cost) and effective (meet the required delivery date) manner. In the early days of their development, the techniques were manual with limited scope of application; however, in recent years, computerization has made them much more effective.

Responsiveness can be shortened with improved knowledge about demand. Better demand forecasting has been a paramount objective for several decades. This effect is characterized by both improved forecasting techniques and by improved information processing. As more is known about demand patterns, especially about their basic features — stability, trends, seasonality, and events — a better forecast can be prepared. However, recent trends in shorter product life cycles and increased variability in demand are proving a difficult adjustment. Companies are finding that better forecasting techniques have inherent limitations; consequently, they are moving toward collaborative efforts with customers to enhance forecasts.

The use of modular components was discussed previously as a way to reduce inventories. It is also proving useful in responding to market demands. If the product can be carried in inventory as a generic model until the order comes in, it can then be finished quickly in a customized form and shipped to the customer. Dell has made this a hallmark of their business. Other companies are making use of this postponement technique to stock a smaller variety of "almost finished" products and then being able to create a wide variety of finished products with the addition of customer-designated accessories.

Reduce Product Development Time

As product life cycles decrease, companies are finding it necessary to develop products faster, such as in automobiles, airplanes, and especially software.

Concurrent engineering is a new approach to product development. Historically, companies used a sequential process to develop new products. Marketing came up with an idea for a new product and passed the idea to Product Engineering, who developed a prototype product. Product Engineering passed the idea along to Process Engineering, who developed a process for making the product. They passed the product and process along to Production, who adapted the recommended process to an existing process and began to make prototype or sample products. Along the way, Accounting got involved to consider product cost and the effect on the income statement. Finance also became interested because of the investment and balance sheet implications. Human Resources found out they had to hire more employees, some with skills not previously needed. Along the way, functional areas had questions that often required changes in the product design. Eventually, after iterations and revisions, a product emerged. In recent years, the process called "concurrent engineering" emerged, which required the use of cross-functional teams in the early stages to resolve questions, select most promising alternatives, and eliminate dead ends in the product development. As a result, it has been possible to reduce the time to develop new products and improve the effectiveness of the new design.

Traditionally, the maker of the product took a make-and-sell approach in which they came up with a product idea, made it at low cost with their specifications for quality, and then tried to figure out how to sell it.[10] Today, businesses are trying to figure out what the customer wants, often through participation from their customers. Then, they develop a product and the corresponding process that meets the customers' requirements. One technique that is becoming more popular is quality function deployment (QFD), in which customer requirements are related to company capabilities and competitor capabilities to decide on the attributes most suitable to the new product design.

Another program that some companies are beginning to use to reduce product development time is outsourcing, in which final assemblers are placing more dependence on their suppliers to participate in the product design process. For example, an automobile manufacturer may rely more heavily on a supplier to develop new braking systems or an engine manufacturer to develop the innovations in engines. This requires a higher level of inter-organizational collaboration and relies heavily on mutual trust among businesses. This arrangement is proving to be a difficult transition for many companies.

Most of the initiatives described here have been developed for application in the continuous process and repetitive manufacturing industries. Batch process manufacturers have tended to lag behind and many of the techniques are not as readily applicable to job shop types of operations. This problem is of special interest as we consider the transfer of knowledge to services because many service operations fit into the classification of batch flow and job shop operations. Table 5-3 offers a summary of the degree of applicability improvement programs can have on improving quality, reducing response times, and reducing product development times.

Service Objectives

Service companies have some objectives that are similar to those of manufacturing organizations; however, they also have some that are different. We will examine those differences more fully in Chapter 6. Where service companies have objectives that are similar to those of manufacturing, the improvement programs that worked well in manufacturing may also work well in services. We will explore some of those programs in the next section.

Table 5-3   Areas of Potential Application of Improvement Programs: Group 2

Type of Improvement Program	Types of Processes and Level of Applicability in Each Type of Improvement Program			
	Continuous	Repetitive	Batch	Job Shop
Improve quality				
Process control	High	High	Limited	Low
Employee empowerment	High	High	Limited	High
Product design	Limited	High	Limited	Low
Reduce response time				
Job scheduling algorithms	High	Limited	High	High
Improved demand forecasting	High	High	High	Low
Modular components ATO	Limited	High	Limited	Low
Reduce product development time				
Concurrent engineering	Low	High	Limited	Low
Quality function deployment	Low	High	Limited	Low
Outsourcing	Limited	High	Limited	Low

Programs That Work in Services

In an effort to achieve the goals described in the previous section, it became useful to develop programs that focused on specific objectives. We described the evolution of improvement programs in Chapter 4; in this section, we will describe several programs that have been successfully implemented in services.

Programs work well in services when the service processes are similar to those for which the programs were developed in manufacturing. Later, we will also describe some of the programs that worked in manufacturing, but did not easily translate to the service side of business.

Theodore Levitt was one of the first to recommend the use of manufacturing techniques in services. He pointed out that even the so-called manufacturing industries contained a wide variety of service activities, such as systems planning, service support, software, repair, maintenance, delivery, collection, and bookkeeping. He emphasized the need for service industries to improve productivity and customer service to remain competitive, and concludes by saying that the use of manufacturing techniques in service operations "can generate liberating new solutions to intractable old problems. It can bring to the increasingly service-dominated economies of the future the same kinds of vaulting advances in productivity and living standards as the newly created goods-producing factory economies brought to the world in the past."[11]

IBM has moved from a manufacturing company to a service company and provides services to purchasing, human relations, and customer relations programs. They have a growing interest in "services science." "The hybrid field seeks to use technology, management, mathematics, and engineering expertise to improve the performance of service businesses like transportation, retailing, and health care — as well as service functions like marketing, design, and customer service that are also crucial in manufacturing industries."[12] Services represent an "understudied field" that is attracting the interest of universities, corporations, and government agencies.

To provide a sampling of the progress that companies are making in adapting manufacturing programs to service applications, we will describe the successes for two types of programs. JIT and lean production represent programs to reduce costs and response times. TQM and Six Sigma have been used primarily to improve quality.

JIT and Lean

JIT is an outgrowth of the Toyota Production System (TPS), which, to some extent, was patterned after the work by Henry Ford at the River Rouge plant. It was designed to reduce inventories and eliminate waste. Lean production is a concept pioneered by Womack, Jones, and Roos following their study of automobile assembly plants throughout the world.[13] Lean production also stressed the need to eliminate waste and to smooth the flow of goods along the value chain. For practical purposes, it has become the successor to JIT in the minds of many writers. In the following paragraphs, we list a few of the successful applications of lean manufacturing to services.

One common theme throughout the lean production literature is the focus on manufacturing companies. Par Ahlstrom investigated why it is difficult to apply lean production to service companies. He found that lean production is applicable to service operations, although there are contingencies to the application. The contingencies stem from the characteristics of services. There are also instances where lean production is perhaps more applicable to services than manufacturing, again due to the nature of services. The conclusions indicate a need to make operations more general, avoiding making clear differentiation between manufacturing and services, and instead focusing on the similarities between the two and thus the possibility to learn from each other.[14] In this light, it is interesting that one of the newsletters from Daniel Womack at the Lean Institute was about the opportunities to apply lean thinking to health care operations[15]:

- Need for an integrated system. Jeff Liker and James Morgan offer ideas on how to adapt the TPS, with its focus on the customer, to services.[16] Although they believe that the lean manufacturing concept can work in services, they also believe that many of the programs started in service sectors have been piecemeal or quick fixes to reduce lead times and costs or to increase quality. Consequently, these programs never create a true learning culture. They stress that it takes a true systems approach that "effectively integrates people, processes, and technology — one that must be adopted as a continual, comprehensive, and coordinated effort for change and learning across the organization."[16]

- Consider service characteristics. Some scholars provide a general approach to applying manufacturing techniques to services. One study points out that, when faced with the need to become more competitive, services are much like manufacturing in that they both use processes that add value to the basic inputs used to create the final product or service. They identify the unique characteristics of services as inseparability of production and consumption, intangibility, perishability, and heterogeneity. When integrated with the JIT themes of total visibility, synchronization and balance, respect for people, flexibility, continuous improvement, responsibility for the environment, simplicity, and holistic approach, they show that these themes can be applied in the service sector.[17]
- State government. In a more specific study, Ron Bane reported on the successful application of lean principles to a state motor vehicle license bureau that was able to reduce waiting times by 70 percent.[18] The study also reports on other applications in the state of California, such as reduction of water main leaks, reduction of check processing times, and reduction in office space steps. The author cautions that successful application requires patience and top management support, and that continuous improvement is a never-ending process; however, he believes that there are "tremendous opportunities for nonmanufacturing organizations that embrace these approaches."[19]
- Bid quotation process. Lean techniques have been applied to the bid quotation process, an especially important service function in job shops or MTO firms. Quotations convey two important bits of information — price and due date. Successful bids must be accurate and prompt, and the use of lean principles helps to achieve those objectives. One study found that application of lean techniques, coupled with the implementation of an ERP system, led to improvements in the following areas:
    - Reduction of paper use via electronic quoting
    - Reduction of waiting time through electronic reminder
    - Elimination of tasks in the quoting process
    - Coordinating with outside vendors
    - Collection of shop floor data for accurate labor costing[20]
- Mass services. Lean techniques can also be applied to "mass servicing." Mass servicing is similar to mass production in that customer service employees are typically semiskilled representatives "trained a mile wide and an inch deep in handling a high volume of calls for all customer inquires."[21] The

traditional mass servicing model consists of a generically trained representative who rushes the customer off the phone to increase the number of calls handled per hour and reduces the unit costs; however, it almost guarantees errors and the need for callbacks. The proposed approach — customer-centric work cells — provides a simplified flow of the customer's order or request by grouping all knowledge and tasks associated with servicing the customer within the work cell. Employees are trained to perform multiple tasks (order entry, order fulfillment, service, and invoicing) so that the customer's request can be serviced without undue handoffs — "one call gets all."[22]

- Local government. Sandra Furterer and Ahmad Elshennawy studied the use of lean practices in local government.[23] They looked at how the application of such lean concepts as kanban and visual control, waste identification and elimination, and one-piece flow can be used effectively to reduce the average processing times from 40 to 90 percent in processes such as payroll and pension reporting, purchasing/accounts payable, accounts receivable, and monthly reconciliation. They also found that a combination of lean with Six Sigma worked well in increasing both productivity and quality of providing financial services.
- In association with Six Sigma. Allen Pannell found that lean and Six Sigma are "powerful tools that complement — not compete with — each other." He explains that they are both management philosophies and sets of tools. Although they have many similarities, they also have some key differences in their primary focus. Both focus on reducing variation in the process. Although Six Sigma focuses on the use of statistical analysis to identify and reduce variation, lean takes more of a macro approach to reduce lead times and inventory. He cautions that "the challenge is not to look for a quick answer. Changing attitudes, thinking differently, and being willing to embrace the future takes leadership, time, and effort. It will pay off for companies that want to be in business 100 years from now."[24]
- Customer service. Siemens used lean methods to transform its culture to one that promotes active involvement, at all levels of the organization, in continuous improvement activities. The study examines how the company made major improvements in customer service and won a ten-year contract to transform and manage the customer operations of National Savings & Investments, a major retail financial services organization.[25]

The above examples show that JIT and lean concepts have been applied to service operations successfully. However, there is no universal agreement that the transfer of manufacturing methods to service is good. David Bowen and William Youngdahl caution that this industrialization of service is inappropriate because of the unpredictable nature of customer demand for services; however, these same writers advocate the transfer of service principles to manufacturing, a topic to be discussed more fully in Chapter 6.[26]

TQM and Six Sigma

Total Quality Management grew out of the statistical process-control work that originated in the Western Electric organization in the 1920s. Widely popular in the late 1980s and early 1990s, TQM programs were used extensively both in manufacturing and service companies. Although there were a number of successful programs, there were also a number of unsuccessful programs, often the result of inappropriate application or lack of focus by the implementers. In an attempt to revive the quality movement, Six Sigma was developed, originating with Motorola in the mid-1980s. It retained the primary focus of TQM, but placed more requirements on the participants in the form of specific objectives (Six Sigma quality), an infrastructure (black and green belts), more training of participants, a formal approach to the improvement process including define, measure, analyze, improve, and control steps (DMAIC), and, in general, the need for a company to be serious about its implementation. Some of the service areas in which TQM or Six Sigma were successfully implemented include the following examples:

- Universities. Commenting on a 2001 Six Sigma survey in *Quality Digest,* Ron Bane reported that, although most of the respondents were in manufacturing, many were in nonmanufacturing areas, such as document control, shipping, sales/marketing, customer service, and engineering. He also found Six Sigma being used in universities to reduce student scheduling errors.[27]
- Health care. Caldwell, Brexler, and Gillem examined why many Six Sigma efforts in health care fail. They found that, although hospitals use processes, which is the necessary perspective heralded by Six Sigma authorities, those processes have a major difference when looking at the role of the physician. They studied both successful and unsuccessful implementations and found that the level of commitment by the physician was essential for success. They found that physicians resist participating in the program because the change places more burden on their processes — consuming more time, increasing complexity, and providing less service to them or their patients. They suggest that, in order to engage physicians, health care providers should:
  - Seek to fully understand physician needs in general and within the specific process to be changed during the prework, define and improve phases of a DMAIC project, and learn the degree of support required.
  - Seek to build trust. This probably sounds simple, but trust between the physician and the hospital executive has dramatically eroded over the past decade.
  - Educate physicians in all aspects of health care management, financial management, regulatory environment, and competitive pressures with an aim to establish a true visioning partnership about the future.

- Seek win–win projects. That is, find projects that will delight physicians, usually by improving the efficiency in physician–hospital interfacing processes.
- Negotiate a quid pro quo, in which the organization provides a concession in some other area in exchange for physician agreement to embrace the desired change.
- Seek physician influences (referrers, physicians with high credibility) to lead the way, instead of hospital managers and executives.
- Integrate improvement work casually into existing physician committee and task force structures by discussion or nonaction items with action items.
- Consider incentives, but beware. Once offered, they become expected.
- Seek nonphysician care givers to execute the change or influence physicians to embrace the change.

They caution that, in health care situations, process innovation cannot be realized without physician engagement.[28]

- Need for committed employees. Members of the Gallup organization developed a concept they called Human Sigma, which they related to Six Sigma. They viewed Six Sigma processes as data driven, rational, and analytic, and looked at their application to the employee–customer encounter at the retail level. They first found that emotionally satisfied customers were less likely to change to another store and bought more than rationally satisfied customers. They also found that those who were only rationally satisfied, as opposed to emotionally satisfied, were, in some ways, no different from dissatisfied customers. The Gallup studies also found that, although an average performance measure may be satisfactory for a company, there was always considerable variability at the local level. This variability fit a normal curve, indicating an unmanaged situation. They found that employees with a higher level of commitment to the firm were more productive, had lower turnover, and were more effective than less committed employees in managing the customer encounter. To improve local performance, Gallup found that stores needed to look at both the level of customer engagement and employee engagement. These two attributes had a major effect on a store's financial performance. To implement rational programs, such as Six Sigma, in measuring and managing the employee–customer interface, companies need to take customers' emotions into account. Companies must also work to gain greater commitment from employees.[29]
- Local government. In addition to their study mentioned earlier, researchers Furterer and Elshennawy also looked at the effect of implementing lean Six Sigma programs in local government.[30] The authors applied lean Six Sigma tools and principles to the financial administration processes in a 7000-citi-

zen municipality. The financial processes included payroll, purchasing, and accounting functions of accounts payable, accounts receivable, and monthly reconciliations. The program achieved increased output, reduced processing times, and reduced errors.
- Hospital processes. In their research study, James Harrington and Brett Trusko reported that there is no other industry more in need of reinventing itself than the health care industry.[31] They offer some of the following reasons:
  1. It has been estimated that globally somewhere between 1,500,000 to 2,200,000 people die because of health care errors every year.
  2. Health care errors kill up to 100,000 people per year in the United States alone.
  3. One person in the United States dies every eight minutes because of nosocomial infection and 95 percent of those deaths are preventable.
  4. Few health care entities are ISO 9000-certified.

They report that an ever-growing number of health care organizations are using Six Sigma to improve processes from admitting to discharge and all the administrative and clinical processes in between. This adoption is driven by several factors including the need to improve the organization's bottom line, eliminate medical errors, and position themselves for an imminent global consumer-centered health care revolution.

- TQM and change management. Implementation of quality improvement programs such as TQM has major change-management implications. One multi-year study of 20 service companies found that the issues of unrealistic expectations of employee commitment, absence of process focus, lack of organization around information flow, holes in education and training, and failure to create a continuous improvement culture could contribute significantly to the lack of success of the programs. Unsuccessful companies tended to have a lack of commitment to the program; successful companies provided strong leadership based on strategic and tactical planning.[32]
- Intensive care. Eva Lindberg and Urban Rosenqvist conducted a four-year study on the implementation of TQM in an intensive care unit (ICU).[33] They found that the TQM program was implemented during the introduction of two other major programs — a change in employee working hours and scheduling, and a change in the leadership of the ICU. The increased ambiguity of the situation made it difficult to evaluate the true impact of the TQM program. The authors, in speaking of the quality program, report that, on the surface, the managers and staff are working in accordance with the quality system. However, they fear that the quality document has "promoted a mechanically functioning organization, similar to the attributes associated with mass production." They conclude that there is a need for a

quality system that can avoid the negative effects by reducing the ambiguity in the organization.
- Retail. Fred Patton described the use of Six Sigma in a retail setting.[34] He believes that quality professionals understand how Six Sigma principles can be applied to any organization concerned with serving its customers as effectively as possible. The key is in convincing the service employees that there is a concept that was developed in manufacturing that can be adapted to a service setting. He states, "Getting service organization people to attend training is more painful for everyone involved than having your wisdom teeth extracted." Although there are obstacles, Patton is an advocate for applying Six Sigma concepts in service settings.
- Similarity of manufacturing and service applications. Some studies examined the difference between manufacturing and service firms with respect to the implementation of TQM practices, and the relationship of these practices to quality performance. In a survey of 194 Australian firms about equally divided between manufacturing and service firms, he found that there was no significant difference in the level of most of the TQM practices and quality performance between the two sectors. He believes that continuing fragmentation of industry sectors may be an area for further study and speculates that, in manufacturing sectors, process management may be the key determinant in achieving a high-quality product, and in the service industry, people management may turn out to be the significant predictor.[35]
- Sarbanes–Oxley. One of the more innovative applications of Six Sigma was in developing a seven-step process to help "get and keep processes under control so management can comply with SOX 404 reporting requirements more cost effectively."[36] The authors go on to caution that Six Sigma is only a tool and is most effective when used correctly.
- Decision support systems. David Rylander and Tina Provost studied how Six Sigma principles could be combined with technology, such as a decision support system (DSS) to enhance the customer service area.[37] Good front-line employees can reduce the negative effect of product problems; however, poor front-line employees can negate the positive effects of good products. They stress the need for employees to have good information to aid them in their customer interface dealings.

Programs More Difficult to Adapt to Service Operations

Programs such as JIT/lean and TQM/Six Sigma have found extensive application in service industries. Although each program required some adaptation, particularly in the employee acceptance area, the evidence supports the basic assumption that techniques from these programs can be beneficial in a variety of service applications.

Product Costing

Stan Brignall studied the application of product costing techniques such as standard costs and activity-based costing (ABC) in service industries.[38] He also studied how performance measurement systems (PMS) developed in manufacturing could be extended to service sectors. As a reference, he used the service classification system developed by Rhian Silvestro and colleagues that included professional services, service shops, and mass services.[39] He observed that, as services move from professional services to service shops to mass services, traceability of costs decrease and the level of overhead allocation increases. As a result, the concept of standard costs, with their corresponding variance analysis, may have some applicability in mass services, but less in service shops and probably not at all in professional services. Performance measurement systems that tie shop-floor performance to overall company performance may be appropriate for many manufacturing companies (although less so for job shops), it is more difficult in service environments where there is more variability in the processes and quality of results.

Activity-Based Costing

Mostaque Hussain and Angappa Gunasekaran studied the applicability of ABC to service organizations.[40] Although they conclude that ABC can be useful in services, the results of their survey of banks indicate that ABC is not yet widely used. ABC was designed to more precisely determine the product costs in manufacturing. Products are easily identified and the unit costs do not have great variability for the same product. Services are more difficult to classify, making it more difficult to define a specific unit to be costed. The cost per unit can also vary considerably because, in many cases, the customer is involved in the service and individual customers have different needs that can influence the time required to complete the unit of service. ABC can also be used for performance measurement in manufacturing; however, it is oriented toward financial measures, and services are finding that they need nonfinancial measures to measure performance best. The net result is that ABC has potential for services, but needs considerable adaptation to be effective.

Materials Requirements Planning

MRP has been a fixture in the manufacturing toolkit for a half century. Because of its popularity and intricate structure, many researchers have tried to figure out how to apply it to services. There have been a number of clever applications. It could be used in surgical operations where the surgery to be performed is the independent demand item and the dependent items needed include the surgeons, the staff, the equipment, the supplies, and so on. MRP could also be used in a university to plan for the needs of students. The graduating student is the dependent demand item and the courses needed to complete the degree are the dependent demand items. Although these and other applications sound interesting, MRP is not likely to be used in service applications, especially with the advent of lean production.

Enterprise Resources Planning

ERP systems have been largely associated with manufacturing and have yet to be widely adopted in service organizations. Although there is a need, many service companies need to consider how to select a system that provides what they need without buying all of the extras that were designed into the manufacturing ERP systems. A study of professional firms reports that because an ERP system often is expensive and complex to deploy, it is important to understand the motivation for acquiring such a system. Programs that are project-oriented and have tools for tracking billable hours and producing what-if projections are the most practical in a service environment.

Performance Measures

The performance measures used in manufacturing often are based on specific units of output. It is not always easy to select a unit of output in the service industries because of the variability of the unit. For example, a unit of output may be a patient at a hospital. However, depending on the type of service and the characteristics of the patient, the time required per patient may or may not be a meaningful performance measure. In measuring quality, manufacturing companies use percent defective or parts per million, considering that the unit is either good or bad. This binary classification does not work as well in services, where the quality is more apt to vary along a continuum from very bad to very good, so another measure has to be adopted.

Automation

Automation is a common theme in manufacturing process improvement. It can provide both cost reductions and quality improvements. Automation is used extensively to reduce costs in service applications, such as in the banking industry with automatic teller machines (ATM) and online banking or in retail stores with self-checkout stations. However, quality in services is often associated with customer service, and the move toward automated services may not be compatible with the customer's perception of high quality. On the other hand, the use of online buying — books at Amazon.com where the Website knows who you are if you have been there before (cookies permitting, of course) — is considered to be a higher level of service than going to a bookstore with its limited selection of books and clerks who may not recognize you as a loyal customer.

Resource Utilization

Manufacturing businesses have developed the utilization of resources — employees, equipment, facilities, and systems — to a real science. They work to utilize the assets effectively and efficiently. They have the luxury of building inventory ahead of demand to facilitate the use of level production plans, although this is changing some because of the pressures to keep inventories low. Service companies cannot build inventory; therefore, they have limited control over their utilization of equipment and facilities. They do have some discretion with respect to scheduling employees, and attempt to match capacity with demand whenever possible.

Keys to Extending Manufacturing Techniques to Services

A number of writers insist that the way to facilitate the movement of manufacturing techniques is to think in terms of processes. A manufacturing process involves the Input–Transformation–Output Model, in which materials (steel and wood) are transformed (manufactured) into outputs (automobiles and furniture). Using this model in services would make it possible to view an untreated patient as an input, the hospital surgery as the transformation process, and the treated patient as the output. Ron Snee and Roger Hoerl explain this in some detail in their book *Six Sigma Beyond the Factory Floor*.[41]

There is also a need to recognize the differences and similarities between manufacturing and services. One way to approach this goal is to think in terms of subgroups within manufacturing and services. For example, there is a great deal of difference between continuous flow processes and job shops. However, job shops are more similar to professional service firms than continuous flow manufacturing firms. Airlines share much in common with repetitive manufacturing processes. Online B2C systems like Amazon have commonalities with continuous flow systems. The point is that some service configurations are more similar to manufacturing units than they are to other service arrangements.

David Bowen and William Youngdahl remind us that manufacturing is also changing.[42] We are learning to adapt some of the improvement programs that worked in stable, repetitive environments to rapidly changing, variable environments. As a result, we are moving closer to the type of programs that are needed in the more dynamic service sector. Robert Johnston suggests that the movement of manufacturing techniques to services has gone through several stages:

Service awakening: Initial interest in services
Breaking free from product-based roots: Awareness of the importance of services
The service management era: Cross-disciplinary research, especially among marketing, operations and human resources
Return to roots: Refocus towards traditional operational issues and approaches

He believes that there is potential for applying the basic manufacturing techniques to service operations.[43]

Conclusions

This chapter provided a look at areas in which programs developed primarily in manufacturing have been extended to service industries. We described the types of manufacturing as job shop, batch, repetitive and continuous. Many of the programs were developed in the repetitive and continuous flow processes where cost reduction and quality improvement were needed. JIT and lean manufacturing are programs that have been very successful in reducing manufacturing costs as well as providing other benefits that result from a smoother flow of products. TQM and Six Sigma have been successful as programs that improve quality. These programs are the ones most successfully applied to service operations.

Even though there are numerous examples of successful applications of manufacturing programs to services, their application has been spotty. When programs such as lean manufacturing were used in manufacturing, they spread across industries, such as automobile manufacturing or appliance manufacturing. This has not occurred in services, where the applications have been more company-oriented than industry-oriented. This lack of migration could be because it requires adaptation of these programs across industries while, in manufacturing, there are more similarities in processes across industries. It could also be that service industries have had other priorities and have developed their own programs to meet these needs, as we will see in the next chapter.

A characteristic of service industries is their concern with providing customer service. However, this strength can become a weakness. Some service industries have overshot their customer needs and are susceptible to new competitors who offer less service to customers at a lower cost.[44] Two examples stand out. The airline industry has been revolutionized by low-cost airlines, to the point that they have transformed the industry. Another example is Wal-Mart in retailing. Their "everyday low price" approach has also been revolutionary, even though it has numerous critics. Low costs have a place in service operations, and many of the techniques developed in manufacturing to lower costs work in services.

However, other service industries, such as health care, need to reduce costs and improve quality. Although there is a need to use the concepts developed in lean thinking and Six Sigma programs, these applications need to be applied with discretion. An additional area they need to address more closely is resource utilization, as in the deployment of expensive medical equipment, such as MRI scanners. Their approach to quality has been to add features, such as additional testing, to help diagnose and treat ailments; however, cost containment continues to be a problem.

Manufacturing has been outstanding in improving internal operations. The need is to focus more on the customer, a strength of services industries, as we will see in Chapter 6.

References

1. Bowen, D.E. and Youngdahl, W.E., "Lean" service: In defense of a production-line approach, *International Journal of Service Industry Management*, 9, 3, 207, 1998.
2. Duclos, L.K., Siha, S.M. and Lummus, R.B., JIT in services: A review of current practices and future directions for research, *International Journal of Service Industry Management*, 6, 5, 36, 1995.
3. Snee, R.D. and Hoerl, R.W., *Six Sigma Beyond the Factory Floor*, Pearson Prentice Hall, Upper Saddle River, NJ, 2005.
4. Hayes, R.H. and Wheelwright, S.C., Link manufacturing process and product life cycles, *Harvard Business Review*, 57, 1, 133, 1979.
5. Chase, R.B., Jacobs, F.R., and Aquilano, N.J., *Operations Management for Competitive Advantage* (Tenth Edition), McGraw Hill-Irwin, Boston, 2004.
6. Schmenner, R.W., *Plant and Service Tours in Operations Management* (Fifth Edition), Prentice Hall, Upper Saddle River, NJ, 1998.
7. Hounshell, D.A., *From the American System to Mass Production, 1800–1932*, The Johns Hopkins University Press, Baltimore, 1984.
8. Hounshell, D.A., *From the American System to Mass Production, 1800–1932*, The Johns Hopkins University Press, Baltimore, 1984.
9. Wren, D. A., *The History of Management Thought* (Fifth Edition), John Wiley & Sons, Inc., Hoboken, NJ, 2005.
10. Haeckel, S.H., *Adaptive Enterprise Creating and Leading Sense-and-Respond Organizations*, Harvard Business School Press, Boston, 1999.
11. Levitt, T., The industrialization of service, *Harvard Business Review*, 54, 5, 63, 1976.
12. Lohr, S., Academia dissects the service sector, but is it a science? *New York Times*, April 18, 2006, p. C.1.
13. Womack, J.P., Jones, D.T., and Roos, D., *The Machine That Changed the World, The Story of Lean Production*, Harper Perennial, New York, 1990.

14. Ahlstrom, P., Lean service operations: Translating lean production principles to service operations, *International Journal of Services Technology and Management*, 5, 5–6, 545, 2004.
15. Womack, J.P., Jones, D.T., and Roos, D., *The Machine That Changed the World, The Story of Lean Production*, Harper Perennial, New York, 1990.
16. Liker, J.K. and Morgan, J.M., The Toyota way in services: The case of lean product development, *The Academy of Management Perspectives*, 20, 2, 5, 2006.
17. Canel, C., Rosen, D., and Anderson, E.A., Just-in-Time is not just for manufacturing: A service perspective, *Industrial Management + Data Systems*, 100, 2, 51, 2000.
18. Bane, R., Leading edge quality approaches in non-manufacturing organizations, *Quality Congress. ASQ's Annual Quality Congress Proceedings*, 245, 2002.
19. Bane, R., Leading edge quality approaches in non-manufacturing organizations, *Quality Congress. ASQ's Annual Quality Congress Proceedings*, 245, 2002.
20. Buzby, C.M., Gerstenfeld, A., Voss, L.E., and Zeng, A.Z., Using lean principles to streamline the quotation process: A case study, *Industrial Management + Data Systems*, 102, 8/9, 513, 2002.
21. Erlich, B.H., Service with a smile, *Industrial Engineer*, 38, 8, 40, 2006.
22. Erlich, B.H., Service with a smile, *Industrial Engineer*, 38, 8, 40, 2006.
23. Furterer, S. and Elshennawy, A.K., Implementation of TQM and lean Six Sigma tools in local government: A framework and a case study, *Total Quality Management & Business Excellence*, 16, 10, 1179, 2005.
24. Pannell, A., Happy together, *Industrial Engineer*, 38, 3, 46, 2006.
25. Pollitt, D., Lean delivers savings for Siemens Business Services, *Training & Management Development Methods*, 19, 3, 223, 2005.
26. Bowen, D.E. and Youngdahl, W.E., "Lean" service: In defense of a production-line approach, *International Journal of Service Industry Management*, 9, 3, 207, 1998.
27. Bane, R., Leading edge quality approaches in non-manufacturing organizations, *Quality Congress. ASQ's Annual Quality Congress Proceedings*, 245, 2002.
28. Caldwell, C., Brexler, J., and Gillem, T., Engaging physicians in lean Six Sigma, *Quality Progress*, 38, 11, 42, 2005.
29. Fleming, J.H., Coffman, C., and Harter, J.K., Manage your Human Sigma, *Harvard Business Review*, 83, 7, 106, 2005.
30. Furterer, S. and Elshennawy, A.K., Implementation of TQM and lean Six Sigma tools in local government: A framework and a case study, *Total Quality Management & Business Excellence*, 16, 10, 1179, 2005.
31. Harrington, H.J. and Trusko, B., Six Sigma: An aspirin for health care, *International Journal of Health Care Quality Assurance*, 18, 6/7, 487, 2005.
32. Hug, Z., Managing change: A barrier to TQM implementation in service industries, *Managing Service Quality*, 15, 5, 452, 2005.
33. Lindberg, E. and Rosenqvist, U., Implementing TQM in the health care service: A four-year following-up of production, organizational climate and staff wellbeing, *International Journal of Health Care Quality Assurance,* 18, 4/5, 370, 2005.
34. Patton, F., Does Six Sigma work in service industries? *Quality Progress*, 38, 9, 55, 2005.
35. Prajogo, D.I., The comparative analysis of TQM practices and quality performance between manufacturing and service firms, *International Journal of Service Industry Management*, 16, 3/4, 217, 2005.

36. Juras, P.E., Martin, D.R., and Aldhizer, G.R., Adapting Six Sigma to the SOX 404 compliance beast, *Strategic Finance*, 88, 9, 36, 2007.
37. Rylander, D.H. and Provost, T., Improving the odds: Combining Six Sigma and online market research for better customer service, *S.A.M. Advanced Management Journal*, 71, 1, 13, 2006.
38. Brignall, S., A contingent rationale for cost system design in services, *Management Accounting Research*, 8, 2, 325, 1997.
39. Silvestro, R., Fitzgerald, L., Johnston, R., and Voss, C., Towards a classification of service processes, *International Journal of Service Industry Management*, 3, 3, 62, 1992.
40. Hussain, M.M. and Gunasekaran, A., Activity-based cost management in financial services industry, *Managing Service Quality*, 11, 3, 21113, 2001.
41. Snee, R.D. and Hoerl, R.W., *Six Sigma Beyond the Factory Floor*, Pearson Prentice Hall, Upper Saddle River, NJ, 2005.
42. Bowen, D.E. and Youngdahl, W.E., "Lean" service: In defense of a production-line approach, *International Journal of Service Industry Management*, 9, 3, 207, 1998.
43. Johnston, R., Service operations management: Return to roots, *International Journal of Operations & Production Management*, 25, 12, 1278, 2005.
44. Christensen, C.M., Anthony, S.D., and Roth, E.A., *Seeing What's Next, Using the Theories of Innovation to Predict Industry Change*, Harvard Business School Press, Boston, 2004.

Chapter 6
How Service Techniques Are Being Extended to Manufacturing

John walked on. He had never had such a strange dream. It wasn't unpleasant; in fact, he was mildly excited by the experience (and John was not an excitable person) and was looking forward to the next stage.

As he approached, he could see a bustle of activity, much like at the last stage. It wasn't quite as well organized, but what it lacked in a structured approach, it made up for in enthusiasm. As usual, John asked the nearest person, "What are you doing?" A bright-eyed young man turned and, with a burst of obvious exultation, exclaimed, "We are sending a shipment of our products to Manufacturing. They have been so good to us over the years in sharing their products with us that we want to return the favor and share some of our products with them. You see, we are ruled by a special breed of people. We call them 'Customers.' Although we depend on them for our existence, we find them very hard to care for and it is difficult to keep them from deserting us and moving to another area. We can't survive if we don't have customers and supplies from Manufacturing. We have always tried to be the buffer between the customers and the Manufacturing towns. We believed that would make it easier for Manufacturing to make their products without having to deal with the customers. Now, we are finding that some customers would be happier if they could have some input into the products that the Manufacturing towns make, instead of just accepting whatever goods we get for them. We have learned a lot about how to deal with customers and we want to share that with Manufacturing. They may not understand yet how difficult it is but, if we can work together with them, we believe that we can keep the customers around for a long time."

This stage was making perfect sense to John Merchant. He asked the young man another question: "If the customers want to talk directly to Manufacturing, do you worry that they will not need you anymore?" John thought he had asked an intelligent question. But deep in his mind, he knew it was also a realistic question; he wondered if he was losing some of his customers to Manufacturing. At this point, the young man stopped his work and looked at John directly. "Yes, that does concern us. In fact, we think some of our customers are buying their products directly from the manufacturer through something called the Internet. Still, we seem to be very busy, but something peculiar is happening over here in the town of Services. It is as if we are becoming more like the town of Manufacturing and they are becoming more like us. As a result, both of our towns are actually prospering quite well. We learn from them, and they learn from us."

John picked up the pamphlet from the stage. It was entitled "From Services to Manufacturing." He was getting more interested, and, at the same time, he was finding some meaning to his experiences. Sometimes they didn't make a lot of sense, but John was beginning to see that if he could somehow apply some of this to his business, he would be better off.

A Pause to Reflect: John awoke and sat up in bed. Whew! That was a strange dream, but he was sure that if he could remember it, he might find some good ideas for this business. First, he remembered that he was in the retail business. He had customers and they were sometimes difficult to please. He depended on manufacturing to provide the goods to sell to his customers. From the first stage, he learned that there were a number of interesting exhibits in the hallway. At the next stage (Chapter 2), he learned that, although there were lots of things to do in a business, a few things were the most important, even critical to the survival and success of the business. The phrase "critical success factors" seemed to be repeating itself in his mind. From the next stage (Chapter 3), he had learned that this thing called a supply chain connected the supplier to the customer and was one of the ways to improve the quality of the products and services that he could provide to the customer. At the next stage (Chapter 4), he had learned that there were three types of business activities: sustaining or day-to-day activities, problem-solving activities, and, finally, improvement activities. He was convinced he was doing a good job on the first two activities, but far from doing anything that could be considered actual improvement activities. He also recognized that if he didn't do a better job, he might lose his customers to some other competitor or to the manufacturer.

What he had seen in the last two stages (Chapter 5 and Chapter 6) began to make sense. Both of the groups were talking about improvement programs. Manufacturing had developed a lot of ways to reduce costs, improve quality, and increase flexibility in supplying customers, the wholesalers, and retailers. At the same time, the service industries had become very good at developing improvements in the customer service areas. Finally, they were beginning to work together and with their customers to make even greater improvements. John wanted a piece of this action, too. He knew that if he could somehow take on an improvement program, his business might do better.

John was now ready to get up, go to work, and get started on implementing some improvement programs. He knew he could find out a great deal about the specific programs and how to adapt them to his business. If he couldn't figure it out for himself, he could read a book, network with some of his peers at other businesses that had tried these programs, or hire some consultants to help him do it. But as the day progressed, something was nagging at him. What was it he had heard at the last professional society meeting? The speaker cautioned that the new technology and methodologies that constituted these improvement programs were good, but they were not the complete answer. Businesses also had to realize that it would be necessary to change their infrastructure and fit the improvement program with the culture of the company.

John knew some things about the new technologies in his industry; he was less familiar with infrastructure issues, and this thing called culture was altogether new to him. He remembered he had taken a sociology course in college. Was that what culture was all about? If so, he knew he was in trouble; that was the one class he kept oversleeping through during that semester! That night, John once again settled into a restless sleep.

Introduction

In Chapter 5, we described the movement of continuous improvement programs from the manufacturing sector to the services sector. In this chapter, we will describe the movement of continuous improvement programs from services to manufacturing. For a long time, services were considered a necessary by-product of the more important agriculture and manufacturing sectors of the economy. Many authors even claimed that true wealth could be created only through the production of goods. In recent years, however, there has been a change in this thinking. Today, most business authorities agree that services are an integral and increasingly important part of the global economy, at least in industrialized nations. In other words, services have become the "tail that wags the dog." Service businesses were dependent on the manufacturing sector for its improvement ideas for a long time. However, in recent years, a number of new concepts and techniques have been developed in the services sector and are being extended back into the manufacturing sector. Before we look at some of these improvement programs, we will review the reasons that services have achieved their increased level of dominance.

The Rise of Services as a Part of the Economy

Services are sectors of the economy whose time has come. The movement from agriculture to manufacturing to services is a natural progression for most countries. In its simplest form, the progression is from providing survival (food) to providing increased standard of living (manufactured goods) to increased enjoyment (services). Automation and improved processes leading to giant increases in productivity reduced the need for workers in agriculture and provided the opportunity for increased manufacturing of goods that could further increase productivity (tools and machines) or provide a higher standard of living (transportation and appliances). As increased productivity reduced the need for manufacturing employees, many moved into the services area, some because it was an attractive change and others because they had no other alternative.

The Swing of Power from Manufacturing to Retail

The growth of services has also brought a shift in economic power. Until the latter part of the 20th century, manufacturing companies tended to have leverage over service companies because of their size. Although manufacturers were the dominant businesses in the past, retailers, such as Wal-Mart and Home Depot, software developers, such as Microsoft, or telecommunications companies, such as AT&T, have become more significant participants in the business world. As retailers have increased in size and purchasing power, they have imposed service requirements on manufacturers. Retail clothing stores want the manufacturer to make the apparel "floor ready" by sorting by category and attaching prices and security tags. Grocery stores want service merchandisers, in such areas as bakery goods or health and beauty aids, to keep the store's inventory stocked with the right products and an attractive appearance. Restaurant chains want frequent and reliable merchandise deliveries to keep their inventories low and fresh. Home improvement chains want attractive packaging and liberal return policies, especially for seasonal goods.

In the United States, services account for over 75 percent of the work force and 90 percent of the new jobs being created. A similar pattern exists in other industrialized countries. As a result, the focus of most companies, whether manufacturing or service, is on the customer. Improvement programs, even those designed to reduce costs or improve product quality, must consider the implications for the customer. As this occurs, some programs can help manufacturing companies provide better customer service; however, they may also increase the cost of providing the product/service bundle.

Maturity of the Customer as a Shopper

In the past, shoppers were less informed about the product or service offered by a company. As a result, they were often satisfied with a basic product with little in the way of a complementary service package. Today, shoppers are more knowledgeable about almost all products, thanks to the Internet, where they can conduct product and price comparisons. As they find an added feature on one product and another feature on a second product, they become more demanding of the product they buy. To enhance their products, manufacturers must often add services. As we described in Chapter 3, customers are not only the recipients of goods and services, they may also be involved in the design of the goods and services they buy.

Increasing Complexity of the Marketplace

The marketplace is changing. Customers have become more discriminating and desire greater variety in the products and services from which they can choose. Manufacturers have created greater variety in products; however, so have their competitors. To gain a competitive advantage, many companies are turning to a greater level of services. This adds complexity so that the marketplace has taken on a different configuration than in the past. Markets no longer distribute themselves in a normal curve; they tend to have at least two major segments — one with customers who continue to want low-cost products and services, and a second market segment with customers who want customized products and are willing to pay and wait for that customization. We will describe that more in the section on mass customization.

Need for Manufacturing Companies to Add Services

Manufacturing companies are finding that they need to add services to complement the goods that they produce. Customers are demanding services to accompany the products they buy. For personal computers, they want help desks and security software downloads. In ordering after-dinner mints, they want the manufacturer to put their names on the package. For home assembly of furniture, they expect that the directions will be complete and easily understood. James Brian Quinn was one of the early writers to project that successful companies in the future would gain their competitive advantage from having service-based "core competencies" to complement their superior products. He further predicted that physical facilities — manufacturing plants and other capital-intensive facilities — would not likely be enough to provide the competitive edge that companies would need. He explained, "Such physical properties are too easily cloned. From now on, intelligent enterprises will derive sustainable advantage from knowledge and service-based activities that leverage intellectual assets."[1]

One way in which manufacturers can provide an almost seamless link between their product and the accompanying services is by embedding the service as part of an electronic network that can detect potential problems and, in some cases, take automatic action to prevent or ameliorate the problem. Several strategies can be employed in this approach:

- Embedded innovator. This is the most product-centric model. Customers may still perceive the physical product as the source of primary value, and they will expect to continue receiving the support services they have in the past (installation, warranties, maintenance contracts, and so on). Example: Heidelberg, a maker of high-end printing presses, is an embedded innovator.
- Solutionist. A single product is still the dominant gateway to a business opportunity, but the scope of high-value activities with the product is broader. Example: GE with its MRI scanner illustrates the solutionist.
- Aggregator. A number of entities are involved in providing a collection of services and the aggregator is the firm that controls the actual data collection and central-processing power. Aggregators are primarily product companies that do not integrate all aspects of the services related to their product. Example: Gardner Denver makes air compressors and monitors the compressed air processes to provide information to manufacturers of air coolers, filters, and dryers.
- Synergists. Firms that participate in the system by providing a specific device or service to the system. Example: Phillips provides lighting ballasts and controls.[2]

All of these added features require manufacturers to consider their customers and how best to please them, concerns that service companies have always dealt with. Although this extension of products and services is innovative, it also increases the complexity of a business, perhaps to the point where it is difficult to separate innovation from complexity.[3]

Jay Galbraith cautions that blending products and services is not easy. He proposes that companies need to change from offering a product to providing solutions. When they do attempt this change, they may have trouble because they have not restructured their organizations to become customer-centric.[4]

Another possible complication in adding services to a goods-producing company is in the decision-making process. Whereas decisions for a goods-producing company can usually be based on quantitative measures, providing customer service often requires a trade-off between increased quantitative costs versus less tangible benefits.[5]

Move from Product-Centric to Customer-Centric

Another factor that is providing the need to move service industry thinking back to manufacturers is the change in orientation from a product-centric to a customer-centric environment. This change is more than just a need to beef up the product with some selected services or just to involve the customer in some of the discussions on product design. It is a new way of thinking about the goals and strategies of a manufacturing company. Barbara Bund describes this as an "outside-in" way of looking at the business, meaning through the eyes of the customer.[6] Stephen Haeckel describes it as a "sense and respond" approach rather than a "make and sell" approach.[7]

Table 6-1 compares some of the characteristics of product-centric companies with customer-centric companies. Many companies are experiencing difficulty in moving through this transition. It is both an organizational problem and a people problem. How do you begin? How do you measure progress in becoming customer-centric? Some researchers warn that many companies become complacent and underestimate the long-term need for building a more customer-centric business.[8] They point out that most established companies have the means to renew growth effectively by creating by-products from the company's core business. Others caution that it is important to match its organizational structure and behavioral norms when switching to a customer-centric business strategy.[9] Historically, businesses have focused on the lifetime value of a product and sometimes failed to measure the lifetime value of a customer.[10] Managerial Comment 6-1 describes steps in becoming a customer-engaged organization.

Managerial Comment 6-1.
Steps to a Customer-Engaged Organization

1. Reframe the task at hand. Move from "creating a marketing program" to "creating an organization where everyone is focused on and energized about the mainstream of the business — delivering value to customers." In other words, you are working in the world of organizational and cultural change.

Table 6-1 Comparison of Product-Centric and Customer-Centric Strategies

Organizational Factors	Company Strategies	
	Product-Centric	Customer-Centric
Goal of the company	Make the best product for the customer	Provide the best solution for the customer
Mental process	Divergent thinking: How many possible uses of this product?	Convergent thinking: What combination of products is best for the customer?
Organizational structure	Based on product segments	Based on customer segments
Most important process	New product development	Customer relationship management
Performance measures	Tangible: Number of new products, percent of market share	Intangible: Customer satisfaction, lifetime value of a customer, retention
Organizational culture	Open to new product and process ideas; internally focused	Searching for more ways to satisfy customers; externally focused
Main offerings	Specific and cutting-edge products	Personalized package of services
Human resource philosophy	Power to employees who develop products	Power to employees with in-depth knowledge of customers' business

Source: Adapted from Galbraith, J.R., Organizing to deliver solutions, *Organizational Dynamics,* 31, 2, 194, 2002.[4]

2. Mobilize the leadership team. Have your leadership team work together to develop a set of "customer value hypotheses" — real issues, propositions, concerns, and other customer-related matters that speak to what makes customers choose you rather than your competition. Then, have the leadership team charter a functionally and hierarchically diverse team — a diagonal slice of the organization whose task is to conduct face-to-face value conversations with your customers.

3. Engage customers in value conversations. "Conversations" is exactly the right word here. It suggests the open discussion that enables your people to

understand more profoundly — both intellectually and viscerally — what is important about your customers' businesses. These kinds of probing, open-ended questions are certainly nothing new to marketing professionals. But remember who it is that is engaging customers in these value conversations: representatives of all functions throughout your organization. It will be an energizing revelation to them.

4. Create a customer value guide. The customer value guide (CVG) is a synthesis of all the value conversations. Its purpose is to provide a "North Star" by which everyone can navigate through the maze of decisions that are a part of work. The CVG is a series of brief, compelling statements that capture and convey the essence of what represents value to your customers.

5. Conduct customer-engagement workshops with all employees. These workshops serve as the visceral link that connects everyone in the organization with customers. Ultimately, they are intended to ensure that all employees can answer two fundamental questions:
   - What makes our customers decide to buy from us rather than from our competition?
   - How does the way I do my job each day affect the likelihood of customers choosing us over the competition?

6. Integrate customer value insights into the organization. Customer engagement has to become part of "the way we work around here." This has significant implications for activities such as business planning; organization design; performance appraisal; performance improvement infrastructure (e.g., TQM, Six Sigma, re-engineering, CRM [customer relationship management]); processes, systems, and structures; and keeping the customer connections "alive" for all employees.

Source: Adapted from Band, W. and Guaspari, J., Creating the customer-engaged organization, *Marketing Management*, 12, 4, 34, 2003.[11]

Services as a Separate New Business Segment

Manufacturers may find that the movement of the economy toward services will provide an opportunity to view services as new "product" opportunities. If manufacturers view the life cycle of their products more comprehensively, they may find opportunities.

There are several service possibilities to complement a company's existing product line:

- Financing the purchase of their product
- Installing the product
- Modifying other products or processes to work with the product
- Maintaining the product and selling replacement parts
- Training personnel to use the product
- Disposing of product waste
- Disposing of the product

Glen Allmendinger and Ralph Lombreglia state that any industrial manufacturer that has not awakened to the fact that it must become a service business is in serious peril today. Soon, it will not be enough for a company to offer services; it will have to provide "smart services." Smart services are services that help a customer to know ahead of time that a machine is about to fail, that a customer's supply of consumables is about to be depleted, and so on. It is a way to help the customer use the product more effectively and efficiently, usually by providing a connected device that sends feedback to the customer for manual intervention, or senses a problem and automatically takes a corrective or preventive action.[12]

Manufacturers add services to generate additional revenue. IBM was originally a manufacturing company with practically all of its revenue coming from the computers it produced. Today, it is known more as a service company because the majority of its revenue originates from service lines of business.

Economic Growth and the Need for Added Services

As economic conditions improved in the industrialized countries, consumers had time and money to pursue added interests, especially in leisure activities. Travel and tourism were available to an increasing percentage of the world's population. Sports activities, such as skiing and golf, grew rapidly. Professional sports — football, baseball, and tennis — attracted huge crowds. In other words, businesses had to develop new ways (services) for individuals to spend their money.

The Movement from Make-to-Stock (MTS) to Make-to-Order (MTO)

As customers become more discriminating, they want more choices not only in the products they buy, but also in the services that accompany those products. One recent movement is toward customization of orders. If manufacturers can reduce their response times sufficiently, they can wait until they get the order before they make the product. This requires greater communication and coordination between supplier and customer, both service-type activities.

June Ma and her colleagues studied the effects of MTS versus MTO programs on a supply chain. They found both benefits and problems in making the transition. Specifically, they found that introducing the volatility of MTO demand could cause problems in production planning and execution. They concluded that businesses should take an integrated view of demand management, production planning, transportation service selection, and information flow across the whole supply chain. For best results, businesses should carefully combine their goods-producing processes with their service-providing processes.[13]

The Movement toward Mass Customization

Closely aligned with the MTO movement is the concept of mass customization. The idea behind mass customization is to provide a product or service almost as fast and at a price almost as low as standard products and services that were provided in the past.

B. Joseph Pine popularized the movement toward mass customization. He described the approach as one in which manufacturers could customize their products by making their processes more flexible and capable of transforming their products.[14] Some manufacturers have succeeded in making their processes more flexible and are able to keep inventory in a subassembly stage until they receive the customer order, after which they quickly customize the finished product to meet the customer's requirements. They are essentially moving from a make-to-stock strategy to an assemble-to-order strategy. This product design strategy, which shifts product differentiation closer to the consumer by delaying identity changes (such as assembly or packaging) to the last possible supply chain location is known as postponement.[15] Dell is a prominent user of this strategy.

However, many companies have tried to address the need for mass customization by simply making a wider variety of products. This approach offers more options to smaller market segments, but it is not the same as customizing the product for an individual customer, whether a business or a consumer. Some consider the increased variety as both costly to the manufacturer and stressful to the consumer.[16]

Mass customization increases the complexity and costs of making products. As an offset, mass customization provides a benefit when the customer is involved in the co-design of the product because this simplifies and focuses the entire product line and increases customer loyalty.[17]

The factors described above contributed to the growth in services. But what are services? What are some of the continuous improvement programs that were developed in services and are applicable to manufacturing? We will examine both of these questions in the following sections.

What are Services?

The word "services" has many meanings. Roger Schmenner suggests that services are better defined by what they are not rather than what they are. He explains that "the U.S. government defines service employment as a residual, namely, as nonfarming, nonmanufacturing employment."[18] Heskett also concludes that services can be classified as any non-goods-producing business. The goods-producing industries are mining, agriculture, and manufacturing; all else are services.[19] In a more recent book, the authors explain, "Economically, the term services is often defined not by what it is, but by what it is not. Historically, economic reports identify activities as 'service producing' that are not 'goods producing,' which includes manufacturing and construction, and are not 'extraction,' such as agriculture, forestry, fishing, and mining."[20] We will use the same rationale in describing services. In Chapter 3, we described the Input–Transformation–Output (ITO) Model and explained that the model could be used to describe either goods-producing or service-producing.

We will use Figure 6-1 to show the relationships among various types of goods and service producers. We will also identify some of the industries that fall within each of these service types. In Chapter 11, we will describe more fully the relationship between and among these business areas. In this chapter, we will focus primarily on the improvement programs developed in services that are now spreading into the manufacturing and extractive industries. Managerial Comment 6-2 reviews the basic attributes of service.

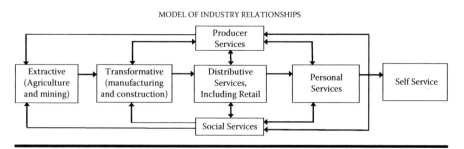

Figure 6-1 Industry classifications.

Managerial Comment 6-2.
Service

In economics and marketing, a service is the nonmaterial equivalent of a good. Service provision has been defined as an economic activity that does not result in ownership, and this is what differentiates it from providing physical goods. It is claimed to be a process that creates benefits by facilitating a change in customers, a change in their physical possessions, or a change in their intangible assets.

By supplying some level of skill, ingenuity, and experience, providers of a service participate in an economy without the restrictions of carrying stock (inventory) or the need to concern themselves with bulky raw materials. On the other hand, their investment in expertise does require marketing and upgrading in the face of competition that has equally few physical restrictions. Providers of services make up the tertiary sector of industry.

Services can be described in terms of their main attributes:

Intangibility: They cannot be seen, handled, smelled, etc. There is no need for storage. Because services are difficult to conceptualize, marketing them requires creative visualization to evoke effectively a concrete image in the customer's mind. From the customer's point of view, this attribute makes it difficult to evaluate or compare services prior to experiencing the service.

Perishability: Unsold service time is "lost," that is, it cannot be regained. It is a lost economic opportunity. For example, a doctor who is booked for only two hours a day cannot work those hours at a later time — she has lost her economic opportunity. Other service examples are airplane seats (once the plane departs, those empty seats cannot be sold) and theatre seats (sales end at a certain point).

Lack of transportability: Services tend to be consumed at the point of "production" (although this does not apply to outsourced business services).

Lack of homogeneity: Services are typically modified for each client or each new situation (customized). Mass production of services is very difficult. This can be seen as a problem of inconsistent quality. Both inputs and outputs to the processes involved providing services are highly variable, as are the relationships between these processes, making it difficult to maintain consistent quality.

Labor intensity: Services usually involve considerable human activity, rather than a precisely determined process. Human resource management is important. The human factor is often the key success factor in service industries. It is difficult to achieve economies of scale or gain dominant market share.

Demand fluctuations: It can be difficult to forecast demand (which is also true of many goods). Demand can vary by season, time of day, business cycle, etc.

Buyer involvement: Most service provision requires a high degree of interaction between client and service provider.

Client-based relationships: Based on creating long-term business relationships. Accountants, attorneys, and financial advisers maintain long-term relationships with their clients for decades. These repeat consumers refer friends and family, helping to create a client-based relationship.

Source: From Wikipedia, the free encyclopedia.

Figure 6-1 shows the relationships among types of industries. At the left, we see the extractive industries of agriculture and mining. These industries provide the raw materials for the transformative industries of manufacturing and construction. Everything else in the diagram can be considered as services. Transformative industries depend on distributive services to move the goods from the factory to the retail stores or directly to the consumer, the personal services area. This horizontal flow completes the supply chain (described in Chapter 3). In addition, we have two other major service areas: producer or business services (the services provided to the goods-producing businesses by other businesses) and social services (provided by government or other nonprofit entities). We will explain each of these service areas briefly below and more fully in Chapter 11.

Distributive Services

There are two principal kinds of services in this category — transportation, as in the movement of goods, and interim storage, such as in regional distribution centers. The functions are becoming more costly as supply chains lengthen, and more critical as customer service demands increase. Distributive services also include the retail outlets that are primarily concerned with making products such as food, clothing, and appliances available to consumers.

Personal Services

There are two main areas of personal service. One is the retail outlets that provide some products, but are primarily thought of as service outlets, such as restaurants, hotels, movie theaters, and theme parks. The other kind of personal service is represented by a one-to-one service, such as that provided by doctors, lawyers, teachers, automobile mechanics, hair stylists, travel agents, stockbrokers, and physical fitness trainers.

Self-Service

There is a special kind of service that is increasing in popularity. It is self-service, where the consumer provides part of the service package by becoming an active participant in the service process, such as in pumping gas at a service station, using the self-checkout station at a grocery store, or buying products online. Technology provides the capability to make self-service possible and the pressure to reduce costs make it attractive. In addition, busy life-styles encourage consumers to participate willingly, even eagerly.

Producer (Business) Services

Producer services are those services provided to businesses by other businesses. Some examples of this type of service include banks, law firms, insurance companies, consultants, public accounting firms, and architectural firms. In addition, other third-party businesses may provide services that were previously handled internally by the manufacturing companies such as food service, janitorial services, and information processing.

Social Services

Governments or nonprofit agencies provide social services. These are services that are not purchased directly, such as producer services, but which may be of great value to a business. Such services include employee training, economic benefits to attract industry, business activity information, and infrastructure, such as roads and utilities access. They may be beneficial; however, they may also impose constraints on a business if they involve mandates to reduce hazardous emissions or increase financial reporting requirements.

Knowledge Transfer From Services to Manufacturing

In Chapter 5, we described how service companies could improve by using improvement programs that had been developed in manufacturing. In this chapter, we will describe programs that have been developed in services and can be used in manufacturing. In some cases, manufacturing companies will want to use these concepts because they will provide a benefit. In other cases, manufacturing companies will have to adopt the programs to remain competitive.

Areas of Manufacturing Expertise

Manufacturing designed and implemented programs focused on cost reduction, product quality and reliability improvement. They tended to be isolated or decoupled from direct customer contact, both physically and systemically. Physically, most of the customer contact resided in the sales and marketing departments. Systemically, manufacturers tended to produce standard products to a demand forecast in a make-to-stock mode. To utilize their costly investment in plant and equipment efficiently, they used a level production strategy. When demand was less than production, they built inventory, and worked down the inventory when demand exceeded production. They designed production jobs to be specialized to improve productivity. They managed the supply side of the demand/supply relationship. To a large extent, they operated in a closed-system environment, shielding their technical core from outside interference. In doing this, they also built a collection of service-type operations — payroll, accounting, engineering, human resources — that were primarily concerned with supporting the manufacturing operations and only incidentally concerned with servicing the customer. Although this promoted efficiency, it left a lot to be desired when it came to customer service.

On the other hand, companies in service industries designed and implemented programs that were primarily focused on improving customer service. Because they did not have the luxury of building inventory to carry them through the peaks and valleys of demand, they had to develop the capability to handle demand variability by designing volume-flexible capacities. Because employees came in direct contact with customers, service companies had to empower employees to handle a variety of situations. In so doing, these companies found that the traditional vertical hierarchical organization structure lacked flexibility; consequently, service industries began to develop alternate forms of organizations that were flatter and facilitated horizontal information flow and decision making. Of necessity, service companies operated in an open-system environment. They managed the demand side of the supply/demand relationship.

Areas of Service Expertise

Although service businesses have not been as effective or efficient as manufacturing businesses, they have always had a customer orientation, a quality sometimes lacking in manufacturing. Many of the continuous improvement programs that developed in the service sector provided improved customer service. Some of the areas in which service businesses have special skills include:

- Customer relationship management (CRM)
- Adaptive organizations
- Reducing response times
- Quality as customer perception
- Qualitative goals and performance measures
- Supply chain management
- Flexibility and agility to adapt to changing needs — customization
- Knowledge workers as the ultimate "perfect" state

Services have been concerned with customer service, quality as a customer perception and flexibility of demand management.[21] Although manufacturing companies respond or react to demand by using inventory as a buffer against demand variability, service companies are more inclined to manage demand.

Examples of Programs Developed in Services

It is easy to find articles about how manufacturing has provided concepts and techniques that can be used in service businesses. It is difficult to find the reverse — how service concepts can be used in manufacturing.

In this section, we describe some of the programs that have originated in services. A number of the concepts that originated in services involved the participation of manufacturing as a partner in the process, such as in Quick Response programs. Other programs, such as the use of the Internet to sell direct to consumers, have been adopted by manufacturing companies. Figure 6-2 shows the approximate relationship of programs, both as to origin and with respect to time. The location of the programs is representative, not absolute, because it is impossible to separate the programs into neat categories. The figure suggests that, contrary to the general movement of improvement programs from manufacturing to services, some programs have originated in services.

Program Focus	1975	1980	1985	1990	1995	2000	2005		
\multicolumn{10}{l}{Evolution of Continuous Improvement Programs - Manufacturing and Services Oriented}
Planning	MRP		MRP II		ERP		ERP Exp.		Origin in manufacturing
Execution		CIM		MES	WMS	APS			
Cost reduction		JIT			Lean		Lean SS		
Quality of goods	SPC			TQM		Six Sigma			
Measurement - tangibles		ABC	ABM		BSC				
Integration				S&OP	SCM	SCM Exp.	SCM-SOA		Combined
Measurement - intangibles					BSC				
Quality of services				TQM		Six Sigma			
Customer						CRM			Origin in services
Response time				QRS	ECR	VMI	CPFR		
Flexibility						Mass Cust	Agile		
Communications		EDI			I-EDI	B2B	B2C		
	1975	1980	1985	1990	1995	2000	2005		

Figure 6-2 Focus of improvement programs.

Customer Relationship Management (CRM)

CRM has evolved from a data-collection technique to a relationship-building strategy. It is a program with value for both the services and manufacturing sectors.[22] CRM builds on the concept that it is better — more profitable — for a company to retain customers than it is to keep trying to find new customers. There is a widespread feeling that loyal customers buy more and may cost less to service than occasional customers do.[23] CRM programs help to develop relationships with customers that cause the customer to want to continue to do business with the company. These programs provide a framework for developing business processes and their supporting infrastructure to improve service delivery. Initially, they focus on automating internal functions, such as sales, marketing, and customer service. However, the hardware and software have to be supplemented by the employees and business processes to implement the CRM system successfully.[24]

Initially, CRM programs focused on obtaining information about customers, such as location, income levels, and buying habits. Companies used this information to plan some of their product and service strategies. Satisfied that there was value in finding out more about their customers, some businesses went even further by asking them directly what they wanted or needed. Over time, suppliers built a relationship with their customers that made it possible to anticipate demand patterns more accurately. With this knowledge, suppliers were able to provide a higher level of customer service.

CRM is one of the more lucrative areas for software developers. In the foreseeable future, the market will be approximately $17 billion for the software and upwards of $100 billion if CRM-related activities, such as consulting and support services, are included.[25] In the following discussion of CRM, we draw heavily from an earlier publication in the *APICS Magazine*.[26]

Definitions

First, about definitions: one study lists 12 different definitions for CRM, ranging from data-driven marketing to an E-commerce application, with numerous variations and extensions in between.[27] Another study found descriptions of CRM as a process, a strategy, a philosophy, a capability, and a technology.[28] A study of acronyms reports that CRM is guilty of "straying acronym syndrome" (SAS) in which the words used vary over time. Agnes Nairn tracked CRM from customer relationship marketing to customer relationship management until 1998; since 1999, PRM, for partnership relationship management, is in vogue, and even more recently e-CRM, for electronic or E-business, has become popular.[29] A final reference describes Operational CRM (OCRM), Analytic CRM (ACRM), Conceptual CRM (CCRM), and Technical CRM (TCRM), and provides an Enterprise Information Roadmap (EIR) to show the relationships among the different versions.[30]

We will use the *APICS Dictionary* definition for CRM:

> CRM is a marketing philosophy based on putting the customer first. The collection and analysis of information designed for sales and marketing decision support (as contrasted to enterprise resources planning information) to understand and support existing and potential customer needs. It includes account management, catalog and order entry, payment processing, credits and adjustments, and other functions. *Syn:* customer relations management.[31]

How about the results of CRM implementations? For the most part, users have not been ecstatic about their results. Many have failed to realize the "perfect order" status heralded by CRM proponents. A Meta Group study found a 55 to 75 percent implementation failure rate.[32] Another study reports that 55 percent of all customer relationship management (software solutions) projects do not produce results.[33] In a survey of 451 senior executives, 25 percent reported that these software tools failed to deliver profitable growth and in many cases damaged long-standing customer relationships.[34]

Should we conclude that CRM is ready to join a host of other "management fads" and fade away? Just the opposite. From its origin in the mid-1990s, CRM had a burst of popularity until 2000, when it fell into decline because of limited success and tighter IT budgets. After a low point about 2004, it is suddenly in vogue again. There should be renewed growth, particularly in nontraditional service areas, such as government and education.[35] One reason for the renewed interest is that many supporters conclude that CRM is not just a marketing program; it is an essential part of the modern supply chain.

Background

How did CRM get started? One driver was global competition, which forced many companies to become more customer-oriented as a means of securing a competitive advantage. Progressive businesses want to become "customer-centric." Many expressed this as simply formalizing their ever-present "caring for the customer" position; however, a number of people viewed it as a blending of capabilities not heretofore available. A host of IT technologies made it possible to collect and analyze data about customers, and then to translate that knowledge into meaningful marketing strategies.

Another driver was the need for ERP vendors to find new products to sell. After the dizzying burst of ERP implementations leading up to Y2K, it became apparent that the next wave of software implementations was in supply chain management (SCM). SCM offered a bonanza of applications that included CRM, along with other processes that needed software and implementation consulting help. Although specialist vendors such as Seibel developed CRM software, the major ERP vendors (SAP and Oracle) are selling extensions of their ERP packages that include CRM software. Even Microsoft wants a share of the market, with the introduction of Microsoft's Dynamic CRM Live in 2007.[36]

What Does It Do?

What should the CRM program do? Shaun Doyle provided a detailed checklist of activities in CRM.[37] Raymond Ling and David Yen developed an analysis framework and implementation strategies for CRM. They also describe the evolution from the direct sales of a bygone era to mass marketing in the 1960s to target marketing in the mid-1980s to CRM in the 1990s.[38] Adrian Payne and Pennie Frow provided a conceptual framework for CRM strategy consisting of the strategy development process, value creation process, multi-channel integration process, information management process, and the performance assessment process.[39] CRM programs have the following major components:

- CRM collects information about customers, primarily from sales transactions, but also from a number of other "touch points," such as complaints or inquiries. E-CRM is popular as a means of recording Website contacts, searches, and other nonsales activities. For example, Amazon.com recommends books based on your previous searches and requests that you provide reviews of books that you have purchased.

- From the collected data, a CRM program organizes the customer base into segments or groups of similar customers. The groups may organize around age, income level, location, books searched, or whatever the marketing group deems useful. The ultimate objective is to get to a group of one (an individual customer), if that is practical.
- The marketing organization designs a program to appeal to the groups described above. Although the primary emphasis of a CRM program is customer retention, marketers are not above designing sales programs that will attract new customers. One example describes some typical programs as customer service, frequency/loyalty programs, customization, rewards programs, and community building.[40]
- Marketing implements the programs designed to enhance the relationship with existing customers. Sometimes sales people may feel they are asked to change from a "hunt and kill" mode to a "tend the farm" mode.[41] It means that progress in these programs will be monitored closely and results will be measured.
- The marketing literature does not say much about "how" the customer is better served. It stresses the need for customer service, but does not always expand on its content. Sometimes, "inventory" slips into a diagram, but there is not much space devoted to its role. Operations has a key role; we will describe it later.
- The CRM program develops a set of metrics to measure the results and to revise, modify, discontinue, and reverse the marketing initiatives that have been introduced.

Benefits

The primary benefit of a CRM program is increased customer retention. There is some indication that customer retention has greater value than customer acquisition and that businesses should expect to see improved financial results from their CRM program.[42–44]

In addition to the tangible benefits, there are intangible benefits. The relationships with customers should be more open and effective. Hopefully, things will go smoothly; however, if there are incidents concerning customers, the closer relationships with them should help in their resolution.

Internally, the need to develop cross-functional programs should increase the collaboration among internal functions. Even without considering the operations functions, such as purchasing, production, distribution, and inventory management, there is a need to get sales, marketing, and IT people more comfortable with one another. Agnes Nairn highlighted the difficulty in communication between the "emotions-driven sales force and the clinical binary-driven IT expert" as difficult unless intelligently managed.[45]

Problems

Why have CRM programs not been more successful? It would be easy to blame the software vendors for "over-hyping" their product, but that does not excuse the companies that bought the software from doing their part. One book describes 53 myths attributed to customer loyalty programs. These myths include a number of promises that have not been met satisfactorily in practice.[46]

One of the leading problems has been the myopic view of CRM in considering it an IT technology and not a strategic process. It is not sufficient to just create a database of customers, no matter how cleverly designed. Another limited-scope problem has been to consider CRM as only a marketing program. Although marketing is the driver, cross-functional support is needed from the rest of the organization.

Sometimes, businesses design the software around what they think they can do (their capability) rather than around the needs of the customer. As a result, the program may be efficient at performing tasks that the customer does not care about or respond to. Some companies fail to get the support within their organization for the CRM program. Marketing personnel may not like the closer monitoring, finance may not agree with the deployment of resources, operations may feel left out, and IT may resent the cavalier attitude toward their innovative systems design. Companies may attempt to do too much too soon. Most researchers advocate a selective approach to implementation of CRM.

Finally, customers may be "turned off" instead of "turned on." Most customers want some but not excessive attention. Although subtlety is not generally associated with marketing efforts, it may be a trait to cultivate in CRM.

Relation to the Supply Chain

Although the idea may be in conflict with the marketing world, Douglas Lambert and Terrance Pohlen in the logistics field have proposed that CRM is only one of eight processes in the world of supply chain management. Other processes are customer service management, demand management, order fulfillment, manufacturing flow management, supplier relationship management, product development and commercialization, and returns management.[47]

CRM's Future

CRM research supports the conclusion that it is a management "fashion" and not a "fad." It should have great appeal in service industries, such as banks,[48] public accounting,[49] retailing,[50] insurance,[51] and increased use of Web services.[52]

However, CRM is a long way from achieving its potential. Some companies reap substantial benefits by assimilating the principles of CRM into their normal business practices. Others will go through the motions, stall, and move into a newer program that promises to be easier to implement and provide even greater benefits. Unfortunately, still others will never know that CRM was available.

Response Time Reduction

A variation of the CRM program involves programs that provide the customer with products and services faster. For example, a retail store sells stock items. When the items are out of stock, the customer may go to a competitor to get the same product. Unfortunately, the customer may like the competitor and return there the next time. In response to this type of dilemma, programs such as Efficient Consumer Response (ECR) in the grocery industry and Collaborative Planning Forecasting and Replenishment (CPFR) quickly replenish stock at the retail store. These programs also help to reduce excess inventory of slow or nonselling items that require markdowns to move them.[53]

Quick Response Systems (QRS)

This section on response time reduction programs comes from an earlier article by one of the authors.[54] In the early 1980s, a group of manufacturers and retailers in the textile and apparel industries initiated a program that would reduce the number of stockouts by providing a closer matching of demand and supply. The program was to provide a way to reduce lead times for stock replenishment orders from manufacturer to retailer and to reduce the lead times for introducing new products. There were a number of components to the program, but the essential ingredients were a willingness and a capability to communicate actual demand data more quickly from the retailer to the supplier who could have time to prepare for orders from their downstream customers. Another goal was to reduce the bullwhip effect by placing orders on a more regular and predictable schedule.

The potential was high, so how did it do? Alan Hunter reported, "In the ten years since its formulation, Quick Response has made only limited progress despite its well-demonstrated benefits to the apparel industry."[55] He listed the expected benefits as reductions in pipeline inventories, greater probability of garment designs and colors being acceptable to the consumer, ability to re-estimate stock keeping unit (SKU) demand at retail, and greater competitiveness for domestic producers facing increased levels of imports.

He also described some of the problems that have delayed the widespread adoption of QR:

- Naivety — participants did not realize the magnitude of the task
- Difficulty in creating "partnerships" — the retailers received the benefits while the suppliers incurred the costs

- Structural issues, such as the staggering number of unique SKUs (1.2 to 1.4 million at a department store every four months), overwhelming effect of fashion (shelf lives are decreasing), and the makeup of the pipeline (retailers and textile companies dominate, apparel manufacturers are small)
- Technical problems, including inadequate accuracy of bar codes, inadequate storage, and manipulation of inventory and sales data, and lack of standards in information transmission (EDI, electronic data interchange)

Finally, Hunter outlined the factors necessary for growth of Quick Response as the need for UPC/EAN compliance and standardization, the clarification and acceptance of the role of value-added networks (VANs), and the need to find a way to extend EDI to smaller manufacturers.

Kurt Salmon Associates expanded the scope of Quick Response programs to include product development and product sourcing as well as product distribution. Despite its slow beginning, QRS programs provided a model for other industries to follow and for subsequent programs to build upon.[56]

Continuous Replenishment Programs (CRP)

Closely allied with QRS, CRP programs encouraged automatic replenishment ordering so that the customers would automatically place an order when their inventory management system indicated a need for a reorder.[57]

Efficient Consumer Response (ECR)

Encouraged by the positive results of the QRS programs, in 1992 several grocery executives formed a voluntary group, known as the Efficient Consumer Response Working Group, and commissioned a study by Kurt Salmon Associates to identify opportunities for more efficient, improved practices in the grocery industry. The consultants returned in early 1993, claiming that the industry could reduce inventory costs by 10 percent, or $30 billion.[58] In addition to efficient replenishment, this group added the requirement of category management, consisting of efficient new product introduction, efficient store assortment, and efficient promotion. The program included collection of demand (sales) data with point-of-sales (POS) terminals, and feedback of this data to suppliers with EDI. Suppliers could then avail themselves of a variety of techniques, such as cross-docking, to move the product more quickly to the customer.

Vendor-Managed Inventory (VMI)

Some retailers decided that, inasmuch as the suppliers had the consumer demand data, they (suppliers) could assume the responsibility of managing their (retailers) inventory. Although the idea of suppliers managing a retailer's inventory was not new — rack jobbers and service merchandisers had performed this function in the health and beauty aids categories years before — it did have the added element of rapid feedback of demand information. One study by Andres Angulo and colleagues revealed the information technology challenges, especially the effects of information delay and accuracy.[59] Another study has concluded that, even for products with stable demand, a partial improvement of demand visibility can increase production and inventory control efficiency.[60]

Sales and Operations Planning (S&OP)

This program has been around for at least 30 years and was originally intended to get marketing and manufacturing to collaborate on a production schedule within a company. S&OP represents a way to get companies to talk with one another and smooth the flow of goods along the supply chain. In a recent survey, it was identified as the No. 2 initiative of global companies, following strategic sourcing of direct materials.[61]

Collaborative Planning, Forecasting, and Replenishment (CPFR)

In 1997, voluntary inter-industry commerce standards (VICS) created a subcommittee to develop CPFR as an industry standard. The following year, VICS issued the first document on CPFR: "VICS CPFR Guidelines," which has been constantly updated since then.[62] Although QRS and ECR provided the flow of demand information from the retailer upstream to suppliers, it was the responsibility of the supplier to anticipate demand and the retailer (except in VMI) to actually do the ordering. CPFR attempted to eliminate the disconnect by advocating that both the retailer and the supplier collaborate to plan a joint demand forecast and replenishment schedule.

Several issues that were addressed for the first time with CPFR included:

- The influence of promotions in the creation of the sales forecast (and its influence on inventory management policy)
- The influence of changing demand patterns in the creation of the sales forecast (and its influence on inventory management policy)
- The common practice of holding high inventory levels to guarantee product availability on the shelves

- The lack of coordination between the store, the purchasing process, and logistics planning for retailers
- The lack of general synchronization (or coordination) in the manufacturer's functional departments (sales/commercial, distribution, and production planning)
- The multiple forecasts developed within the same company (marketing, financing, purchasing, and logistics)

The report also provided an excellent summary of benefits, barriers to implementation, and enablers of the CPFR process, and concluded that trust and good information technology must be present for success.[63]

Supply Chain Management (SCM)

A SCM program includes all of the features of the programs described above plus other attributes that we will describe more fully in Chapter 10. An integrated supply chain is the ultimate in linking entities through communications and collaboration.

Present Status

How are companies doing in their quest to reduce response times? A simplistic answer is that the programs work, but that few companies are getting the full benefit of the programs because of incomplete implementations. There are three key components to any program of this type: technology, infrastructure, and change management. The technology, primarily in the form of electronic data collection, with POS terminals, and data transmission, with EDI or the Internet, is good and getting better. The infrastructures — organization, systems, and functional relationships — are inconsistent among companies and even industries, but improving. Managing change is a continuing problem, particularly in the area of company culture and trust. How much do you trust? Which customers/suppliers do you trust? Almost every study finds trust is not only essential for success, but also often lacking in many programs.

Coleen Crum and George Palmatier list the following reasons why demand collaboration programs have not realized their potential:

- The pace of adopting new ways of doing business is slow
- Demand information supplied by customers is not put to use in trading partners' own demand, supply, logistics, and corporate planning in an integrated manner
- Demand management and supply management processes are not integrated, and sales and operations planning is not utilized to synchronize demand and supply
- There is a lack of trust among trading partners to share pertinent information and collaborate on decision making

- There is a desire to partner, but not commit to executing the communicated plans
- A common view exists that demand collaboration is a technology solution and that the current technology is too complex.[64]

Another study reports that, although most companies think of CPFR as primarily a technology, it is the business process supported by the internal culture that makes CPFR successful.[65] Computer Sciences Corp., in conjunction with *Supply Chain Management Review*, recently concluded its second annual Global Survey of Supply Chain Progress. Their findings suggest that some companies understand the advantage of leveraging the buy across a more strategic supply base, and others are content to pursue more limited, tactical improvements. Despite the progress needed in some areas, most companies pursuing the various supply chain initiatives are generally happy with the results. However, only 28 percent said that an awareness of the need to increase customer satisfaction ratings was a main factor in the success of their initiatives.[66]

One seemingly simple problem is the need for consistency in product data identification and transmission. UCCnet, a nonprofit unit of the Uniform Code Council standards organization, established a global online registry that requires product data with as many as 151 attributes, or descriptors, about 40 of which are mandatory. One well-known company found that it was transmitting information by phone, E-mail, fax, CDs, EDI, PDFs, spreadsheets, Websites, and printed price pages. The same survey found that the percentage of companies that felt they have the business processes in place to take full advantage of real-time information varied from 20 percent in the construction and engineering industry to slightly over 70 percent in the logistics and transportation industry, with the overall average of all industries around 45 percent.[67]

Another nagging question is the relative benefits between retailers and suppliers. Corsten and Kumar studied the question of whether or not collaborative relationships with large retailers benefit suppliers. They found that, although suppliers benefit in the economic sense and in capability learning, they believe that suppliers bear more of the burden and receive less of the benefits they deserve.[68]

Some activities are increasing the response times, most notably the offshore outsourcing movement. The route to cheaper supplies from overseas sources may be attractive, but supplier lead times are elongating, a step backward from the improvements in domestic operations from years of lean implementations. Some companies may find the payoff worth it. Others may find themselves with cheaper raw materials and components, but fewer customers. The problem is that it just takes longer to get product from another country. The costs may be lower when outsourcing, but the response times and uncertainty of supply are higher. Balancing the trade-offs between supplier responsiveness and order costs requires a long-term strategic outlook.

Future

What does the future hold? No doubt, the pressure to clean out excess inventories in supply chains and gain a better matching of supply with demand will continue. Individual programs, such as those described above, are losing their focus, as they become a part of more general programs such as supply chain management and integrated demand-driven collaboration systems. A survey of over 60 retailers in the United States, Europe, and Asia found that the persistent problem of out-of-stock to be their biggest inventory challenge.[69] Another study suggests that retail exchanges, stemming from improved IT, may provide part of the answer.[70]

Whatever the program is called, companies must integrate their communication systems and develop sufficient trust with one another to collaborate effectively and to gain the benefits of their efforts. If they do, they can hope to succeed. If they do not, they face an uncertain future.

Flexibility

The service sector has been progressive in increasing their flexibility in several areas — product and service offerings, organization structure, demand management, facility location, and employee versatility.

Product and Service Flexibility: Mass Customization

Although the idea of mass customization is still emerging, it is beginning to become a more common theme in many companies. Just as in product or service design, the customer becomes the focal point in that mass customization means building a product or providing a service exactly the way the customer wants it. Service companies can achieve some level of mass customization by empowering employees to make the modifications needed such as "hold the mustard" in a deli, to "a little more off the side" in a barber shop, to "put more emphasis on growth stocks" in a stockbroker's office, to "I want to carry 15 hours this semester" at a university. Manufacturers often must make major changes in their production processes to achieve mass customization; however, they are being driven to do this by the service industries. B. Joseph Pine outlines the future of mass customization in Managerial Comment 6-3.

Managerial Comment 6-3.
Mass Customization

"Mass customization is a new way of viewing business competition, one that makes the identification and fulfillment of the wants and needs of individual customers paramount without sacrificing efficiency, effectiveness, and low costs. It is a new mental model of how business success can be achieved, one that subsumes many of the 'silver bullets' of prevailing management advice such as time-based competition, lean production, and micromarketing. Further, the development of Mass Customization as a paradigm of management explains why product (and service) life cycles are decreasing, why development and production cycle times must follow, why businesses are re-engineering into networked organizations."[71]

Table 6-2 summarizes some of the factors that influence operational effectiveness in mass customization.

Horizontal Communication and Organization Structure

The traditional vertical hierarchical organization structure — the command and control concept developed by the military establishment — was good in maintaining structure and discipline from top to bottom of an organization. However, it was slow in moving information and decisions in a horizontal direction — towards the customer. This type of organization was effective in managing the normal day-to-day operation of a stable demand business. However, it is deficient when it comes to managing a business in a more dynamic demand business. It is also not very effective in planning and managing improvement programs or projects. Service organizations have been active in developing more flexible organization forms than in manufacturing because services have several unique needs:

- They often have smaller operating units that are widely dispersed — chain restaurants, insurance agents, and stockbrokers. They need decentralized authority to handle day-to-day operations.
- They are closer to the customer and need empowered employees to deal with a variety of transactions and problems.
- The smaller units cannot afford to support a specialized staff; therefore, staff support is located at the central control point and unit operations are self-supporting with general purpose employees.

Table 6-2 Operational Effectiveness in Mass Customization

Mass Customization Practice	Strategic Priorities Affected	Mechanisms of Enablers of Strategic Advantage
Modular design	Customization, cost, speed	Complexity reduction, ease of planning, economies of scale, demand variance reduction, and lead-time reduction through delayed differentiation
Delayed design, fabrication, assembly (postponement)	Cost, speed, customization	Reduction in demand variability, safety stocks, cycle times, and lead-times
Cellular or flexible processes	Cost, quality, speed	Cell efficiency, recurrent problem-solving skills, process stability and flexibility, reduced setup costs and times, quick learning efficiencies, complexity reduction, planning efficiencies
Customer co-design	Customers' satisfaction	Product fit, customer perception of design ownership, reduced cost of warranty, returns, and product liability, reduced risk of obsolescence

Source: Copyright © 2007. Reprinted with permission of the Institute of Industrial Engineers from September 2006, *Industrial Engineer*. All rights reserved.

- Service operations do not have the tradition of functional staffs that became prominent in manufacturing during the scientific management era and continue today.

One of the most recognized authorities on both service operations and their flexible organization forms is James Brian Quinn. In Managerial Comment 6-4, he describes some of the types of organizations and their common characteristics and problems. Quinn provides detailed descriptions and diagrams for each of the organization forms in his book *Intelligent Enterprise: A Knowledge and Service Based Paradigm for Industry*[72] and, with others, in a separate journal article.[73]

Managerial Comment 6-4. Flexible Organization Forms

Service and service technologies have opened a wide variety of new organizational options for managing intellect on a much more disaggregated basis. The infinitely flat, spider's web, and inverted forms embrace the most common structures. But there are many other configurations, the most significant being "starburst," "cluster," and "voluntary" organizations. All these radical organizations enjoy certain common characteristics and similar problems:

- All tend to be flatter than their hierarchical predecessors. All assume a constant dynamic rather than stasis in their structures. And all support greater empowerment of people closer to the customer contact level.
- All develop around a maintainable set of core service competencies at their center. These competencies — typically consisting of special depth in some unique technologies, knowledge bases, skills, or organizational–motivational systems — are then radiated outward toward customers by information technologies or through organizational dynamics that enable mixed clusters of people, with greater knowledge access, to work directly with customers.
- All recognize intellect, motivation, and knowledge bases as the most highly leverageable assets of a company. They use the new organization forms and their supporting systems as improved ways to attract, keep, and leverage key people as knowledge resources.
- Successful companies use their new organizations as both aggressive and defensive strategies. Most companies look to these forms initially to lower costs, decrease cycle times, and ensure more precise implementation of strategy. At the same time, they allow the companies to bring more energy, focus, and responsiveness to bear on customers than ever before. But companies have also found that these organizations make the firm a more exciting, personally challenging, and satisfying place to work.

Like earlier decentralization or SBU concepts, some of these new organization modes have often been touted as cures for almost any management ill. They are not. Each form is useful in certain situations, and not in others. But more importantly, each requires a carefully developed infrastructure of culture, measurements, style, and rewards to support it. When properly installed, these disaggregated organizations can be awesomely effective in harnessing intellectual resources for certain purposes. When improperly supported or adapted, they can be less effective than old-fashioned hierarchies.

Source: Adapted from Quinn, J.B., *Intelligent Enterprise: A Knowledge and Service Based Paradigm for Industry,* The Free Press, New York, 1992.[74]

Demand Management

Service industries have learned to live with variable demand patterns, whether daily or weekly at a restaurant, and seasonal in a variety of businesses. Consequently, they have learned to vary their capacity, especially in the form of employees, by developing scheduling techniques to best match the number of employees to the demand highs and lows. In some cases, where the employees represent a scarce and costly resource, such as among physicians, lawyers, and other professional people, they use an appointment system to control the demand to fit the availability of their capacity. These and other demand management techniques can be extended to a number of manufacturing applications.

There are a number of approaches that have been suggested to match supply better with demand. Christopher Lovelock and Robert Young were among the first to suggest that customers could help reduce the variation in demand by participating in the service process. "Service managers, utilizing demand management, can use marketing tools to encourage consumers to modify their behavior so that services can be delivered in a more productive and economically efficient manner."[75]

Kenneth Klassen and Thomas Rohleder provide a thorough review of the demand management literature. They describe the desirability of simultaneously planning demand and capacity by involving both marketing and operations in the planning process, especially in a service environment where it is not possible to build inventory as a buffer against demand fluctuations.[76]

Demand management involves sales forecasting, usually a responsibility of sales and marketing functions. However, a company's position in the supply chain affects their role in sales forecasting. The business closest to the consumer — the retailer — can forecast independent demand. Once the independent demand is determined, all demand upstream in the supply chain is dependent on that demand forecast. "Recognizing the differences between independent, dependent, and derived demand, recognizing which type of demand affects a company, and developing techniques, systems, and processes to deal with that company's type of demand can have a profound impact on supply chain costs and customer service levels. To realize the benefits that are possible from managing demand in a supply chain, communication among the enterprise must be open, coordination in the form of sales and operations planning processes must be implemented, and a culture of collaboration must be established."[77]

Demand management extends beyond day-to-day operations. When combined with logistics and new product development, demand management can contribute to improving the financial performance of a firm.[78]

The health care industry is one that could benefit from better demand management. One study in an acute care hospital found that a demand management program better utilized bed and clinic capacities, reduced patient length of stay and treatment costs, and provided input for nursing staffing.[79]

Shielding strategies work best when demand uncertainty is high and flexibility is low. Pricing and rationing changes offer a strategy available under shielding. Absorbing strategies are recommended when both demand uncertainty and flexibility is low. Using time buffers and slack capacity buffers is recommended with this strategy. Containing strategies are offered when demand uncertainty is low and flexibility is high. Focusing on work force flexibility and information technology is recommended. Finally, mitigating strategies are recommended when both demand uncertainty and flexibility is high.

Location near the Market versus Lowest Cost

Manufacturing companies build large plants to be low cost and efficient. Service companies have to build small outlets to be close to their markets, such as retail grocery stores, restaurants, hospitals, and branch banks. Consequently, service companies have learned how to manage smaller units, often through process designs that are modular and scaled to fit the situation. Manufacturers need to be more responsive to the customer, both in faster response and more appropriate products. To achieve these objectives, they have to locate closer to their markets and use smaller facilities.

Closely allied with the concept of smaller facilities is the need to manage multiple locations or branch operations within the total company. Some of this can be done with standardized processes for use at the branch, such as customer contact services, with centralized processes maintained at the corporate office, such as procurement. In a retail specialty store, such as The Gap, the in-store personnel handle the direct sales to the customer; the corporate office handles stock replenishment functions, based on sales information collected at the point-of-sale registers. Even with specific guidelines, empowered employees are needed to carry on the branch operations.

Evolution from Job Specialization to Self-Directed Teams

Services have generally been more oriented toward broader job designs, as opposed to the specialization of manufacturing. This stems from the variability of services, the need for lower-level decision making and the smaller size of the operating units. As a result, the human resources function in service businesses has been responsible for designing programs that move from training employees to do simple tasks, to education programs that enhance the capabilities of the employees to perform multiple tasks, and to work effectively in cross-functional relationships. Because of the increased capability of its work force, a business can gain a competitive advantage through improved utilization of its human resources. Although operations management can provide contextual insights to human resources, human resources can provide behavioral insights to operations management.[80]

Replacement of Inventory with Information

For years, the availability of inventory has covered a multitude of sins in manufacturing. Today, largely through the development of information, inventory is being replaced. In services, we have seen money, as a medium of exchange, change its form from cash to checks to online payments to electronic funds transfer. With each step in the progression, there is less inventory, but more information. In manufacturing, better demand information has made it possible to reduce the amount of inventory needed to satisfy demand. An even more tangible example is the replacement of newspapers with electronic news transmissions. The computer download capability could mean that, some time in the future, electronic storage and accessibility will replace books, although this carries with it the need for readers to adapt from reading books to reading computer screens.

Inter-Organizational Communications

Another area in which service companies have taken the lead is in the area of inter-organizational communications (IOC). Although EDI originated almost three decades ago, the use of the Internet for IOC is increasing rapidly and will probably become the primary tool in the future.

Traditional EDI

Traditional electronic data interchange is the paperless (electronic) exchange of trading documents, such as purchase orders, shipment authorizations, advanced shipment notices, invoices, and using standardized document formats. It became available to companies in the early 1970s as a means of transferring information electronically from one business to another. One of the more popular applications was in order processing. A business could place an order with a supplier, who would acknowledge the order, and ship the order with an invoice. The receiving business could prepare a receiving report, match it with their purchase order and the invoice, and authorize payment through their bank, which would send the payment to the supplier's bank. This process eliminated paperwork because all of the transactions and accompanying documents described above were electronic. EDI reduced costs of order processing by as much as 90 percent, reduced errors, speeded up deliveries, and reduced the time required of employees to identify errors or track orders. The biggest problem was that it was expensive to implement and operate. As a result, only a small fraction (less than 5 percent in 1995) of the businesses adopted EDI.[81] Many of those that used EDI did so reluctantly because their major customers demanded it. Because it was a one-to-one kind of communication, it was secure. However, it required that each customer–supplier relationship be set up individually.

Large companies implemented EDI and found it worth the investment. Some of the major applications included global communications, financial funds transfers, health care claims processing, and manufacturing and retailing.[82] Small companies could not achieve the volume necessary to make it a worthwhile investment. If they had to use EDI to do business, they usually did it through value-added networks (VANs), entities that facilitated the transfer of information between customer and supplier. Using a VAN eliminated a portion of the initial investment cost, but did not change the high transaction costs. Sawabini reported, "Where EDI fails, it is because (1) it's cost prohibitive and too complex for smaller suppliers, and (2) it offers few bottom-line benefits for suppliers."[83] In the early days, users viewed EDI as an inter-organizational system (IOS). In recent years, as the concept of an IOS has enlarged, writers now view EDI as an element of an IOS.

In summary, EDI was good for a few companies, but a burden for many others. Today, competitive pressures make the need for rapid information transfer imperative. Is the Internet a viable alternative? Many say yes; however, a few are more cautious.

Internet EDI

Internet EDI looks attractive because of the lower costs potential. The initial investment is lower and the transactions costs are lower.[84] In addition to the lower costs, other reasons why the use of the Internet is attractive include:

- Publicly accessible network with few geographical constraints
- Offers the potential to reach the widest possible number of trading partners
- Powerful tools facilitating inter-organizational systems are becoming available
- Consistent with the growing interest of business in increasing electronic services
- Can complement or replace current EDI strategies
- Leads to an electronic commerce strategy[85]

The Internet also offers close to real-time transactions because there is no longer the need to go through VANs that use the batch-and-forward method of transmitting information. But how do companies ensure the secure transmission of information? If a satisfactory solution to this issue becomes available, Internet EDI could become a major growth application in the next few years.

Research tells us to expect that Internet EDI will become more widely used than traditional EDI because of its lower costs. At the same time, traditional EDI will also become more widely used because of its advantage over paper processing. Both forms of electronic data interchange will begin to realize their potential in the development of inter-organizational systems. With the ever-expanding global business community, it is no longer possible to use hard copy documents in most business transactions. The speed and efficiency of electronic information is compelling businesses to make the transition to paperless systems. Although Internet EDI may become the standard at some point in the future, traditional EDI will remain a significant factor for several years because it is a proven method and many companies already have an investment in their present systems.

Like many innovations, the use of the Internet as a replacement for traditional EDI has many supporters, some of whom have a stake in seeing it succeed. Technology opportunities abound. For those that understand the technology, it is like being in a candy store trying to decide which treat to choose. For those that do not understand, it is a jungle with danger lurking behind every acronym. A number of articles and Websites provide more complete descriptions of the technology involved.[86-88] Although the future looks promising, Internet EDI is not without some hurdles, some of which we describe below.

There is still a need to develop global standards for the formatting and transmission of data. Applicability Statement 2 (AS2) is gaining attention because Wal-Mart advocates it.[89] AS2 is a draft specification from the Internet Engineering Task Force (IETF) that describes how to create a connection and securely transport an EDI file over the Internet. AS2 provides security for the transport of HTTP packets through digital signatures and data encryption. AS2 also provides for nonrepudiation — proof that a transaction was preformed at a certain time and by legitimate parties — using receipts. The AS2 specification was developed by the IETF's EDI over the Internet (EDIINT) Working Group, which was formed in 1996 to create a set of secure standards for sending EDI data over the Internet. AS2 is also being used in an intranet or a LAN within an organization.[90] Benefits of AS2 include confirmed data delivery, encryption of digital signatures, and automatic data transfer. In addition, AS2 "sets the stage for broader enterprise collaboration, supporting strategic trading relationships with a wider set of key partners through the interoperation of business processes and shared visibility into underlying data."[91]

Extensible Markup Language (XML) appears to hold promise; however, it is not completely ready. "XML is not, however, a panacea for electronic commerce. Support for, and use of, XML is growing rapidly, but all the necessary standards and agreements are not yet in place to enable XML-based application to work seamlessly with all other applications and systems."[92]

Additional actions become necessary to provide greater security for data transmission over the Internet. One example is the use of firewalls at both the sender and receiver, and a virtual private network (VPN) between the two. VPNs use the Internet and its standardized protocols to communicate information by combining the global connectivity of the Internet with the security of a closed, private network.[93]

Internet EDI must overcome the problem that restricted the growth of traditional EDI — dominance by a few large companies in a supply chain. When there is an imbalance among supply chain members, not all participants share equally. Smaller firms may have to adopt innovations to maintain supplier relationships with large customers, but without realizing any of the savings provided by them.[94]

Any inter-organizational system is more than technology. It also includes organization and people. Trust between participants is necessary if the relationship is to be an effective one, supporting collaboration among entities. Even if the cost and security issues can be resolved, the trust issue may remain. Company culture and trust are significant issues when it comes to the use of EDI in either the traditional or Internet format. To move from coordination to collaboration, trust is required.[95]

EDI, either traditional or Internet, cannot stand alone. To be completely effective, a company should integrate it with its mainstream information systems. This requires the capability to interface two disparate systems or to modify the EDI system to allow integration of the data into the main systems, such as Enterprise Resource Planning (ERP), Supplier Relationship Management (SRM), or Customer Relationship Management (CRM).

Traditional EDI is a proven method of transmitting data between entities. Where there is sufficient volume, it is of significant benefit. Unfortunately, only a limited number of companies have the requisite volume. A promising supplement is Internet EDI, when the level of security can be raised to acceptable levels. Other approaches include the use of portals — access point (or front doors) through which a business partner accesses secured, proprietary information from an organization. These portals may be distribution portals (single supplier, multiple buyers), procurement portals (single customer, multiple suppliers), or trading exchanges (balance between suppliers and buyers).[96] Until some approach dominates, a company may use a hybrid approach, such as to continue to use traditional EDI for the more sensitive data applications and move more deliberately into Internet EDI with less sensitive data, using trading portals where they offer benefits. As with any worthwhile innovation, the widespread diffusion of electronic data exchange will take time and will likely involve changes in direction or emphasis during its life cycle.

Other Service Developments

In addition to the major areas described above, services have also pioneered in a number of other ways.

Working in an Open-System Environment

Service industries are exposed to an open-system environment more than the manufacturing industries. Service industries such as retail, insurance, health care, banking, and hospitality are highly visible to the consumer and other elements in the economic world. Consequently, they are less able to shield their operations from the effect of external entities.

Product Development as a Result of Customer Inputs

Historically, product ideas were passed along from sales and marketing, who may have interpreted what they thought was the voice of the customer, to engineering and operations, which designed and made the product. Often there was a series of modifications to the design so that the resulting product was not what the customer wanted. Some companies have found that direct customer involvement in the design of products and services is beneficial and increases the likelihood that the product will be a successful entry in the marketplace.

Quality as Customers' Perceptions, Not Just Conformance to Specification

For years, manufacturers have measured product quality by comparing the end product with the product specification. If the product met the specification, it was considered to be of good quality, whether or not it was suitable for the customer. Although Juran and Gryna introduced the concept of "fitness for use,"[97] the quality measures still tended to be quantitative and measurable without regard for the customer. Services have changed this. The service industries must consider customer perceptions when assessing their service quality. The hamburger that was made to specification may not please the customer because the order taker was not courteous or there were food scraps on the floor. Designing products to meet customer perceptions, where each customer has slightly different perceptions, is a new way of life for manufacturers.

Managing the Customer Encounter

Customer encounter implies a face-to-face meeting between employee and customer. It is a common occurrence in services. Much of the success of such an encounter depends on the knowledge, ability and motivation of the employee. Customers look for a combination of service content, quality of service, and humane treatment. To achieve this combination of outcomes, companies can use an approach that weaves together a consistent method for assessing the employee–customer encounter and a disciplined process for managing and improving it.[98]

Nonquantitative Performance Measurement

Manufacturing has always been thought of as tangible, but services have been assigned the attribute of intangibility. Whether or not this differentiation is valid, services do have the need to use some nonquantitative measures, whether it has to do with customer perceptions of quality or the decision-making capability of empowered employees. Manufacturing is facing the need to become more adept in measuring nonqualitative variables. For example, what is the risk cost of outsourcing to an offshore supplier? What is the impact of a packaging change on the customer's propensity to buy that product? What is the cost of a back order? Although services may not have developed foolproof ways to handle this dilemma, they are forcing manufacturers to address the issue.[99]

The Use of Business Intelligence

Business Intelligence is a term that is usually associated with the increased understanding of the available information, developed from collected data, about a company's business. It tends to be about the demand side of the business, rather than the supply side, so we associate it with the external factors that affect the business. It probably originated with the point-of-sales data collection and has been enhanced with the collection of additional data from surveys and Websites. The additional information processing capabilities makes it feasible to begin to analyze, interpret, and act on this information. An Aberdeen report by Greg Belkin examined the status of business intelligence in retailing. He found that the top factors driving the implementation of Business Intelligence included the need for more rapid response to consumer demand, the need to become more operationally efficient, and the need to manage demand across multiple channels. He also determined that some of the most pressing challenges for retailers involved the quality of their data, e.g., the data is not good enough, it is not well organized, and it is not customer specific enough to generate valid analysis and actions.[100]

Summary

The boundary between manufacturing and services is vanishing.[101] Progressive companies are building a bridge over the chasm that separated "hard goods" from "soft services." For years, we have read how programs that originated in the manufacturing area are being adapted for use in services applications. Authors suggest that MRP has application in academic institutions, JIT works in retail, banks use TQM, lean manufacturing revolutionizes distribution, and the health care industry benefits from Six Sigma.

Although most of the articles describe using manufacturing techniques in services, an increasing number report how concepts and techniques that originated in service industries are adopted in manufacturing. Although manufacturing industries concentrated on reducing costs and improving quality, service industries focused on improving customer relationships and service. Certainly, service businesses want to reduce costs and improve quality; that is why they adapt manufacturing techniques to fit their needs. Beyond that, they understand that businesses exist because of their ability to find customers and satisfy their needs. As Peter Drucker described it, "With respect to the definition of business purpose and business mission, there is only one such focus, one starting point. It is the customer. The customer defines the business. Businesses exist for one purpose only — to find customers and provide service to them."[102] Companies must look at their business from the outside in — from the perspective of the customer. Barbara Bund builds on Drucker's basic ideas and includes a number of case histories of businesses that have succeeded, in part because they have taken the "outside-in" look at their companies. She uses service companies such as FedEx, eBay, and Costco to illustrate the customer-centric thinking. She also considers GE and Dell as outside-in thinkers, but only because they think of themselves as service businesses.[103] Low cost is not enough for survival in the future; manufacturing companies must integrate service strategies in their planning.

Does this mean that manufacturing companies are not interested in their customers? Of course not; however, most manufacturing customers are other businesses, not individual consumers. The service industries deal with other businesses, but they also focus on individual consumers. This understanding requires that they have a more diverse customer base and deal with a greater variety of needs and wants. There are no mass markets anymore. Although we have not yet reached single-customer markets, we are certainly into niche markets and the niches are getting smaller.

Differences between Manufacturing and Services

A key difference between manufacturing and services is that manufacturing companies provide a tangible product and service companies provide a less tangible service. Almost all companies provide a combination of goods and services; however, a manufacturing company provides a product as its focus, such as an automobile, and surrounds the product with a service package that may include such things as post-sales service or financing. Service companies provide a service — a complete dining experience — that includes tangible goods such as food. The difference is in the primary focus. There are "defensible differences" between manufacturing and services. There are differences in production strategies, production processes, and performance measures.[104] However, the fact that manufacturing and services are different does not mean that they are incompatible. On the contrary, they are an important complement to each other. As Chase and Garvin put it, "Today's flexible factories will be tomorrow's service factories."[105] They emphasize that manufacturing companies must bundle services with their products; in addition, they must anticipate and respond to an increasing range of customer needs. Bund refers to the "augmented product" — the basic product plus the support, brand name, documentation, and all the other attributes that customers think matter. In this section, we will show how knowledge transfer between manufacturing and service businesses is helping each to be more competitive.[106]

From Services to Manufacturing

What have the service industries been doing that should be of value to manufacturing? Many of these ideas or concepts are not neatly packaged like JIT or Six Sigma, so they are not always obvious. It is easy to find articles about how manufacturing has provided concepts and techniques that can be used in service businesses. It is difficult to find the reverse — how service concepts can be used in manufacturing. We will describe a few of the ideas that originated in services.

Service businesses have not been viewed as effective or efficient as manufacturing businesses. However, they have always had a customer orientation, a quality sometimes lacking in manufacturing. One of the programs that is being actively pursued by a number of companies is customer relationship management (CRM). As CRM becomes more fully defined, understood, and applied, this concept — which heavily depends on knowledge workers — will be assimilated more effectively into manufacturing businesses.

Historically, quality in manufacturing has been measured by how well the product conformed to the specifications. Although Joseph Juran and Frank Gryna pointed out "quality is customer satisfaction" or "fitness for use," an idea that encouraged manufacturers to consider customer needs, it has been convenient to measure product performance against tangible and quantifiable specifications.[107] On the other hand, services have always been somewhat at a disadvantage because customers do not always agree on what is good quality. How do you measure such desired attributes as reliability, responsiveness, competence, access, courtesy, communication, credibility, security, and understanding the customer?[108] As a result, services have always needed to recognize that quality is more than conformance to specifications; it includes customers' perceptions. As manufacturers become more involved in providing complementary services, they must also consider individual customer perceptions of quality.

As a corollary to the intangibility of quality, there is a need to develop qualitative goals and performance measures. How do we "improve customer service" or "enhance the acceptance of our product in the marketplace"? Sometimes the answers can be meaningful, even if not precise. Feedback information comes from surveys, focus groups, extended interviews, complaint letters, and personal observation. As a result, service businesses search for meaning in a labyrinth of ambiguity. Although manufacturers may be uncomfortable with this uncertainty, they have to learn how to deal with it in a meaningful way. The number of customer complaints may not sound as tangible as number of defective units, yet it is a measure of performance. Bowen and Ford advise that "assessing organizational effectiveness and efficiency for an intangible product relies on subjective assessment by the customer; traditional objective measurements need to be supplemented with subjective measures for assessing service experiences."[109]

Manufacturers have subscribed to the notion that bigger is better in building facilities. They locate plants to minimize costs and worry about how to get the product to the market later. Service businesses have no choice; they must locate near their market. This move has forced them to learn how to manage smaller units. As manufacturing companies focus on reducing lead times, they will have to consider moving closer to the market and learn how to manage smaller, more flexible units.

Organization structures are also changing — from hierarchical to horizontal or flat. Quinn describes how "service firms — because of their communications–information intensity and localized delivery needs — have generally pioneered" some of the more innovative organization forms such as the matrix, spider web, or inverted pyramid forms.[110] New organization structures enable empowered employees to respond to the customer faster and more effectively.

Collaborative Efforts

Some concepts and programs are difficult to classify as manufacturing or services originated, such as mass customization. Both manufacturing and service businesses are active in this area. Mass customization represents the convergence of basic manufacturing and service principles, rather than a transfer from one to another. It combines the individual customization associated with service and the efficient volume associated with manufacturing. Mass customization is service-oriented because it is responsive to the customer. However, manufacturing makes it happen.

Quick Response programs also seem to be a collaborative effort. Retail organizations recognized that if they were to have the right goods available at the right time in the right quantities, they had to give up on the idea of accurately forecasting and rely more on rapid replenishment of goods from the manufacturer. Services had the need; manufacturing provided part of the answer.

For decades, manufacturers worked to eliminate the worker through automation. To date, we have not achieved that ultimate state. On the other hand, the labor-intensive services sector has been learning how to capitalize on the versatility and knowledge of the worker.[111] We are nearing the time when the effort will be on how to optimize the worker–equipment interface. How do we design equipment to aid the workers, not to replace them?

Conclusion

Does it matter where the ideas originate? Most improvement programs require a cross-functional approach. Whether in manufacturing or services, sales and marketing provide the customer contact; operations and finance provide the analysis and execution. The world is changing and competing successfully is becoming more difficult. Businesses need to take advantage of good ideas, regardless of the source. Manufacturing and services can learn from each other; many companies have already started.

References

1. Quinn, J.B., *Intelligent Enterprise: A Knowledge and Service-Based Paradigm for Industry,* The Free Press, New York, 1992.
2. Allmendinger, G. and Lombreglia, R., Four strategies for the age of smart services, *Harvard Business Review,* 83, 10, 131, 2005.
3. Gottfredson, M. and Aspinall, K., Innovation vs. complexity: What is too much of a good thing? *Harvard Business Review,* 83, 11, 62, 2005.
4. Galbraith, J.R., Organizing to deliver solutions, *Organizational Dynamics,* 31, 2, 194, 2002.
5. White, R.E. and Pearson, J.N., JIT, system integration and customer service, *Organizational Dynamics, International Journal of Physical Distribution & Logistics Management,* 31, 5, 313, 2001.
6. Bund, B., *The Outside-In Corporation: How to Build a Customer-Centric Organization for Breakthrough Results,* McGraw Hill, New York, 2006.
7. Haeckel, S.H., *Adaptive Enterprise Creating and Leading Sense-and-Respond Organizations,* Harvard Business School Press, Boston, 1999.
8. Slywotzky, A. and Wise, R., The dangers of product-driven success: What's the next growth act? *The Journal of Business Strategy,* 24, 2, 16, 2003.
9. Olson, E.M., Slater, S.F., and Hult, G.T.M., The importance of structure and process to strategy implementation, *Business Horizons,* 48, 1, 47, 2005.

10. Jain, D. and Singh, S.S., Customer lifetime value research in marketing: A review and future directions, *Journal of Interactive Marketing,* 16, 2, 34, 2002.
11. Band, W. and Guaspari, J., Creating the customer-engaged organization, *Marketing Management,* 12, 4, 34, 2003.
12. Allmendinger, G. and Lombreglia, R., Four strategies for the age of smart services, *Harvard Business Review,* 83, 10, 131, 2005
13. Ma, J., Nozick, L.K., Tew, J.D., and Truss, L.T., Modeling the effect of custom and stock orders on supply-chain performance, *Production Planning & Control,* 15, 3, 282, 2004.
14. Pine, B.J., II, *Mass Customization, The New Frontier in Business Competition,* Harvard Business School Press, Boston, 1993.
15. Blackstone, J.H. and Cox, J.F., III, *APICS Dictionary* (Eleventh Edition), 2004, APICS — The Educational Society for Resource Management.
16. Schwartz, B., *The Paradox of Choice: Why More Is Less,* Harper Collins, New York, 2004.
17. Piller, F. and Kumar, A., For each, their own: The strategic imperative of mass customization, *Industrial Engineer,* 38, 9, 40, 2006.
18. Schmenner, R.W., *Service Operations Management,* Pearson Custom Publishing, Boston, 1995.
19. Heskett, J.L., *Managing in the Service Economy,* Harvard Business School Press, Boston, 1986.
20. Metters, R.M., King-Metters, K., and Pullman, M., *Successful Service Operations Management,* Thomson South-Western, Mason, OH, 2003.
21. Klassen, K.J. and Rohleder, T.R., Combining operations and marketing to manage capacity and demand in services, *The Service Industries Journal,* 21, 2, 1, 2001.
22. Legarreta, J.M.B. and Miguel, C.E., Collaborative relationship bundling: A new angle on services marketing, *International Journal of Service Industry Management,* 15, 3/4, 264, 2004.
23. Jain, D. and Singh, S.S., Customer lifetime value research in marketing: A review and future directions, *Journal of Interactive Marketing,* 16, 2, 34, 2002.
24. Kapanen, R., Customer relationship management and service delivery, *International Journal of Services Technology and Management,* 5, 1, 42, 2004.
25. Zablah, A.R., Bellinger, D.N., and Johnston, W.J., An evaluation of divergent perspectives on customer relationship management: Towards a common understanding of an emerging phenomenon, *Industrial Marketing Management,* 33, 6, 475, 2004.
26. Crandall, R.E., A fresh face for CRM: Looking beyond marketing, *APICS Magazine,* 16, 9, 20, 2006.
27. Payne, A. and Frow, P., Customer relationship management: From strategy to implementation, *Journal of Marketing Management,* 22, 1/2, 135, 2006.
28. Zablah, A.R., Bellinger, D.N., and Johnston, W.J., An evaluation of divergent perspectives on customer relationship management: Towards a common understanding of an emerging phenomenon, *Industrial Marketing Management,* 33, 6, 475, 2004.
29. Nairn, A., CRM: Helpful or Full of Hype? *Journal of Database Management,* 9, 4, 376, 2002.
30. Chan, J.O., Towards a unified view of customer relationship management, *Journal of American Academy of Business,* 6, 1, 32, 2005.

31. Blackstone, J.H. and Cox, J.F., III, *APICS Dictionary* (Eleventh Edition), 2004, APICS — The Educational Society for Resource Management.
32. Chan, J.O., Towards a unified view of customer relationship management, *Journal of American Academy of Business*, 6, 1, 32, 2005.
33. Lambert, D.M., Garcia-Dastugue, S.J., and Croxton, K.L., An evaluation of process-oriented supply chain management frameworks, *Journal of Business Logistics*, 26, 1, 25, 2005.
34. Lambert, D.M., Garcia-Dastugue, S.J., and Croxton, K.L., An evaluation of process-oriented supply chain management frameworks, *Journal of Business Logistics*, 26, 1, 25, 2005.
35. Compton, J., It's not business as usual, *Customer Relationship Management*, 8, 12, 32, 2004.
36. Bois, R., Live, From Boston! It's Microsoft CRM, www.amrresearch.com, July 12, 2006.
37. Doyle, S., A sample road map for analytical CRM, *Journal of Database Marketing & Customer Strategy Management*, 12, 4, 362, 2005.
38. Ling, R. and Yen, D.C., Customer relationship management: An analysis framework and implementation strategies, *The Journal of Computer Information Systems*, 41, 3, 82, 2001.
39. Payne, A. and Frow, P., Customer relationship management: From strategy to implementation, *Journal of Marketing Management*, 22, 1/2, 135, 2006.
40. Winer, R.S., A framework for Customer Relationship Management, *California Management Review*, 43, 4, 89, 2001.
41. Nairn, A., CRM: Helpful or full of hype? *Journal of Database Management*, 9, 4, 376, 2002.
42. Winer, R.S., A framework for *Customer Relationship Management, California Management Review*, 43, 4, 89, 2001.
43. Lambert, D.M., Garcia-Dastugue, S.J. and Croxton, K. L., An evaluation of process-oriented supply chain management frameworks, *Journal of Business Logistics*, 26, 1, 25, 2005.
44. Kennedy, M.E. and King, A.M., Using customer relationship management to increase profits, *Strategic Finance*, 85, 9, 36, 2004.
45. Nairn, A., CRM: Helpful or full of hype? *Journal of Database Management*, 9, 4, 376, 2002.
46. Keiningham, T.L., Vavra, T.G., Aksoy, L., and Wallard, H., *Loyalty Myths: Hyped Strategies That Will Put You out of Business — and Proven Tactics That Really Work*, John Wiley & Sons, Inc., Hoboken, NJ, 2005.
47. Lambert, D.M. and Pohlen, T.L., Supply chain metrics, *International Journal of Logistics Management*, 12, 1, 1, 2001.
48. Giltner, R. and Ciolli, R., Re-think customer segmentation for CRM results, *The Journal of Bank Cost & Management Accounting*, 13, 2, 3, 2000.
49. Hayes, M., Get close to your clients, *Journal of Accountancy*, 201, 6, 49. 2006.
50. Feinberg, R.A., Kadam, R., Hokama, L., and Kim, I., The state of electronic customer relationship management in retailing, *International Journal of Retail & Distribution Management*, 30, 10, 470, 2002.
51. West, J., Customer relationship management and you, *IIE Solutions*, 33, 4, 34. 2001.

52. Shah, J.R. and Murtaza, M.B., Effective customer relationship management through Web services, *The Journal of Computer Information Systems*, 46, 1, 98, 2005.
53. Ko, E. and Kincade, D.H., The impact of Quick Response technologies on retail store attributes, *International Journal of Retail & Distribution Management*, 25, 2, 90, 1997.
54. Crandall, R.E., Beating impossible deadlines, *APICS Magazine*, 16, 6, 20, 2006.
55. Hunter, N.A. and Valentino, P., Quick Response — Ten years later, *International Journal of Clothing Science and Technology*, 7, 4, 30, 1995.
56. Kurt Salmon Associates, Quick Response mandates today, *Online Viewpoint*, March 1997.
57. Lummus, R.R. and Vokurka, R.J., Defining supply chain management: A historical perspective and practical guidelines, *Industrial Management + Data Systems*, 99, 1, 11, 1999.
58. Frankel, R., Goldsby, T.J., and Whipple, J.M., Grocery industry collaboration in the wake of ECR, *International Journal of Logistics Management*, 13, 1, 57, 2002.
59. Angulo, A., Nachtmann, H., and Waller, M.A., Supply chain information sharing in a vendor managed inventory partnership, *Journal of Business Logistics*, 25, 1, 101, 2004.
60. Smaros, J., Lehtonen, J.-M., Appelqvist, P., and Holstrom, J., The impact of increasing demand visibility on production and inventory control efficiency, *Organizational Dynamics, International Journal of Physical Distribution & Logistics Management*, 33, 4, 336, 2003.
61. Poirier, C.C. and Quinn, F.J., How are we doing? A survey of supply chain progress, *Supply Chain Management Review*, 8, 8, 24–31, 2004.
62. www.cpfr.org, VICS 2000.
63. Barratt, M. and Oliverira, A., Exploring the experiences of collaborative planning initiatives, *Organizational Dynamics, International Journal of Physical Distribution & Logistics Management*, 31, 4, 266, 2001.
64. Crum, C. and Palmatier, G.F., Demand collaboration: What's holding us back? *Supply Chain Management Review*, 8, 1, 54, 2004.
65. Harrington, L.H., 9 Steps to Success with CPFR, *Transportation & Distribution*, 44, 4, 50, 2003.
66. Poirier, C.C. and Quinn, F.J., How are we doing? A survey of supply chain progress, *Supply Chain Management Review*, 8, 8, 24–31, 2004.
67. Sullivan, L. and Bacheldor, B., Slow to Sync, *Information Week*, 992, 55, 2004.
68. Corsten, D. and Kumar, N., Do suppliers benefit from collaborative relationships with large retailers? An empirical investigation of Efficient Consumer Response adoption, *Journal of Marketing*, 69, 3, 80, 2005.
69. Rosenblum, P., The Business Benefits of Advanced Planning and Replenishment, Aberdeen Group, 2005, white paper.
70. Sparks, L. and Wagner, B.A., Retail exchanges: A research agenda, *Supply Chain Management*, 8, 3/4, 201, 2003.
71. Pine, B.J., II, *Mass Customization, The New Frontier in Business Competition*, Harvard Business School Press, Boston, 1993.
72. Quinn, J.B., *Intelligent Enterprise: A Knowledge and Service Based Paradigm for Industry*, The Free Press, New York, 1992.
73. Quinn, J.B., Anderson, P., and Finkelstein, S., Leveraging intellect, *The Academy of Management Executive*, 10, 3, 7, 1996.

74. Quinn, J.B., *Intelligent Enterprise: A Knowledge and Service Based Paradigm for Industry*, The Free Press, New York, 1992.
75. Lovelock, C.H. and Young, R.F., Look to consumers to increase productivity, *Harvard Business Review*, 57, 3, 168, 1979.
76. Klassen, K.J. and Rohleder, T.R., Combining operations and marketing to manage capacity and demand in services, *The Service Industries Journal*, 21, 2, 1, 2001.
77. Mentzer, J.T., A telling fortune, *Industrial Engineer*, 38, 4, 42, 2006.
78. Morash, E.A., Droge, C., and Vickery, S., Boundary-spanning interfaces between logistics, production, marketing, and new product development, *Organizational Dynamics, International Journal of Physical Distribution & Logistics Management*, 27, 5/6, 350, 1997.
79. Rhyne, D.M., The impact of demand management on service system performance, *The Service Industries Journal*, 8, 4, 446, 1988.
80. Boudreau, J., Hopp, W., McClain, J.O., and Thomas, L.J., On the interface between operations and human resource management, *Manufacturing & Service Operations Management*, 5, 3, 179, 2003.
81. Lankford, W.M. and Johnson, J.E., EDI via the Internet, *Information Management & Computer Security*, 8, 1, 27, 2000.
82. Kalakota, R. and Whinston, A.B., *Frontiers of Electronic Commerce*, Addison-Wesley, Reading, MA, 1996.
83. Sawabini, S., EDI and the Internet, *The Journal of Business Strategy*, 22, 1, 41, 2001.
84. Angeles, R., Revisiting the role of Internet-EDI in the current electronic commerce scene, *Information Management*, 13, 1, 45, 2000.
85. Senn, J.A., Expanding the reach of electronic commerce, *Information Systems Management*, 15, 3, 7, 1998.
86. Kalakota, R. and Whinston, A.B., *Frontiers of Electronic Commerce*, Addison-Wesley, Reading, MA, 1996.
87. Jessup, L.M. and Valacich, J.S., *Information Systems Today* (Second Edition), Pearson Education, Inc., Prentice Hall, Upper Saddle River, NJ, 2006.
88. www.ediuniversity.com.
89. Bury, S., Piggly Wiggly's doing it, *Manufacturing Business Technology*, 23, 2, 42, 2005.
90. Bednarz, A., Internet EDI: Blending old and new, *Network World*, 21, 8, 29, 2004.
91. Anonymous, AS2 is A-OK at Wal-Mart, *Chain Store Age*, 80, 2, 38, 2004.
92. Jessup, L.M. and Valacich, J.S., *Information Systems Today* (Second Edition), Pearson Education, Inc., Prentice Hall, Upper Saddle River, NJ, 2006.
93. Jessup, L.M. and Valacich, J.S., *Information Systems Today* (Second Edition), Pearson Education, Inc., Prentice Hall, Upper Saddle River, NJ, 2006.
94. Grossman, M., The role of trust and collaboration in the Internet-enabled supply chain, *Journal of American Academy of Business*, 5, 1/2, 391, 2004.
95. Ruppel, C., An information systems perspective of supply chain tool compatibility: The roles of technology fit and relationships, *Business Process Management Journal*, 10, 3, 311, 2004.
96. Jessup, L.M. and Valacich, J.S., *Information Systems Today* (Second Edition), Pearson Education, Inc., Prentice Hall, Upper Saddle River, NJ, 2006.
97. Juran, J.M. and Gryna, F.M., *Quality Planning and Analysis, From Product Development through Use*, McGraw-Hill, New York, 1993.

98. Fleming, J.H., Coffman, C., and Harter, J.K., Manage your Human Sigma, *Harvard Business Review*, 83, 7, 106, 2005.
99. Evans, J.R., An exploratory study of performance measurement systems and relationships with performance results, *Journal of Operations Management*, 22, 3, 219, 2004.
100. Belkin, G., Business Intelligence in Retail: Bringing Cohesion to a Fragmented Enterprise: A Benchmark Report, Aberdeen Group, June 2006.
101. Crandall, R.E., Looking to service industries, new resources for learning a thing or two, *APICS Magazine*, 16, 10, 21, 2006.
102. Drucker, P.F., *Management: Tasks, Responsibilities, Practices,* Harper & Row, New York, 1974.
103. Bund, B., *The Outside-In Corporation: How to Build a Customer-Centric Organization for Breakthrough Results,* McGraw Hill, New York, 2006.
104. Bowen, J. and Ford, R.C., Managing service organizations: Does having a "thing" make a difference? *Journal of Management*, 28, 3, 447, 2002.
105. Chase, R.B. and Garvin, D.A., The service factory, *Harvard Business Review*, 67, 4, 61, 1989.
106. Bund, B., *The Outside-In Corporation: How to Build a Customer-Centric Organization for Breakthrough Results,* McGraw Hill, New York, 2006.
107. Juran, J.M. and Gryna, F.M., *Quality Planning and Analysis, From Product Development through Use,* McGraw-Hill, New York, 1993.
108. Parasuranam, A., Zeithaml, V.A., and Berry, L.L., *Delivering Quality Service and Balancing Customer Expectations,* Free Press, New York, 1990.
109. Bowen, J. and Ford, R.C., Managing service organizations: Does having a "thing" make a difference? *Journal of Management*, 28, 3, 447, 2002.
110. Quinn, J.B., *Intelligent Enterprise: A Knowledge and Service Based Paradigm for Industry,* The Free Press, New York, 1992.
111. Rayport, J.F. and Jaworski, B. J., *Best Face Forward: Why Companies Must Improve Their Service Interfaces with Customers,* Harvard Business School Press, Boston, 2005.
112. Piller, F. and Kumar, A., For each, their own: The strategic imperative of mass customization, *Industrial Engineer*, 38, 9, 40, 2006.

Chapter 7

The Role of Technology in Continuous Improvement

John began to dream again. Surprisingly, he found himself in the same great hallway. Imagine that, continuing the same dream from the night before! He moved along to the next stage and found what appeared to be a family on a picnic in a lovely park. There were children of various ages, running, throwing balls, yelling, turning somersaults, talking on cell phones, measuring how far they could jump with a GPS locator device, and in general, having a good time, at least insofar as their parents allowed. There was another group that seemed to be even more intolerant than the parents. They were the attendants of the park who had posted a lot of signs to prohibit walking on the grass or picking flowers or walking dogs and a lot of other restrictions. Some of the signs even said to turn your cell phone off before entering any of the buildings! Another sign said no pictures could be taken past this door. It was as if the park attendants wanted to keep things from changing at all.

At the entrance to the park, he noticed a newsstand with a number of pamphlets. They were titled "Technology," "Infrastructure," and "Culture." He picked up one of each and carefully placed them in his tote bag along with the pamphlets he had collected earlier.

John walked up to one of the children who had stopped long enough to catch her breath. "Are you having fun?" he asked. She looked up rather wistfully and said, "We could have a lot more fun if our parents would let us. They don't understand that times have changed and that we don't enjoy the same things they did when they were young. They try to restrict what we do and we have so many new games and gadgets we could play with or things we could build or other kids we could do something with. We don't just want to play, we want to create new things. We could talk to other kids on the other side of the park with these little things (at this point, she produced what looked like a cell phone, and also like a small computer). Why, I could even talk to other kids in other parks, all over the world, if they would just let us."

John was amazed. These children were certainly different from his generation. "What is your name?" The girl replied, "My name is EDI; I like to send electronic messages. That boy in the green shirt is CAD. He loves to draw and design things. The girl next to him is CAM. She takes CAD's drawings and builds the things he has designed. We have other friends we call by their nicknames like Lean, Six Sigma, BPR, and ECR. That small child over there bouncing up and down like a yo-yo is Internet. She is growing up quickly and, although she has just started walking, she can get around really fast. We have so much energy that we need to burn off. I know it sometimes distresses our parents because they want us to be orderly and listen to them talk about what they did when they were children, but we are of the technology generation and we have things to do before we get older and have children of our own."

The girl made a lot of good points, John thought. He wondered why the parents couldn't see that their children were so talented and needed some freedom of expression and action. Maybe he should ask them.

Managerial Comment 7-1.
Humanness

"When we think of the 21st century, we think technology: space travel, biotechnology, robots. But the face of the future is more complex than the technology we use to envision it.

"The most exciting breakthroughs of the 21st century will occur not because of technology but because of an expanding concept of what it means to be human."[1]

The Role of Technology in Continuous Improvement ■ 197

In this chapter, we will try to describe the role of technology in implementing continuous improvement programs. We say "try" because the task is daunting, if not overwhelming. To cover the entire topic is impossible; to select what is relevant to our discussion is challenging. Consequently, we will be selective in what we present and supplement it with a number of references that you can refer to if you want additional information. Technology is important; however, it is only one of the three essential elements in making continuous improvement programs successful. In Chapter 8, we will describe the need for building the infrastructure necessary to support the improvement programs, and in Chapter 9, we will discuss the importance of organizational culture in the change management process.

Paul Kidd points out the importance of recognizing the limitations of technology. Without technology, there would be no manufacturing, but manufacturing is more than technology. It is also about people and their relationship with the technical resources of an organization. Manufacturing is about organization, people, technology, management accounting, business strategy, and so on. It is also about the connections between all these dimensions. In the past, we have tended to ignore not only the connections, but also some of the dimensions. We have placed too much faith in technology, using technology to compensate for inadequacies elsewhere, and trying to solve all problems as though they were technical problems.[2]

What do you think of when you hear the word "technology"? It probably depends on your background and interests, but chances are that some of the thoughts that you have will include machines, information systems, computers, vehicles, electronic communications, magnetic resonance imaging (MRI) devices, virtual reality, and so on. All of these examples and more are examples of technology. So are such things as enterprise resources planning (ERP) software, self-directed work teams, and supply chains, just to name a few that may not at first come to mind. If it is a new way of thinking about or doing something, it probably falls under the umbrella of technology. Technology is also about ideas, people, and how to blend them together.

In many improvement efforts, the use of technology outpaces the redesign of the company's infrastructure and we will discuss this more thoroughly in Chapter 8. For example, the new technology may require a change in organization structure to facilitate horizontal communications with customers and suppliers. The company may also need the effective participation of its employees to accept and implement the technology changes, a subject we will discuss in Chapter 9. The new technology may require the involvement of self-directed teams. If the company is still using extreme job specialization, the employees will not be intellectually or emotionally prepared for the change.

In a typical improvement program, technology often precedes changes in infrastructure and cultures. Technology innovation is a proactive exercise; the changing of infrastructures and cultures are reactive exercises. Consequently, they lag the development of technology. For example, new technologies include EDI and Six Sigma. However, successful implementation requires changes in infrastructure and culture. Often, the technical part of the improvement program is implemented before the changes in infrastructure and culture. Because they lag, the changes in infrastructure and culture are often implemented under significant time pressures, which may mean they are done piecemeal or inappropriately.

One area of growing importance is inter-company communication, leading to collaboration. This goal is now possible because of new developments in technology. Although information and communication technology are important to collaborative networks, technology can never overcome a lack of trust and ineffective goal setting between key cross-company partners.[3]

Technology is often linked with innovation.[4] Innovation implies change, so the introduction of new technologies in continuous improvement programs is accepted as a natural and worthwhile progression. However, infrastructure and culture convey the impression of permanence and, as a result, can be impervious to change.

Although there is often a resistance to new technologies, the evolution of technology has a persistent forward movement that makes it almost inexorable that there will be a continuous wave of new technologies. Resistance may delay the adoption of new technologies, but seldom prevent a worthwhile technology from finding its niche in business practice.

Organizations develop new technologies, such as the introduction of new clothing styles, because they want to and can. On the other hand, some new technologies, such as a new vaccine or a new management program, are developed as a result of identifying a need.

Technology is anything that enables a person or an entity to do something differently, hopefully better than before. In this sense, it is a driver of change. Perhaps that explains some of the resistance to the introduction of new technology; it is the introduction of change. It may cause a change in how we do things, such as the use of PCs instead of typewriters. It may change the way we think, such as when we set goals of zero defects instead of a range of acceptable defects. It can even change the way we manage, such as with self-directed teams instead of through a "show and tell" use of job specialization.

In Chapter 3, we described the Input–Transformation–Output (ITO) Model and its role in developing the supply chain. In the ITO model, technology relates primarily to the transformation process; consequently, we will describe the role of technology in the evolution of both manufacturing and service transformation processes. We will show how technology enhances the effectiveness of the resources used in the transformation processes — employees, equipment, facilities, systems, and information. In later sections, we will describe the implications of technology in strategic planning, and describe some of the issues concerned with technology and new technology implementation.

What is technology? Many people conjure up visions of space ships and automated production lines when answering this question. Those are examples of high technology, but the fact is, a handsaw or a hammer is also considered technology. So is the thought process you use to solve a math problem. Lee Krajewski and Larry

Ritzman define technology as "any manual, automated, or mental process used to transform inputs into products or services."[5] Operations managers are concerned with the level or degree of technology used and the corresponding capital intensity of the operations process.

The fast pace of technology change and the many technologies available can tempt managers to upgrade process technology without thoroughly analyzing their actual needs. Changes in technology affect all aspects of an organization — human, technical, and financial — and are usually long lasting. Because of the high stakes involved in the technology game, it is important that decisions about upgrading technology are based on the needs of the organization and customers, and that these decisions are supported by an appropriate analysis of needs. Changing the technology of a service delivery system for reasons other than productivity, service quality, or delivery time improvements can be a serious management mistake.

Technology can be related to the ITO Model described in Chapter 3. Technology can be an input (data), a transformation agent (database), or an output (information). An extension of this is to view the input as information, the transformation process as learning, and the output as knowledge. It can be a driver of innovation (the computer used in system design). Technology has many faces. We will concentrate primarily on its role in improvement programs in this chapter.

Definitions

The *APICS Dictionary* defines technologies as the terms, concepts, philosophies, hardware, software, and other attributes used in a field, industrial sector, or business function.[6] This definition is so broad that we will have to look further.

The *American Heritage Dictionary*'s definition of technology is almost as broad. It says that technology is

a. The application of science, esp. to industrial or commercial objectives, or
b. The entire body of methods and materials used to achieve such objectives.[7]

The definition from www.wipedia.com makes the following distinctions:

- *Science* is the formal process of investigating natural phenomena. It produces information and knowledge about the world.
- *Engineering* is the goal-oriented process of designing and building tools and systems to exploit natural phenomena for a practical human means. *Engineers* work within the constraints of natural laws and societal needs to create technology.

- *Technology* is the consequence of these two processes and societal requests. Most commonly, the term "technology" is used as the name of all engineering products.[8]

Technologies can be described as "hard" — cellular telephony, railway signaling, or electricity generation. There are also "soft" technologies — management competencies or governmental policy development processes. Some define technology as the construction and use of machines, systems, or engineering. Others, the "socio-technologists," take a broader view and consider technology to be meaningful only when it becomes a social fact.[9]

From the above discussion, we conclude that technology has a wide variety of meanings. In the following sections, we will examine subsets of technology to show how technology relates to continuous improvement programs. To focus our discussion, we will consider technology as consisting of two major categories — tangible products or "hard" technologies and concepts, or "soft" technologies.

The Role of Technology in Continuous Improvement

Advances in technology are commonplace today, so we seldom stop to think about what impact technology has had and will have on our lives. Every economy in the industrialized world is experiencing the effect of this phenomenal explosion of technological advances. New technology offers new opportunities and new threats to individuals and businesses alike. Those who seize the opportunities will move forward, those who fail will probably not be around to compete in the next decade.

The primary use of technology in business and industry is to improve competitiveness — improve productivity, quality, and service delivery. In the previous section, we stated that technology alone is not the key to productivity improvement. The key to redefining the tasks includes clarifying what needs to be done and eliminating what does not need to be done. This is a good beginning because many organizations apply advanced technology to tasks that do nothing to add value for the customer. When considering the application of technology in operations, rule 1 is to streamline the tasks before applying higher technology. Do not automate tasks that do not need to be done. A corresponding fact is technology will not correct poor management. Our discussions in this chapter assume that rule 1 is in effect and that you understand this fact of technological life.

Technology can be separated into two categories: information technology and process technology. Both are key resources for any organization, and computer technology plays a significant role in both. In this chapter, we will discuss the functions of equipment and process technology in organizations, and some of the decision areas involving technology. We will explore the criteria used in decision making, the steps in adding technology to the process, and key issues in the use of technology.

Technology and the Infrastructure

For purposes of discussion, we refer to the structure of an organization and the systems used in managing the firm as the infrastructure. Does the introduction of higher technology in a firm influence the infrastructure, and if so, how? Certainly, with more and higher technology, there is a corresponding increase in the rate of capital intensity, and this creates a new awareness of the cost of capital. Applying technology for productivity improvements implies changes, perhaps unwelcome changes, in work tasks. Wickham Skinner and Kishore Chakraborty discuss the impact on an organization's structure and business strategies as "changes in the structure of jobs, changes in the content of jobs, and the need to redefine the environment at work."[10]

Even though it is difficult to generalize about technology's impact on the infrastructure throughout the various service industries, there are some common threads among industries. Initially, the introduction of new equipment and process technology increased the responsibilities of managers and tended to increase middle management ranks. Staff specialists were hired to support the needs of this growing cadre of technology specialists and managers. Ironically, in recent times, the introduction of powerful information and communications technology has reversed that trend. If information can be made more meaningful and widely distributed quickly, fewer middle managers are required. Upper-level managers now have access to management information that not long ago was the territory of middle managers. This trend is creating flatter organizations and facilitating easier information flow along the supply chain.

Technology has implications in other ways in an organization. Many employees have to learn new skills, and absorb the duties of those who are re-engineered out of the firm. There is a potential for new technology to affect the quality of working life, both positively and negatively. On the positive side, word processors and spreadsheet technology have expanded the abilities of and made work easier for the users. On the negative side, new technology, especially in the form of computer hardware and software upgrades, may cause stress among employees to stay current with the latest developments. These primarily affect job design and work assignments.

Another infrastructure area that is affected is the system used in managing the organization. Again, introducing new technology has the potential to affect the operating system either positively or negatively. When management recognizes that individual technologies cannot operate independently, new technology that can integrate functions will have a higher positive impact.

There is a trend in the manufacturing sector referred to as "a focused factory" approach. In this approach, a given factory focuses on producing a single product or family of products. The advantages of this approach include narrowing the number of demands that will result in higher productivity, fewer levels of management to coordinate, teams are more effective in solving problems, communications are easier and less likely to be misunderstood, and there is oftentimes a closer-knit culture. This trend is also evident in some service firms. Fast food restaurants, business schools with specialized majors, niche killer retailers, and specialty medical practices are a few examples. This trend necessitates organization changes, often requiring a decentralization of management responsibilities.

There is a trend to outsource more of the tasks performed by organizations. This will disrupt the existing organizational structure and systems, and require that companies develop new organizational forms to accommodate the new arrangements. This practice of outsourcing may prove to have more impact on the infrastructure than technology changes.

Technology and Organizational Culture

The successful implementation of continuous improvement programs involves knowledge transfer and subsequent knowledge management. Companies must adapt their organization structure and systems to accommodate the new improvement program; in addition, they must also adapt their corporate culture to fit the new program, or adapt the program to fit their culture.

What is corporate culture? It is the belief system of the organization. It results in a feeling among employees about how they should act in carrying out the company's work. It is not the same as the vision, mission, strategies, and plans of a business. Companies could have very similar strategies, but differing cultures. Culture is the cumulative result of blending strong leadership, environmental influences, personal characteristics of the region in which the company operates, and the manner in which the company competes in its marketplace. Cultures form over long periods of time and are highly influenced by the founder(s) of the business.

Corporate culture should be carefully assessed when implementing new technology, such as a continuous improvement program. Although technology may be the driver of the program, it will not be successful unless it is properly assimilated into the culture of the business. Some cultures actively accept and endorse change; other cultures resist and even oppose change.

That corporate cultures exist is a reality. To be able to implement new technology, companies must successfully deal with their corporate cultures. There are two diametrically opposite approaches — change the culture to accommodate the proposed changes or modify the program to fit the culture. The following examples will illustrate this dilemma.

One approach is to accept the culture as it is and adapt the technology to fit the culture. Consider a business that wants to implement a quality improvement program. The corporate culture is heavily oriented toward low cost as the competitive advantage. If the quality improvement program were presented as a way to reduce costs of scrap and rework, it would probably be more acceptable in the existing culture.

Another approach is to attempt to change the corporate culture. Consider the same quality improvement program. If presented as a way to encourage greater employee involvement and job enrichment, employees would have to reorient their approach to their work as well as changing their relationships with their supervisors. Some employees would readily change; others would change more slowly; and some would resist the change. Working through this transition takes careful planning and recognition of individual differences.

A third alternative is to replace completely the existing culture with a new one. This is a radical approach that usually causes major disruptions. The quality program is intended not just to improve the quality of existing products; it is a movement to replace the entire product line with a new emphasis on high quality instead of low cost. Such a change within the existing employees, both at the production level and manager level, requires extensive preparation. Some employees, especially at the middle management level, may not be able to adjust and will have to be replaced or moved to other positions. In the most extreme cases, it may be more advantageous to move the operation to a new site, hire new employees, and build a new corporate culture at that location. IBM did something like this when deciding to start its PC operation in a remote location in Florida rather than risk its rejection in the existing mainframe culture at established facilities.

John Kotter and James Heskett were among the first to study the impact of corporate cultures on a firm's financial performance. They found that firms that included all of the key managerial constituencies and leadership from managers at all levels outperformed firms that had less well-developed cultures. They also concluded that cultures would inhibit firms from adopting strategic or tactical changes, a serious negative effect in a rapidly changing world. Corporate cultures formed over a long period are most difficult to change. Although difficult to change, corporate cultures can enhance performance with good leadership.[11]

Corporate cultures also have an effect on knowledge management (KM) practices. Standard KM tools may be used differently by diverse organizational cultures and, as a result, achieve different outcomes. Management's choices are to change how the different subcultures use the tools or encourage uses that fit the cultures. One study concluded that it was best to "avoid KM initiatives to achieve uniform targeted outcomes across the firm. Rather, facilitate KM initiatives that appeal to a wider range of more favorable outcomes."[12] Another study found that one of four enablers in developing a KM system is the forming of a culture of sharing, a major shift from the present thinking of many individuals that knowledge is power and not to be shared with others.[13]

Enterprise resource planning (ERP) systems have been one of the major technological developments in the past decade. Studies indicate that many companies are less than completely satisfied with the results of their ERP implementations despite the potential benefits. Although some of the less than favorable results may be a result of the technology, researchers are finding that "managers as well as scientists in many cases attribute the problems associated with the development and implementation of ERP to aspects of organizational culture. However, in general culture has not yet been a topic of explicit concern in studies of ERP."[14] Because

ERP systems are designed to standardize processes and centralize control, they tear at the framework of existing cultures, especially among multi-division companies. They also attack the job content of the individual employee at all levels of an organization, affecting the worker's skills and attitudes.

Many companies are moving toward globalization of their businesses. When a business moves into another geographic area or country, it must confront the issue of national cultures. In some cases, the cultures are markedly different and the literature is full of missteps that companies have made in violating the cultural norms of the country in which they operate. The culture of an individual country may be well entrenched; however, the globalization of business is having an effect. Although businesses may adapt to different cultures, the cultures are also adapting to global management practices. "Thus, as much as the managerial values are changing, so much so the organizations are cross-verging with different cultures, and a new cultural community is emerging in global business."[15]

As part of the dynamic changes in businesses, there is a revolution in entrepreneurship as large corporations are finding it difficult to keep up with the innovations required, in part because of the deeply entrenched cultures and their resistance to change. Consequently, many of the innovations are being developed by start-up companies, or as separate entities under the umbrella of the parent company. The R&D Limited Partnership model also offers a route for large companies to evolve sequentially into these new ways of operating, without threatening their entrenched bureaucracies. Major cultural shifts are required, but these can be managed in incremental steps.[16]

Traditional software development relied on the fundamental assumption that systems can be fully defined ahead of time. It used a mechanistic command-and-control form of organization that was formal and included a project manager and specialized team members. It involved formal communication and developed extensive documentation, or implicit knowledge. A newer approach to software development involves the use of agile methodologies where the basic assumption is that high-quality, adaptive small teams using an iterative design and test approach that relies on rapid feedback and change can develop software. It is people-centric and builds tacit knowledge using self-organizing teams with informal communication and heavy reliance on customer input. An organic organizational form encourages flexible and collaborative interaction in the development process. Bureaucratic and formal organizations with entrenched cultures are not as likely to be successful in using agile methodologies as organizations that support change and innovation.[17]

Corporate culture is a key ingredient in the successful implementation of technology innovations. We will discuss corporate culture more fully in Chapter 9.

Technology Transfer

Technology is a powerful aid to businesses; however, before they can take advantage of it, businesses must know that the technology exists and they must learn how to use it. In general, knowledge first resides in individuals, as tacit knowledge. Individuals can learn in several ways:

- Read about the technology in a publication
- Attend a workshop or some other educational program
- Work as an aide to someone who has deeper knowledge about the technology

As individuals learn, they can pass that knowledge along to others in the organization either informally or through training classes. As a greater number of individuals in an organization learn about the technology, they prepare memos, manuals, and other documents that help to establish a store of knowledge within the organization, called explicit knowledge.

It may be possible to learn enough about a simple technology by reading about it. However, as the complexity of the technology increases, most businesses find that one of the following scenarios is advantageous:

- Have selected employees attend training sessions
- Hire a person who has had experience with the technology in another company
- Hire a third party to help them introduce the technology into the organization

Continuous improvement programs are generally of sufficient complexity that businesses find it necessary to use more formal methods to acquire knowledge about the new technology.

Thomas Davenport and Laurence Prusak prepared one of the more popular books about knowledge management. It included chapters on knowledge generation, knowledge codification and coordination, knowledge transfer, knowledge roles and skills, technologies for knowledge management, along with some examples of knowledge management projects.[18]

Technology for Process Improvement

An almost universal expectation of technology is that it will be used to improve the transformation processes used by businesses. Technology enables companies to be more effective and efficient. Although the introduction of new technology is often disruptive, the availability of new technology eventually prevails if it has basic value to the operation of a business.

Agriculture, Mining, Construction, and Manufacturing: Goods Producers

The goods-producing industries have been revolutionized by hard technology. Reapers and baling machines in agriculture, automated cutting machines and conveyors in mining, factory-built trusses and high-rise cranes in construction, and flexible manufacturing systems (FMS) and robots in manufacturing are just a few examples of how hard technologies have enabled these industries to reduce costs, improve quality, and provide other benefits.

Table 7-1 is a representative list of technologies that have enhanced production and productivity in goods-producing industries. These industries have also improved their processes with management concepts, or soft technologies. In Chapter 5, we described a number of management programs that originated in manufacturing. In this section, we will briefly describe some of the major concepts that made it possible to apply technology to the manufacturing sector. Table 7-2 provides a sampling of these concepts.

There is an abundance of literature about technologies that apply to the goods-producing area. We include just a sampling of the sources in the following section.

Lean production was pioneered by James Womack and Daniel Jones.[19] Since then, there has been a steady stream of books and articles about lean production. It has replaced JIT as the program of choice when companies want to streamline their operations to reduce costs and customer response times.

Mohsen Attaran and Sharmin Attaran studied the re-engineering movement. They found that re-engineering initiatives have produced a range of results. They believe that information technology, with a major emphasis on the use of the Internet, will play an increasingly important role in re-energizing business process improvement efforts.[20]

Manufacturing technology innovations continue to be developed. One study found that early adopters gained financial advantage because they were able to combine the technology, such as CAD and cellular manufacturing, with knowledge transfer capabilities and culture adjustments to assimilate the technology into their operations successfully.[21]

RFID is a new technology that is still in its early stages. Although it has great promise, high tag costs, evolving reading reliability, and the need for industry standards is making the move to RFID tags a slow and somewhat painful process. As with most technology innovations, there is a need to develop the infrastructure to facilitate the change and the culture to accept it.[22] Ariel Markelevich and Ronald Bell studied the effect that RFID will have on accounting practices. They expect that accounting will have to make significant changes to capitalize on the added information provided by RFID.[23]

Supply chains will benefit from new technologies. In addition to new forms of materials handling, there will be new inter-organizational communications systems that will more closely link companies with one another. We will describe this more

Table 7-1 Process Technologies in Manufacturing

Type of Technology	Description
Automated storage and retrieval system (ASRS)	A high-density rack inventory storage system with vehicles automatically loading and unloading the racks.
Automated guided vehicle system (AGVS)	A transportation network that automatically positions materials handling devices at predetermined destinations without operator assistance.
Automatic identification system (AIS)	A system for transforming data into electronic form with the use of bar codes or RFID tags.
Computer-aided inspection and test (CAIT)	The use of computer technology in the inspection and testing of manufactured products.
Computer numerical control (CNC)	A technique in which a machine tool controller uses a computer or microprocessor to store and execute numerical instructions.
Electronic data interchange (EDI)	The paperless (electronic) exchange of trading documents, such as purchase orders, advanced shipping notices, and invoices.
Flexible manufacturing system (FMS)	A group of numerically controlled machine tools connected by a central computer system and an automated transport system.
Process control	The use of information technology to control a physical process such as temperatures, pressures, and quantities.
Rapid prototyping	The transformation of product designs into physical prototypes through cross-functional teams, data sharing, and computer technology.
Robots	A flexible machine with the ability to hold, move, or grab items. Electronic impulses activate motors and switches.
Servo system	A control mechanism linking a system's input and output, designed to feed back data on system output to regulate operation of the system.
Vision systems	Using video cameras and computer technology in inspection roles to assure that cans of soup are full.

Source: Adapted from Blackstone, J.H. and Cox, J.F., III, *APICS Dictionary* (Eleventh Edition), APICS, Alexandria, VA, 2005.[48]

Table 7-2 Soft or Conceptual Technologies

Concept	Description
Advanced planning and scheduling (APS)	Techniques that deal with analysis and planning of logistics and manufacturing over the short, intermediate, and long-term periods.
Business-to-business commerce (B2B)	Business being conducted over the Internet between businesses through electronic communications networks.
Cellular manufacturing	A concept that produces families of parts within a single line or cell of machines controlled by operators who work within the line or cell.
Collaborative planning, forecasting, and replenishment (CPFR)	A collaboration process whereby supply chain trading partners can jointly plan key activities required for effective supply chain operations.
Design for manufacturability	Simplification of parts, products, and processes to improve quality and reduce manufacturing costs.
Design for manufacture and assembly (DFMA)	A product development approach that involves the manufacturing function in product design to ensure ease of manufacturing and assembly.
Design–measure–analyze–improve–control (DMAIC)	A Six Sigma improvement process with a structured approach to the problem-solving process.
Concurrent engineering or participative design	Involves the participation of all the functional areas of the firm in the product design activity, often including customers and suppliers.
Flow chart	The output of a flowcharting process, a chart that shows the operations, transportation, storages, delays, and inspections related to a process.
Value engineering and/or analysis	A disciplined approach to the elimination of waste from products or processes by deciding if such functions add value to the good or service.
Computer-aided design (CAD)	The use of computers in interactive engineering drawing and storage of designs.
Computer-aided manufacturing (CAM)	The use of computers to program, direct, and control production equipment in the fabrication of manufactured items.

Source: Adapted from Blackstone, J.H. and Cox, J.F., III, *APICS Dictionary* (Eleventh Edition), APICS, Alexandria, VA, 2005.[49]

fully in Chapter 10. One study found that firms have to incorporate new product and process technologies and, at the same time, develop capabilities that allow them to be flexible and agile. To accomplish this, they have to increase their effectiveness, exploit the synergies, and learn throughout the organization.[24]

Research and development (R&D) centers develop much of the new technology for manufacturing and service operations. Consequently, R&D centers have a heavy responsibility to manage an effective technology transfer system. Jeffrey Lind describes a process developed by The Boeing Company that "combines insights from systems engineering, software process improvement, organizational psychology, and anthropology to provide a coherent approach to innovation in a large enterprise."[25]

Simulation is a technology that has been around for several decades, but has probably never reached the level of use that it merits. A study by Sameer Kumar and Promma Phrommathed looked at how process mapping and simulation could be combined with some popular software tools — Visio, Excel, and Arena — to develop a way to optimize the scheduling of a manufacturing process.[26]

Technology has been a major contributor to the success of manufacturing operations, as we saw in Chapter 5. In the next section, we will describe some of the technologies used in service operations.

Services

One of the significant differences between services and manufacturing is that the customer often participates in the service process. This means that the success of technology changes depends on customers and their acceptance or rejection of the change. For example, many people will not give up the personal attention of a bank teller for the convenience of an ATM. Still others refuse to purchase gasoline from a pay-at-the-pump facility because they do not want to learn how to do it. Some callers do not have the patience to go through the 1, 2, 3, or # sequence to leave a voice mail. The point is, the role of the customer in the service process must be considered when making changes in the service delivery system. Customers often rebel and find new sources for services when forced to make a change they resist.

Service operations delivery systems usually consist of two parts: the front end, where there is high customer contact, and the back room, which supports the front-end customer encounter. Bank tellers work in the front end, check clearing is a back-room operation. There appears to be a trend of pushing more tasks into the back-room operations to facilitate productivity improvements. Moving tasks away from the customer contact part of the process offers more opportunities for technology upgrading and automation projects. However, this move does take away some of the long-term value of "personalized service," and you must have understanding customers who are willing to move to lower contact. A real, or perceived, improvement in delivery, timing, or quality, or a price reduction may be necessary to facilitate this move.

Over the past few years, many service industries have taken a different approach to productivity improvement by involving customers more in the service delivery system. ATMs, salad bars, fast food and clean-your-own table restaurants, and self-service gas stations are but a few examples of involving customers more to gain productivity improvements.

Some may not think of service operations as candidates for technology upgrades and automation, but there is a wide range of opportunities for applying automation in the varied services sector. Table 7-3 gives some examples of automation in service industries.

The Japan Industrial Robot Association and the American Robot Industry have classified equipment available for automation. Table 7-4 offers a summary of these classifications with examples of the application in a service business.

In addition to the automated systems listed in Table 7-4, a wide variety of manual machines or manipulators for technology improvements are available to the operations manager.

Table 7-3 Examples of Automation in Services

Industry	Technology
Airlines	Air traffic control systems, auto-pilot systems, reservation systems, self-ticketing, security checks
Health care	MRI and CAT scans, RFID tags, patient monitoring systems, remote diagnostics, online information systems
Financial services	Automatic teller machines (ATMs), credit cards, debit cards, electronic funds transfers, online transactions
Hospitality industry	Online reservation systems, automatic checkout, electronic key/lock systems
Wholesale and retail	Point-of-sale terminals (POS), computerized stock replenishment, self-checkout terminals, E-commerce
Transportation	RFID and GPS tracking, automated toll booths, containerized shipping, security checking
Restaurants	Terminals to place orders with kitchen, POS terminals, computerized employee scheduling
Government services	Online inquiry capability, one-person garbage trucks, automated telephone response during power failure

Table 7-4 Categories of Robots

Type of Robot	Description	Application
Fixed sequence robot	A machine that repetitively performs successive steps of a given operation according to a predetermined sequence, condition, and position, and whose set information cannot be easily changed.	Automated one-person garbage trucks
Variable sequence robot	A machine that repetitively performs successive steps of a given operation according to a predetermined sequence, condition, and position, and whose set information can be easily changed.	Automated teller machine
Playback robot	A machine that can produce from memory operations originally executed under human control.	Telephone answering machine
Intelligent robot	A machine with sensory perception, such as visual or tactile, that can detect changes in the work environment or task itself, and has its own decision-making abilities.	Medical information systems
Totally automated systems	A system of machines that performs all the physical and intellectual tasks required to produce a product or service.	Space shuttle

Source: Adapted from Collier, D., *Service Management: The Automation of Services*, Reston Publishing Company, 1985, p. 20.[50]

Before ending this discussion of equipment and process technology, it is appropriate to begin to examine the human aspect of technology changes. There are advantages and disadvantages to using people and machines for work. In general, people can think creatively and adapt to unexpected situations much better than machines can. On the other hand, machines are better suited for complex or repetitive tasks that require speed and precision. The challenge is to provide the optimum mix of people and technology.

Self-Service

Self-service is a growing part of the service economy. At one time, most retail stores made it difficult for customers to touch the merchandise because grocery items were stored on shelves behind the clerk at the counter, and tools, watches, and other valuables were locked in cases. This practice was the standard applied to reduce theft. Over time, merchandisers decided that customers might be more apt to buy the goods if they could examine them more closely, so they displayed the merchandise in open shelves and counters. This was the beginning of self-service.[27]

Over time, customers have increased the level of their participation in the service process. They collect groceries in a cart and check themselves out through a self-checkout station. They pump their own gasoline and use a credit card to pay for it without ever entering the store. They check their own blood pressure or administer their insulin shots without assistance. Homeowners do more "do it yourself" projects, thanks to improved instructions and training courses provided by the retailer. Self-service reduces the cost of the business and is usually more convenient to the consumer. However, it does require an increasing knowledge on the part of the consumer and increased blending of human and machine skills.

Over time, a number of self-service technologies have caught on as customers conclude that there is a benefit to them and that they can learn to use the technologies without extreme difficulty. One of the newest and most demanding areas in which the self-service concept is being introduced is in E-business.

E-Business

The world is moving rapidly, albeit erratically, toward electronic communications systems. As new technologies evolve, participants increase. This is a case where the technology is definitely the driver, and infrastructures and cultures are scurrying to keep pace.

E-business in the broad sense is conducting business through electronic-aided means, with emphasis on the use of the Internet. Managerial Comment 7-2 includes some more formal definitions from the U.S. Census Bureau. Although E-business is generally considered a broader term that includes E-commerce, it appears that many writers use the two terms interchangeably.

Managerial Comment 7-2.
E-Business: Definitions and Concepts

The three primary components of our electronic economy, and the feature shared by two of them, are defined below. Each definition includes examples of its scope and content. The definitions are intentionally broad to provide an inclusive framework for planning statistical measures, and to allow flexibility to incorporate continuing changes in the electronic economy.

E-business infrastructure is the share of total economic infrastructure used to support electronic business processes and conduct electronic commerce transactions. It includes hardware, software, telecommunications networks, support services, and human capital used in electronic business and commerce. Examples of E-business infrastructure are:

- Computers, routers, and other hardware
- Satellite, wire, and optical communications and network channels
- System and applications software
- Support services, such as Website development and hosting, consulting, electronic payment, and certification services
- Human capital, such as programmers

Electronic business (E-business) is any process that a business organization conducts over a computer-mediated network. Business organizations include any for-profit, governmental, or nonprofit entity. Their processes include production-, customer-, and management-focused or internal business processes. Examples of electronic business processes are:

- Production-focused processes including procurement, ordering, automated stock replenishment, payment processing, and other electronic links with suppliers as well as production control and processes more directly related to the production process
- Customer-focused processes including marketing, electronic selling, processing of customers orders and payments, and customer management and support
- Internal or management-focused processes including automated employee services, training, information sharing, videoconferencing, and recruiting

Electronic commerce (E-commerce) is any transaction completed over a computer-mediated network that involves the transfer of ownership or rights to use goods or services. Transactions occur within selected E-business processes (e.g., selling processes) and are "completed" when agreement is reached between the buyer and seller to transfer the ownership or rights to use goods or services. Completed transactions may have a zero price (e.g., a free software download). Examples of E-commerce transactions are:

- An individual purchases a book on the Internet.
- A government employee reserves a hotel room over the Internet.
- A business calls a toll-free number and orders a computer using the seller's interactive telephone system.
- A business buys office supplies online or through an electronic auction.
- A retailer orders merchandise using an EDI network or a supplier's extranet.
- A manufacturing plant orders electronic components from another plant within the company using the company's intranet.
- An individual withdraws funds from an ATM.

Source: Adapted from Mesenbourg, Thomas L., Measuring Electronic Business: Definitions, Underlying Concepts, and Measurement Plans, U.S. Bureau of the Census, 2007.[28]

E-business requires a great deal of technology, such as described in Managerial Comment 7-2. However, it requires more than technology. Table 7-5 contains a number of characteristics found in successful E-business programs.

Technology for Resource Enhancement

In addition to improving processes, technology can enhance the effectiveness of the resources used in the transformation process.

Human Resources

Technology can affect human resources in a number of ways. In a study of a major aerospace company, Robert Kazanjian and Robert Drazin concluded that there

Table 7-5 Characteristics of Successful E-Commerce Implementations

Leadership	Strategy
Commitment at the top	Well-positioned online brand
Thorough competitive analysis	Online-friendly offerings
Significant financial investment	Reliable customer service
Cultural transformation	Cross-channel coordination
Structure	**Systems**
Internal investment	Modernized internal processes
Integrated management teams	Incentive-laden HR practices
IT know-how from within	Aligned performance measures
Strategic partnerships	Improved customer management

Source: Adapted from Epstein, M.J., *Implementing E-Commerce Strategies, A Guide to Corporate Success after the Dot.com Bust,* Praeger, Westport, CT, 2005, p. 20.[51]

was adequate technology to enable the manufacturing sector to meet the market requirements of flexibility and quality in the face of increased worker demands and resource scarcity. However, they also concluded "successful process innovation is a function of the nature of the innovation itself and the amount of new knowledge to be developed, the appropriate configuration of organizational structure, and the existence of a constellation of critical staffing roles."[29]

Sometimes, advocates of technology have promoted its use as a way to overcome the weaknesses of the worker. In that scenario, the success of a technology depended only on training employees so they could overcome organizational deficiencies, implying that the technology was all right. A different perspective is "fix the process, not the people."[30] This approach recognizes the need to consider organizational structure and culture when implementing new technology.

In the early stages of technology applications in services, it aided employees in performing their work. An electric-powered lift truck in warehouses is an example. As competitive pressures increased, technology improved and replaced employees. Automated materials-handling systems is an example of replacement. Today, with demands for high customization and flexibility, there is a trend toward person/machine combinations as an integral part of the service delivery system. The fixed-sequence robot, one-person garbage truck is an example of an interdependent person/machine combination.

As an Aid to the Employee

Some of the early efforts in productivity improvement in services centered on using technology to make work less difficult and less time consuming. As is the case today, much of this occurred in the back-room operations and offices of service firms. One of the early technology introductions in offices was the electro-mechanical calculator. Previous methods of performing calculation functions were time consuming and difficult, and accuracy was a problem. The evolution to electronic calculators was a further enhancement to the speed and accuracy of calculation functions. Computers, spreadsheets, and other specialized application software have moved calculation functions to higher ground.

The technology of digital document systems offers desktop control over the production, management, and finishing of documents. A person can print, fax, copy, collate, staple, and distribute documents without leaving his or her desk. This technology closed the gap between electronic and paper documents and significantly improved workflow productivity. In general, managers and workers welcomed technology that aided work performance.

Technology can aid employees and the service firm in ways other than as specific work aids. Technology can change the total service delivery system and help create jobs. Consider the use of plastics and refrigeration in the food business. The ability to prepackage, transport, and keep predetermined portions of food has allowed fast-food service businesses to adopt a production line approach to service delivery. The next time you opt for a Big Mac lunch, observe the process of preparing your order. It is an example of a division of labor production line system made possible by the application of technology in a service environment.

Examples of technology that helps people accomplish their work are everywhere in the workplace. Some of the more common are:

- Telephone headsets
- Word processors and spreadsheets
- Electronic meter readers
- Electronic data interchange (EDI)
- Bar codes
- Retail computerized checkouts
- Police car installed computers
- Weather reporting systems
- Magnetically leveled passenger movers
- Mobile phones with text messaging
- Notebook computers with wireless connections
- Intranet and extranet systems

As a Substitute for the Employee

The one overriding characteristic of higher technology and automation is the capability to produce standardized goods and services rapidly and consistently at a reduced cost. In the manufacturing sector, this prompted a drive to integrate computer-assisted machines into larger clusters to assume even greater control of the production process and use fewer people. These systems are flexible manufacturing systems, or FMS. Some of this technology has spilled over into the services sector. Automated guided vehicles (AGV) are flexible materials-handling systems that perform pick and put away operations as well as transport materials from receiving to storage and then from packing to shipping.

In a flawed belief that more computerization is the ultimate answer to success, many U.S. firms have leap-frogged basic FMS into computer-integrated manufacturing (CIM). This powerful technology is very appealing, but for all the wrong reasons. "Its appeal is the lights-out factory, no more labor troubles, and a quick edge on the competition. That's why CIM, touted as the big manufacturing breakthrough of the age, has turned into a disaster at many companies. And the reason is simple: to implement computer-integrated manufacturing requires the kind of sweeping organizational changes that many companies investing in CIM have yet to make."[31]

What are these sweeping organizational changes? Organizational structures, groups, and people need to be formed into teams that operate as an integrated organization. This requires changes in attitude, work practices, and people's skills, and few companies are willing to make such drastic changes. Chapter 8 describes more fully the organizational changes required and much of Chapter 9 is dedicated to people-oriented actions that support the needed changes.

Just as in manufacturing, automation technology is being used in many service companies for the wrong reasons — to replace people in service jobs. This replacement function of technology has created a controversy among some sociologists. Displacement of people frequently hurts the individual worker and creates a burden on society as a whole. Some have gone so far as to predict that the majority of the total work force will eventually be trapped in low-skill, low-pay job positions. Others contend that there are both social and economic obstacles to automation. The economic obstacles are well understood: high tech is high cost. The social obstacles are more complex.

In the past, the rapid growth of services absorbed many of those displaced by automation in other sectors. If there are fewer and fewer jobs in the service sector, can there be continued job growth? If fewer people have jobs, who will bear the increased burden on the Social Security system and other social systems? Can companies find new markets in the global economy to provide jobs for those displaced by automation? Alternatively, is all of this just a technical Chicken Little shouting "the sky is falling"? We do not know the answers to these questions, nor is there consensus among those sociologists who study them. The point is that operations managers' jobs and responsibilities are, and will continue to be, affected by these issues. The next part of this section discusses a trend that may impact using technology to replace people.

As an Integral Part of the Process

There is a growing consumer trend toward less standardized, more customized products and services. This trend will continue, indeed accelerate, as we move into the 21st century. This will require a different kind of systems integration, and it will have a significant impact on service delivery systems.

In CIM, what is integrated? The computers are integrated; the people interface to support the system! In instances where systems have been designed to interface with people, the interface was reactionary to solve a problem with the system or to compensate for the inadequacies of the technology.[32] This approach is compatible with a service environment that is cost-driven with standard services. Today, you must have a system that can deal with more customization and flexibility in service delivery. A system that meets these needs requires a different kind of system design, a design that integrates people and machines rather than interfaces between people and machines. In other words, people characteristics add flexibility and cognitive thinking to a service delivery system. "Flexibility cannot be purchased by acquisitions or merger. Flexibility is something people do. Organizations of people may be combined or disassociated, but flexibility itself is not attained through things, financial or physical."[33]

We will examine some of the forces that influence introducing technology into a service organization and some of the characteristics of people/machine relationships.

Technology applications are either "pulled" or "pushed" into service organizations. Pulled applications occur when market forces demand services that cannot be provided by existing technology. From the advent of credit card use, the approval process was an obstacle to using a credit card. Improvements in this process to the point of electronic approval in seconds were pulled by the market. Pushed applications occur when technology developers create new technology and find or create a market for it. Bar-code systems were developed, but were not widely used in services until a universal product code was developed. Bar codes were pushed into the market and the pushing pulled a universal product code.

In his book, *Things That Make Us Smart,* Donald Norman contends that up until now, most technology development has been a push process. He illustrates this by using the motto of the 1933 World's Fair: *Science Finds, Industry Applies, Man Conforms.* It is Norman's contention that we are approaching the time when there must be human-centered technology: *People Propose, Science Studies, Technology Conforms.*[34] Paul Kidd essentially says the same thing. "Balance (people/machine) is an important conceptual basis for the new paradigm of agile manufacturing. It will help us to address the interconnected elements of our overall manufacturing enterprise. A balancing capability avoids polarization from any perspective."[35] He describes a participatory model for computer-based technologies that support people. Table 7-6 illustrates this concept.

The demand for customization is pulling a change in people/machine relationships, and social changes in the workplace are pushing for a change in service

Table 7-6   Interdisciplinary Design Methodology[52]

One of the primary concepts of agile manufacturing is concerned with the integration of organization, people, and technology. To achieve this integration we need to adopt an interdisciplinary design methodology. The design paradigm that should underlie agile manufacturing is characterized by: • A holistic systems based approach • Concurrent engineering concepts applied to the design of the manufacturing/service enterprise • Application of insights from the organizational and psychological sciences to the design of the technology and the overall manufacturing enterprise • Consideration of organization and people issues at all stages, from formulation of business strategy right through to the design and implementation of systems

Source: Kidd, P.T., *Agile Manufacturing,* Addison-Wesley Publishing Company, 1994.

delivery system design. These forces are pushing toward technology in which people/machine combinations operate as an integral part of the service delivery system. Employees will be less available to compensate for inadequacies of technologies, and will be more a part of a system where they apply their knowledge to operate flexible systems that provide customized services.

Equipment

Technology is used in equipment to enhance the performance of the equipment or to use the equipment as a data collection point.

Enhance the Performance of Equipment

Much of the technology described in Table 7-1 related to improving the precision and quality of output from the equipment resource. Much of the focus in recent years has been in using the computer to enable equipment to do more without the aid of human operators. Although equipment can perform a number of automated functions, there is a growing realization that the best results come from a blend of human and machine skills, not just from machine automation alone.

Provide a Source of Performance Information

Through sensing devices and connected software, equipment can now provide a great deal of information about the amount and quality of the products and services processed. For example, machines can count and report the number of units produced. They can sense the thickness of paper and provide a permanent record of the quality of the paper processed.

Facilities

The degree of technology employed in a service delivery system influences the design and layout of the service facility. However, just as customer contact and customer needs influence the degree of technology used, they equally influence facility design and layout. It is, then, the nature of the basic service and overall operations and marketing strategies that determine the kind of facility required. When selecting a design and layout, organizations must consider a number of factors, including the level of customer contact, volume of demand, range of the types of services offered, degree of personalization of services offered, skills of employees, and costs as well as the technology employed.

The diversity of services makes it difficult to generalize about facilities and technology; however, we can establish two ends of a continuum. There are service facilities with limited or no customer contact where the technology employed can be at the high end. The design of these facilities focuses on technology and optimizing productivity. At the other end, there are service facilities where the entire facility is designed around customers. We previously used the example of the backroom operations of a bank as the former, and the front end of a bank as the latter. Other service facilities combine the characteristics of these two extremes. Again, the degree of customer contact versus technology for productivity depends on the services provided and the tactics employed in providing the services.

Design

The service package is a bundle of goods and services consisting of four features: supporting facility, facilitating goods, explicit services, and implicit services. The makeup of the service package is a function of the competitive service concept and strategies selected, and technology decisions are mostly associated with the supporting facility feature. These decisions are about location and physical size of the facility, layout to optimize the customer encounter and facilitate the flow of goods and information, and equipment selection. The operating capacity of the service process is determined by these decisions, and most of the decisions create "hard" results. That is, once they are made, it is difficult to undo the results of the decisions. Conversely, service process design is never final, for once the operation begins, it is necessary to modify it to fit ever-changing customer needs, and to accommodate technological advances.

- Optimizing the customer encounter. This includes easy and safe access and egress; providing for customer waiting lines and service counters; good direc-

tional signs; attention to customer comfort; and aesthetic factors. Designers must also consider the level of customer involvement in the encounter: Is the customer a part of the process, acted upon, or a passive encounter?
- Facilitating the flow of goods and information. This is a prerequisite to improving human and machine productivity. A necessary activity is to determine optimum flow paths and the relative location of the various areas. Flow charting, load-distance analysis, and closeness ratings are tools that assist in these determinations. Correct closeness relationships are also important to good communications as well as reducing personnel travel times.
- Optimizing facility flexibility. Service organizations frequently find that they must quickly adapt to changes in the nature and quantity of demand. How well they can adapt depends on the degree of flexibility designed into the facility. The design should focus not only for today's needs, but also for possible future needs. Balancing flexibility with practicality and capital intensity is the operations manager's challenge.
- Improving human and machine productivity. The type of operation determines the means to accomplish this objective. A good approach is to find the most costly or most time-consuming tasks and concentrate on improving them. For example, in a retail furniture warehouse, stocking, picking, and transporting are the dominant costs. Automatic materials-handling from the storage racks to the loading dock can improve productivity. This is also an example of the impact of technology choices on facility design.
- A safe and morale-building environment. Certain safety features are a given in today's society, but the effective operations manager will provide beyond the minimum. Security for employees' personal belongings, employee parking, spaces for privacy, adequate (above building code requirements) restroom facilities, and decorating schemes are some of the things to consider.

Location

When you consider the trends in how consumers receive their services, technology plays a significant role in location decisions. Buying goods from catalogs is at an all-time high. Banks provide online financial services. You can purchase goods and services on the Internet. These consumer practices are having an impact on location decisions. Firms that utilize catalogs and communications technology in the service delivery process change their position on the customer contact continuum from high to low. Low customer contact creates the opportunity to locate for cost advantages rather than customer convenience. (Also, owners can live where they want to live.) Lands' End can sell a sweater from Dodgeville, Wisconsin, as effectively and at a lower operating cost as the department store in the mall. Think of the implications of banking via computer. Not only is there an opportunity to reduce personnel costs, but there is also the significant potential to reduce the number of branch offices. Internet banking offers an even less costly and very flexible alternative to branch banking. Imagine a bank with 24-hour service that you can access from anywhere and receive customized services for a low fee.

Layout

There is a strong relationship between facility layout and technology. However, what influences the use of technology in a service facility? Once again, we emphasize that the service concept and strategies selected determine the service delivery process and, therefore, the technology and corresponding layout of the service facility. Regardless of a firm's level of customer contact, managers have some common facility design and layout objectives:

- Optimizing the customer encounter
- Facilitating the flow of goods and information (good communications)
- Optimizing facility flexibility
- Improving human and machine productivity
- Providing a safe and morale-building environment

Information Technology (IT)

Information technology (IT) is the technology of computers, telecommunications, and other devices that integrate data, equipment, personnel, and problem-solving methods in planning and controlling business activities. Information technology provides the means for collecting, storing, encoding, processing, analyzing, transmitting, receiving, and printing test, audio, or video information.[36]

IT can perform three important functions:

1. Control and direct equipment and facilities to perform programmed functions, such as using CNC machines to control machining operations or adjust temperatures in a chemical process.
2. Collect and use information to perform administrative functions without human intervention, such as in placing automatic reorders for stock inventory items or in assigning late payment charges in credit card processing.
3. Collect and report information that requires further analysis and decision making by employees. This generally is an interactive process in which the IT system and the employee interface to facilitate the flow of information.

Table 7-7 describes the links between IT and the business operations. It distinguishes among process, information, services, and technology integration, and identifies some of the alignment requirements and issues confronting each of these categories.

Table 7-7 Issues in Matching IT to Business

Process-Driven Architecture (PDA) Layer	Description	Examples
Process	A structured and measured sequence of activities designed to produce a specific output based on defined input	Process design and modeling Continuous process change Process improvement programs Process management Workflow management
Information	Managing the conditions of effective and efficient information production and delivery	Need for a single view Data warehousing Real-time information needs Consistent meaning of data Multi-channel data delivery Availability of metadata Semantic Web initiatives
Services	Well-defined, self-contained functions fulfilling a particular business need provided by an application or module on request of another application	Standardization of Web services Service-oriented architecture (SOA) Linking software development with application deployment and management mobility Access for different user devices Bridging the difference between business needs and technical capability
Technology integration	Provide a means for gradually migrating to a service-oriented (SOA) and process-driven architecture (PDA)	Using the still useful logic and functions of legacy systems Functional integration: formatting, transformation, routing Technical enterprise application integration (EAI); message-oriented middleware (MOM), message queuing

Source: Adapted from Strnadl, C.F., Aligning business and IT: The process-driven architecture model, *Information Systems Management*, 23, 4, 67, 2006.[53]

Integrated Systems

At some point, companies have to integrate a variety of processes, functions, and techniques into an integrated system. An example is a supply chain. We described this in Chapter 3 and provide a detailed description of an integrated supply chain in Chapter 10. In the following section, we provide a few examples of technology that have helped to integrate systems.

Enterprise Resource Planning (ERP) Systems

ERP systems have been popular since the mid-1990s as an attempt to integrate a large portion of a company's information processing. Earlier, materials requirements planning (MRP) and manufacturing resources planning (MRP II) systems, integrated the manufacturing operations and, to some extent, linked with marketing and financial functions. ERP was designed to finish the integration of all functions within a business. Although there are successes, there is also evidence that some companies have not realized the level of integration promised, despite substantial investments.[37] Companies are also finding that it is difficult to link their ERP systems with their partners along the supply chains because of incompatibility problems. ERP systems are well integrated and represent a large investment, in both time and money, to organizations.[38]

Inter-Organizational Systems (IOS)

One of the recent IT technologies that help to link companies along the supply chain is the inter-organizational system (IOS). The IOS is designed to help organizations communicate with one another using such technologies as EDI, extranets, extensible markup language (XML), and Web services.[39] "An IOS consists of computer and communications infrastructure for managing interdependencies between firms."[40] Some of the benefits of IOS include an increase in the competitiveness of a firm, a reduction in the cost of communicating, the ability to enable tight integration between firms, an atmosphere that facilitates knowledge sharing and trust building, and the opportunity for enhanced innovation. Although there are still compatibility problems, companies are beginning to make progress in connecting with their supply chain partners.[41] Although IOS increases the communications capability of organizations, it does not necessarily provide a way to improve individual business applications.

Service-Oriented Architecture (SOA)

One of the areas that holds promise as an integration enabler of business applications is service-oriented architecture (SOA). SOA is not a single software package or even a neatly packaged collection of software applications. It is an approach — an architecture — to moving a company from a stand-alone entity to a member of a loosely coupled network of companies, such as the partners along the supply chain. Instead of being a tightly integrated system of applications, such as in ERP, SOA offers a way for sharing companies to "mix and match" IT hardware, software, protocols, and standards to fit their mutual needs most directly. There are numerous definitions for SOA; however, there is no standard definition as yet.[42] We will describe SOA further in Chapter 12.

Structure, Trust, and Collaboration

Technology is necessary, but not sufficient in itself, for the success of integrated systems. As we will describe in Chapter 8, adapting the structure of a company to fit the new technology is also a necessary ingredient. Before a company is comfortable sharing information with its customers and suppliers, it must build a trusting relationship with them. Once there is trust, cooperation and collaboration follow.

Criteria Used in Decision Making

In this section, we will discuss criteria to consider when making decisions about strategic needs versus short-term needs, behavioral versus scientific management issues, and costs versus benefits of added technology. All decisions about the level of technology to apply must link to the strategies selected in the strategic planning process described in Chapter 2.

Strategic Needs versus Short-Term Needs

When considering technology issues, managers often encounter conflict between strategic needs and short-term needs. Too many times the urgency of the short-term needs overshadows the long-term necessity of the strategic needs. At times, priorities must be on short-term needs, particularly when customers are involved; however, decisions about short-term items should always take into consideration the effect on the strategic, long-term needs. When these situations arise, you should return to the vision statement and identify the organization's driving force. This central theme should guide current and future decisions about the company. Companies should answer such questions as: if we concentrate our technology resources on short-term needs, what impact will that have in the future on the company's capabilities; the company's products/services/markets; or desired financial results? Examining the impact on the elements listed below usually provides enough information on which to base an enlightened decision about possible trade-offs:

- Value added for the customer
- Service delivery — quality, speed, and timeliness
- The service or product itself
- Price
- The employees
- Other elements specific to the company.

There is an abundance of literature about the strategic role of technology. Some of the tools available to assist in such evaluations are the decision tree technique and the preference matrix. A decision tree is a schematic diagram of the alternatives available, which shows potential effects of the various alternatives. The preference matrix is a simple table that can rank alternatives by numerically rating various elements of two or more alternatives. For a complete explanation of these tools, see any operations management textbook, such as *Operations Management* by Jay Heizer and Barry Render.[43]

Behavioral versus Scientific Management Issues

Introducing technological improvements into a service delivery system changes the manner in which tasks are accomplished. These are often unwelcome changes as they affect the people performing the tasks. As work becomes routine, many jobs move from the more satisfying behavioral-based design to specialized job design. Changes in job structure and job content are disturbing to many employees in an organization and may have an adverse effect on the service delivery system. Just witness cashiers learning how to use a new point-of-sale (POS) register system. Oftentimes, they get frustrated when re-learning the process that they believe has been made more complicated by the introduction of new technology.

Job structure refers to the manner in which the job is designed. The introduction of new equipment and process technology can have a double-edged sword impact on job structure. At the lower levels, job structure tends to become routine, akin to the job specialization design. Job structure in higher-level jobs, such as supervision and crafts, tends to expand. The equipment is usually complex, requiring new skills and knowledge to set up and maintain. Supervisors spend less time working with training and scheduling, and more time communicating. High-tech processes are managed differently from low-tech processes.

Job content refers to the number of and complexity of tasks required to accomplish a job. Job structure and job content are closely related, and the impact of changing technology is similar to that of job structure.

Table 7-4 in this chapter compares various capabilities of people and machines. The level of customer contact is an important factor in job design. In high-contact systems, you are seeking the advantages of people. In low-contact systems, you are seeking the advantages of machines. The selection of technology improvements is influenced by the need for customer contact, the degree to which the characteristics in Table 7-4 must be present, and the judgment of the operations manager.

Costs versus Benefits of Added Technology

Improving technology to gain productivity benefits means increasing capital intensity. Few businesses can afford all of the technology available to them because this would require alternative uses of financial resources. Thus, more technology is not always better. The level of technology that supports both operating and marketing strategies provides the most benefits to the company. To achieve this optimized state, technological change must be managed through planning, analysis, and justification.

The place to make and discover mistakes in technological choices is in the planning stage. When considering the application of technology in operations, it is important to streamline the tasks before applying higher technology. Do not automate tasks that do not need to be done.

One of the first steps in planning for technological change is to define the expectations resulting from the change. Simply, reducing costs does not adequately define an objective for an expensive change. An operations manager must set specific goals, defining not only results expected, but also time schedules and responsibilities. Planning for technological change should include gaining knowledge about the ultimate available technology and determining at what time, if ever, this technology may be of value. Know where the technology can take you, and know if you need to go there. Plan for the long-term; implement incrementally. This strategy combines the big picture of what can be done with the capabilities of the organization and provides a road map for the future.

Justifying higher technology projects cannot usually be accomplished using only traditional financial analysis techniques. Even though financial analysis techniques are a good way to start and are necessary, they do not take into consideration the intangible and qualitative benefits that are possible with technology projects. You must move beyond reducing labor and fixed costs to consider the value you can add for the customer: improved customer service, improved quality, increased speed of delivery, reduced inventory levels with no negative impact on service, and improved service delivery system flexibility.

There are a number of tools available to assist in the analysis and justification of technology projects. The decision tree technique and preference matrix approach previously mentioned are effective tools when evaluating qualitative factors. Break-even analysis is a handy "what if" tool, and the traditional analyses of rate of return, payback time, net present value, and inventory turns are good evaluative tools when combined with the qualitative methods.

We discussed the impact of technology on employees in a previous section. It is, however, important to remember that technology affects jobs at all levels in the business. A new technology can create cultural and organizational explosions.

Some people gain and some people lose. These outcomes can create human relations issues that could sink the project. Pay attention to the people aspects of technological change. Early communications and involvement of as many people as possible in the change process is critical.

Steps in Adding Technology to the Process

The manner in which technology is introduced in a production or service delivery system is equally as important as the technology itself. In fact, there are far too many instances where the appropriate technology was selected for productivity improvements, yet the potential was never reached because of the way the change was implemented. In this section, we will discuss a seven-step process that can enhance the probability of success.

Step 1. Communicate, Communicate, Communicate

This step is actually a corollary to the other six steps. As soon as you know that a technological change is going to be considered, tell your employees. This communication must be clear and it must be complete. The reasons why a change is being considered must be given to all those who may be affected. If reasonably well known, the potential impact on work life should be discussed. If the change will cost some jobs, let people know about the possibility in the beginning. However, do not speculate. If you do not know, say so. Any facts about estimated timing should be communicated. Make a commitment to providing education to those affected. Displaced employees should receive outplacement assistance. Make a commitment to continue communicating at the appropriate times as the project progresses.

Every new technology project needs a champion, someone who can do much of the communicating with others in the company and serve as a liaison with the implementation team. The champion does not need to be the CEO or the operations manager. The champion does need to be politically sensitive and have a keen sense of timing to know when to involve non-team employees in the project. Above all, the champion must exhibit an air of confidence, be persistent when dealing with top management, and practice patience throughout the life of the project.

Step 2. Identify Needs and Opportunities

Many times the need for improvement will be painfully obvious. Market share is declining, profits are drying up, service delivery is lagging, and other such red flags may appear. In other cases, there may be symptoms that are disguising problems and the operation may have to be analyzed to find those problems or identify potential areas for improvement. (For example, high sales volume for short periods can disguise a multitude of problems.) In either case, we recommend a concurrent engineering approach in this and all subsequent steps. Concurrent engineering is a team-based approach in which the team members come from diverse backgrounds, and usually have some interest in the outcomes of the change. In a traditional organization, such a team would typically consist of people from operations, marketing, sales, accounting, advertising, and the like. In a team-oriented organization, members would come from various work teams. It is at this point that top management should express its support of the team and the project, and their support must be evident throughout the life of the project.

If a need is not known, the team must develop a means to identify potential candidates for improvement. Whatever the process, measurement criteria must begin to flow into the process. Identifying the changes to be made and the technology to be applied is an integral part of the process.

Step 3. Evaluate Alternatives and Select the Optimum Alternative

In this step, examine the cost–benefit relationships among various alternatives. From a previous section, you may recall that justifying higher technology projects cannot usually be accomplished using traditional financial analysis techniques alone. Even though financial analysis techniques are a good way to start and are necessary, they do not take into consideration the intangible and qualitative benefits that are possible with technology projects. You must move beyond reducing labor and fixed costs and consider the value you can add for the customer: improved customer service, improved quality, increased speed of delivery, reduced inventory levels with no negative impact on service, and improved service delivery system flexibility. Evaluation should include testing the technology either through simulation in your setting or at another company's location. This is also necessary for proper education and orientation.

Step 4. Educate and Orient

Team members and top management should become familiar with the technology and understand its future applications. Talk with others who have experience with the technology. Observe it in action under operating conditions. Visit other production and service facilities that use the technology. As soon as practical, orient other people who might be affected by the change. Renew the commitment to provide education to those affected by the change.

Step 5. Develop the Implementation Plan

Technological change projects are usually large and complex and require an implementation plan. An implementation plan should not be just a schedule of activities; it must be a comprehensive document that can be used to manage the technology change. Absolute musts are:

- The requirements of the new technology (design specifications), with inputs from users
- The expectations from the results of the technological change, with inputs from users
- How and when the results will be measured
- Training and education required
- All the known activities required to accomplish the expectations of the project, from planning through evaluation
- Start and complete times (in days or weeks, dates can be added later)
- Responsibility for each activity, including outside participants
- The critical path of activities
- Other items specific to the project

Two of the more useful tools that have been developed to assist in project planning and management are PERT (project evaluation and review technique) and CPM (critical path method). PERT was developed to handle uncertainties in activity times. CPM was developed primarily for scheduling and managing projects where activity times were known. There are computerized versions of both, and many of the distinctions between the two have disappeared. A Gantt chart can display the relationships of project activities for day-to-day project management.

Step 6. Implement the Technological Changes

If you have sufficiently practiced Step 1, implementation will go more smoothly; however, "more smoothly" is a relative term. Even diligent planning and communicating does not guarantee problem-free implementation. An accepted rule of thumb is that the higher the technology involved, the higher the probability of problems. However, the real value of having a comprehensive plan becomes evident during implementation. With a thorough plan, it is less difficult to determine the impact that changes have on other elements of the implementation process. There will need to be revisions to the implementation plan because of unexpected occurrences or something not performing properly. Knowing the details of all planned activities and the relationships to the critical path will greatly facilitate rescheduling. A word of caution: Not every problem requires a plan change, and the team should make certain that all plan changes are necessary, rather than convenient.

Implementing technological change in an organization tends to disrupt day-to-day operations. This is another reason to consider implementing the components of the technology in increments, or modules. There is often an opportunity to implement and test modules in an area of the business outside the mainstream of daily activity. This provides a venue for making changes and adjustments while optimizing disruptions to more critical operations. The technology can be transferred to the mainstream after it is performing effectively.

The project team should track all activities against the planned schedule and other measures developed as a part of the plan. Variances should be documented as they occur and accompanied by a complete description of how they were dealt with. Technology performance measures should start as early in the implementation process as practical. Corrective actions and results of each action should be documented.

Step 7. Evaluate Results, Redefine Needs, and Redefine Additional Increments

"Evaluate" refers to both the implementation process and the technology performance. At this point, the base for comparison is always the original expectations used for justification. Technology results may exceed expectations, or they may fall short. Seldom do results exactly match expectations. Actual results compared to the original objectives should be documented and studied to determine what caused the differences. This review can be a learning experience and a very important step because it provides an opportunity to improve the implementation process. Expertise in managing technology change can provide a competitive advantage.

You may discover that implementing the first increment of technology changed the original needs; therefore, it is necessary to review the additional increments to determine if they will meet the redefined needs. Some of the activities scheduled for future increments may not be required, or additional activities may be substituted. Perhaps one or more of the planned future modules can be eliminated. There are many possibilities of changes resulting from the first implementation increment, and you will uncover them only by performing this important evaluation review step. Effective planning and control of time and cost performance is one more tool to use in the global competitive markets of today and of the future. What you learn from the review process can help you with planning and controlling other projects in other areas.

Future Considerations for Technology

We have lived in a world where we could see, touch, and feel tangible things. Change occurred, but it occurred at a rate that allowed us to be relatively certain about what to expect. Today, the scope of change is different. Change happens so rapidly that we cannot "touch" things anymore. The world is constantly changing, so we do not always get a chance to get as comfortable with situations and people as we would like. With this scenario as a backdrop, you can see why predicting the future of technology in services is a difficult endeavor, so we will stay away from predictions. However, technology is the driving force for the major changes already experienced and for experiences in the future.

Some current technology trends and issues appear to be influencing service delivery systems. The trends are less customer contact, more customer participation, increased customization, and more flexibility. At times, there will be conflicts among these four systems parameters; however, that is the environment in which businesses operate.

Technology improves productivity, service quality, and delivery times, but one thing drives technology changes — competition in the marketplace. Because the customer is what a service business is all about, we will start there. As you read the following discussions, put yourself in the shoes of the consumer and think about your first reaction to services delivered in the manner described.

Customer Acceptance

Customers have been exposed to technological changes for centuries; yet, they began accepting less customer contact in the early 1990s. We previously discussed ATMs and pay-at-the-pump gasoline delivery as examples of a less-customer-contact service delivery system. From the literature and news media reports, it appears that these technological changes are accepted by the customer. History indicates that even those who reject technology early on eventually warm up to it. Customers begin to accept more participation in the service delivery process. A move toward more customization requires a more flexible service delivery system. Combining the contents of the previous sentences with technological advances in computers and communications technology yields the following scenario of consumer acceptance of technology in services.

The stage is set for consumers to drive the move from high contact/low involvement to low contact (person-to-person)/high involvement in the service delivery system. Communications and computer technology exist which make it possible for the consumer to never leave home yet individualize products and services that fit with her or his unique requirements. Surrounded by computers, fax machines, cellular phones, and online capability in a location where they choose to live, the electronically empowered consumer can reorganize the business landscape. This new electronic heartland is spreading throughout the globe, especially in the United States.[44]

The issue of customers' acceptance of technology that keeps them away from human contact is really one of conditioning. Those who have grown up during the transition to people-less transactions are conditioned to, and probably like the convenience of, automatic transactions. Older generations most likely still resist much of the technological changes. Perhaps businesses will use incentives to help move customers to the use of technology to improve productivity for both parties. As use of time increases as a competitive priority, it may be the customer pushing the business to use technology to improve customer productivity. We all know what the incentive will be in this instance — "do it or lose my business."

Work Force Acceptance

As the service delivery system integrates technology and people, the stage is set for the people affected to accept the changes as friendly. Some of the future technology will continue to be aids to employees; however, much of it will be technology that assists intellect rather than muscle. People will recognize that knowledge technology is for their benefit, and that having additional knowledge increases their value in the marketplace.

Experience shows that employees at the lower organizational levels have less of a problem accepting new technology and change. There is a natural apprehension at first, particularly if the grapevine rather than management is the means of communication. Once the technology is a reality, they usually accept their new roles easily and seriously and do a good job implementing the change. On the other hand, senior and middle managers have a more difficult time with the change. Their perception is that they have lost some degree of managerial control because of the change, and as a result, feel a dilution of their power and influence.

Economic Feasibility

We previously discussed the necessity of considering both qualitative and quantitative factors when analyzing the economic feasibility of technology projects. Another factor involved in much of today's economic feasibility analysis concerns the customer–vendor relationship. Frequently a technology decision made by a customer affects the economic picture of their vendors. There are also opportunities for vendors' actions affecting the customer's economic picture. In the early 1990s when electronic data interchange (EDI) became a standard rather than a novelty, many companies required their vendors to install EDI systems as a prerequisite for conducting business. Some companies could use EDI for only one customer, so it became a questionable economic benefit for those companies. Large customers with much power pressured vendors to install EDI by canceling orders, or at least threatening to cancel. This shared economic factor will become more of an issue as communications technologies entwine customer–vendor relationships.

The business world is becoming more aware of and sensitive to the social and environmental implications of technological changes. We mentioned in a previous section the potential social implications of displacing people with technology. The cost of retraining or supporting a work force trapped in low-skill, low-paying jobs could become a heavy economic burden to society as a whole. The economic impact of technology will be quite the opposite from that predicted by those who view technology as harmful.

The environmental economic impact has been the subject of numerous debates during the early 1990s, and this trend will most likely continue. Regardless of one's position on the environmental continuum, there is a cost to society for protecting our environment and there is a longer-term cost for not protecting it. Environmental problems are viewed as not just quality of life issues, but by many as survival issues; and there is a significant economic cost to dealing with these issues. Social and environmental issues are economic issues and they will become a more important part of responsible businesses. The push for sustainable development, a strategy that encourages careful economic growth while simultaneously ensuring that finite and renewable resources are available to future generations, is also gaining momentum among businesses. "Green" is now becoming synonymous with good.

Technical Feasibility

The emerging technologies of the digital age have the potential to transform all our lives, to change forever the way we work, buy, learn, and communicate with each other.[45] These are suppositions yet unproven, but look at the technical realities of the past few years. Systems are in place that provide, instantly and simultaneously, all the information about a customer's dealings with a business. Small businesses can compete better with larger, multi-national concerns because of low-cost computer hardware and software. Collaboration among software, hardware, and chip vendors has resulted in previously unimaginable off-the-shelf applications for businesses of every size. Whatever is current is made obsolete not by competitors, but by the company that developed it. The applications on the information superhighway, as powerful as they now are, represent the "ice age" of potential.

With these technological advances in place, it would be difficult to argue against future technology that supports the adage "if man can imagine it, technology can make it possible." Management Comment 7-3 gives a perspective from one of the people who brought us to where we are today, Bill Gates.

Managerial Comment 7-3. Bill Gates' View of Future Technology

"The past 20 years have been an incredible adventure for me. It started on a day when, as a college sophomore, I stood in Harvard square with my friend Paul Allen and pored over the description of a kit computer in *Popular Electronics* magazine. As we read excitedly about the first truly personal computer, Paul and I didn't know exactly how it would be used, but we were sure it would change us and the world of computing. We were right. The personal-computer revolution happened and it has affected millions of lives. It has led us to places we had barely imagined.

"We are all beginning another great journey. We aren't sure where this one will lead us either, but again, I am certain this revolution will touch more lives and take us farther. The major changes coming will be in the way people communicate with each other. The benefits and problems arising from this upcoming communications revolution will be much greater than those brought about by the PC revolution.

"There is never a reliable map for unexplored territory, but we can learn important lessons from the creation of the $120 billion personal-computer industry. The PC — its evolving hardware, business applications, online systems, Internet connections, electronic mail, multi-media titles, authoring tools, and games — is the foundation for the next revolution."[46]

Summary

This chapter began with a look at technology today and ended with a view of where technology may take us in the future. Throughout the chapter, we made the points that technology should improve competitiveness, and an operation should be streamlined before introducing new technology. We defined technology as "any manual, automated, or mental process used to transform inputs into products or services." The use of technology affects the capital intensity of the organization, and an operations manager must be concerned with the level or degree to which technology is applied.

The relationship of customers and technology is affected by the fact that in services, customers often participate in the service delivery system. The front end and back room aspects of services were discussed. It is usually less difficult to apply technology to back-room operations than to the front end where customers may be affected. Involving customers more in the service delivery system is a way to gain productivity improvements. We provided some examples of technology in services.

Technology affects facilities and organization infrastructure. Facility decisions should be based on the basic service and overall operations and marketing strategies. The degree of customer contact has a significant influence on facility design and location. The impact of new technology on the infrastructure is changing. In past years, new technology caused an increase in management and staff specialist ranks. As information becomes more meaningful and distributed faster, fewer middle managers and staff specialists are required. This creates "flatter" organizations as opposed to "fatter" organizations. We emphasized that new technologies must be integrated with all parts of the service delivery system to gain optimum benefits.

We discussed the functions of technology and moved through technology as an aid to people, technology as a substitute for people, and integrating technology and people into the service delivery system. The latter function was identified as the current trend and the approach that can gain the most from new technology.

Decision making is important in considering technology changes. When customer contact is at the low end of the continuum, communications technology makes it possible to locate for cost advantage rather than customer convenience. Customer contact also plays a significant role in facility layout decisions. Regardless of a firm's position on the customer contact continuum, operations managers have five common facility design and layout objectives: optimizing the customer encounter, good communications, optimizing facility flexibility, improving human and machine productivity, and providing a safe and morale-building environment.

Some of the criteria used in making technology decisions were presented. Strategic needs versus short-term needs, job design, and cost versus benefits were discussed in some detail. Strategic needs must be the force that guides current and future decisions about a company. Using the driving force of the strategic plan is a good place to begin the decision-making process when conflicts arise between strategic and short-term needs. Job design decisions depend on a number of factors including job structure and job content. Again, the degree of customer contact influences job design. Because improving technology means increasing capital intensity, costs and benefits must be examined. When justifying higher technology projects, operations managers must examine qualitative as well as quantitative factors.

Steps in adding technology to a service delivery system were discussed. A seven-step process was suggested with the first and all other steps including good and timely communications with all who might be affected by the changes. Concurrent engineering was recommended as an approach to developing and implementing a project plan. The final step identified as the need to review, evaluate, and adjust. This step can also help plan and control future improvement projects, yielding another competitive advantage.

The last section of this chapter presented future considerations for technology in services. The ideas are not predictions, but rather extrapolations of current trends, including the relationship of customer acceptance and technology, work force acceptance, economic feasibility, and technical feasibility. The stage is set for consumers to drive the move from high contact/low involvement in service delivery to low contact/high involvement. People at the lower levels of an organization have less of a problem accepting new technology and change than do senior and middle managers. Loss of control and power of managers were cited as the major reasons. The economic feasibility of future technology was expanded beyond cost–benefit considerations to customer–vendor relationships economics, environmental/technology economics, and social economic issues. These economic issues are becoming as important as cost–benefit issues in the life of operations managers. Finally, we closed the chapter with a bridge to the next chapter, some speculation regarding the feasibility of having technology take businesses to where they want to go.

We began this chapter with a quote from John Naisbitt and Patricia Aburdene about the need for technology to be combined with humanness. We would like to end the chapter with a more contemporary quote from an equally recognized writer — Thomas Friedman — with a similar appeal to blend humanness with technology.

Managerial Comment 7-4. The Technology in Our Future

"There is absolutely no guarantee that everyone will use these new technologies for the benefit of themselves, their countries, or humanity. These are just technologies. Using them does not make you modern, smart, moral, wise, fair, or decent. It just makes you able to communicate, compete, and collaborate farther and faster. In the absence of a world-destabilizing war, every one of these technologies will become cheaper, lighter, smaller, and more personal, mobile, digital, and virtual. Therefore, more and more people will find more and more ways to use them. We can only hope that more people in more places will use them to create, collaborate, and grow their living standards, not the opposite. But it doesn't have to happen."[47]

References

1. Naisbitt, J. and Aburdene, P., *Megatrends 2000, Ten New Directions for the 1990's*, William Morrow and Company, Inc., New York, 1990.
2. Kidd, P.T., *Agile Manufacturing*, Addison-Wesley Publishing Company, 1994.

3. Chapman, R.L. and Corso, M., From continuous improvement to collaborative innovation: The next challenge in supply chain management, *Production Planning & Control,* 16, 4, 339, 2005.
4. Rogers, E.M., *Diffusion of Innovations* (Fifth Edition), Free Press, New York, 2003.
5. Krajewski, L. and Ritzman, L.P., *Operations Management, Strategy and Analysis* (Fourth Edition), Addison-Wesley, 1996.
6. Blackstone, J.H. and Cox, J.F., III, *APICS Dictionary* (Eleventh Edition), APICS, Alexandria, VA, 2005.
7. *American Heritage Dictionary,* Second College Edition, Houghton Mifflin Company, Boston, 1982.
8. www.wikipedia.com
9. Bessant, J. and Francis, D., Transferring soft technologies: Exploring adaptive theory, *The International Journal of Technology Management & Sustainable Development,* 4, 2, 10, 2005.
10. Skinner, W. and Chakraborty, K., *The Impact of New Technology: People and Organizations in Service Industries,* Pergamon Press, 1982.
11. Kotter, J.P. and Heskett, J.L., *Corporate Culture and Performance,* Free Press, New York, 1992.
12. Alavi, M., Kayworth, T.R., and Leidner, D.E., An empirical examination of the influence of organizational culture on knowledge management practices, *Journal of Management Information Systems,* 22, 3, 191, 2006.
13. Yeh, Y.J., Lai, S.Q., and Ho, C.T., Knowledge management enablers: a case study, *Industrial Management + Data Systems,* 106, 6, 793, 2006.
14. Boersma, K. and Kingma, S., Developing a cultural perspective on ERP, *Business Process Management Journal,* 11, 2, 123, 2005.
15. Kanunga, R.P., Cross culture and business practice; are they coterminous or crossverging? *Cross Cultural Management,* 13, 1, 23, 2006.
16. Merrifield, D.B., Obsolescent corporate cultures, *Research Technology Management,* 48, 2, 10, 2005.
17. Nerur, S., Mahapatra, R.K., and Mangalaraj, G., Challenges of migrating to agile methodologies, *Communications of the ACM,* 48, 5, 72, 2005.
18. Davenport, T.H. and Prusak, L., *Working Knowledge: How Organizations Manage What They Know,* Harvard Business School Press, Boston, 1998.
19. Womack, J.P. and Jones, D.T., *Lean Thinking: Banish Waste and Create Wealth in Your Corporation,* Simon & Schuster, New York, 1996.
20. Attaran, M. and Attaran, S., The rebirth of re-engineering: X-engineering, *Business Process Management Journal,* 10, 4, 415, 2004.
21. Marquardt, M., Smith, K., and Brooks, J.L., Integrated performance improvement: Managing change across process, technology, and culture, *Performance Improvement,* 43, 10, 23, 2004.
22. Fontanella, J., The RFID Benchmark Report: Scaling RFID Implementations from Pilot to Production, Aberdeen Group, 2005.
23. Markelevich, A., and Bell, R., RFID: The changes it will bring, *Strategic Finance,* 88, 2, 46, 2006.

24. Hyland, P.W., Soosay, C., and Sloan, R.R., Continuous improvement and learning in the supply chain, *International Journal of Physical Distribution & Logistics Management*, 33, 4, 316, 2003.
25. Lind, J., Boeing's global enterprise technology process, *Research Technology Management*, 49, 5, 36, 2006.
26. Kumar, S. and Phrommathed, P., Improving a manufacturing process by mapping and simulation of critical operations, *Journal of Manufacturing Technology Management*, 17, 1, 104, 2006.
27. Lee, J. and Allaway, A., Effects of personal control on adoption of self-service technology innovations, *The Journal of Services Marketing*, 16, 6, 553, 2002.
28. Mesenbourg, T.L., Measuring Electronic Business: Definitions, Underlying Concepts, and Measurement Plans, U.S. Bureau of the Census, 2007.
29. Kazanjian, R.K. and Drazin, R., Implementing manufacturing innovations: Critical choices of structure and staffing roles, *Human Resource Management*, 25, 3, 385, 1986.
30. Cveykus, R. and Carter, E., Fix the process, not the people, *Strategic Finance*, 88, 1, 26, 2006.
31. Davidow, W.H. and Malone, M.S., *The Virtual Corporation*, Harper Business, 1992.
32. Kidd, P.T., *Agile Manufacturing*, Addison-Wesley Publishing Company, 1994.
33. Davidow, W.H. and Malone, M.S., *The Virtual Corporation*, Harper Business, 1992.
34. Norman, D.A., *Things That Make Us Smart*, Addison-Wesley Publishing Company, 1993.
35. Kidd, P.T., *Agile Manufacturing*, Addison-Wesley Publishing Company, 1994.
36. Blackstone, J. H. and Cox, J. F., III, *APICS Dictionary* (Eleventh Edition), APICS, Alexandria, VA, 2005.
37. Manrodt, K.B., Abbott, J., and Vitasek, K., Understanding the lean supply chain: Beginning the journey, Report on lean practices in the supply chain, APICS, 2005.
38. Crandall, R.E., The epic life of ERP, Where enterprise resources planning has been and where it's going, *APICS Magazine*, 16, 2, 17, 2006.
39. Turban, E., Leidner, D., McLean, E., and Wethebe, J., *Information Technology for Management: Transforming Organizations in the Digital Economy*, John Wiley & Sons, New York, 2006.
40. Chi, L. and Holsapple, C.W., Understanding computer-mediated inter-organizational collaboration: A model and framework, *Journal of Knowledge Management*, 9, 1, 53, 2005.
41. Crandall, R.E., Let's talk, Better communications through IOS, *APICS Magazine*, 17, 6, 20, 2007.
42. Urvasula, S., et al., SOA Practitioners Guide: Why Services-Oriented Architecture?, 2006.
43. Heizer, J. and Render, B., *Operations Management* (Eighth Edition), Prentice Hall, 2006.
44. Naisbitt, J. and Aburdene, P., *Megatrends 2000, Ten New Directions for the 1990's*, William Morrow and Company, Inc., New York, 1990.
45. Gates, B., *The Road Ahead*, Penguin (Non-Classics); Revised edition, 1996.
46. Gates, B., *The Road Ahead*, Penguin (Non-Classics); Revised edition, 1996.
47. Friedman, T.L., *The World Is Flat, A Brief History of the Twenty-First Century*, Farrar, Straus, and Giroux, New York, 2006.

48. Blackstone, J.H. and Cox, J.F., III, *APICS Dictionary* (Eleventh Edition), APICS, Alexandria, VA, 2005.
49. Blackstone, J.H. and Cox, J.F., III, *APICS Dictionary* (Eleventh Edition), APICS, Alexandria, VA, 2005.
50. Collier, D., *Service Management: The Automation of Services,* Reston Publishing Company, 1985, p. 20.
51. Epstein, M.J., *Implementing E-Commerce Strategies, A Guide to Corporate Success after the Dot.com Bust,* Praeger, Westport, CT, 2005, p. 20.
52. Kidd, P.T., *Agile Manufacturing,* Addison-Wesley Publishing Company, 1994.
53. Strnadl, C.F., Aligning business and IT: The process-driven architecture model, *Information Systems Management,* 23, 4, 67, 2006.

Chapter 8

The Role of Infrastructure in Continuous Improvement

John walked up to a pleasant-looking couple who were standing near one of the plastic houses with ladders and slides for the children. The expressions on their faces were a combination of pride in their children's actions and fear that they were going to break some bones. "Are those your children on the slides?" John asked. "Yes," the wife replied. "My name is Organization and my husband's name is Plans. We like to see them have a good time, but we want them to grow up to be good citizens and productive members of our community. We realize we don't always agree with them. We come from the Infrastructure generation. We plan the outings, we organize how each member of the family does their part of keeping the house clean, the lawn cut, the errands run, and all of the other things we do as a family. We try to be tolerant of what the children want and we do adjust somewhat, but they are so different. It is difficult sometimes to find a way to change and keep the family happy so that everyone is satisfied. I noticed you were talking to our daughter, EDI; she no doubt complained to you about how structured we are as parents." "Well, she did say something about her technology toys and how much she enjoys her brother and sister, and all of their friends," John replied, trying to dodge the issue.

John quickly asked, "How do you decide on the structure of your family's life?" At this point, the husband picked up on the conversation. "We have a variety of ways to run a family. We know how our parents operated. Of course, we have changed some things, but we have come to realize that maybe our parents were smarter than we thought they were. We also look at how some of the other families operate. We like to use some of the ideas from those who seem to do well and, of course, we reject some of the things that the dysfunctional families do. We also have to consider how society in general views the family. Although we like to be somewhat independent, we admit that we tend to work within the mainstream of what is acceptable to the general public. We also have to follow the laws of the government and the global community. There are so many conditions and constraints that it is difficult to organize them all into a family's lifestyle."

What a dilemma, John mused. I'm not sure I could handle those children. They have such great potential, but are they ever going to get a chance to show what they can do? The parents want to give their children freedom to express themselves, but how can they do it when they have so many constraints of their own?

In Chapter 7, we described various technologies that are used in pursuing continuous improvement programs. In this chapter, we will describe why it is also necessary to address the infrastructure of an organization to be successful. Determining the infrastructure of an organization does not always receive much attention in business books for practitioners. It is, to be perfectly honest, not the most exciting aspect of running a company. And yet, it is deceptively one of the most important items that need to be "right" for the business to be successful.

What Is Infrastructure?

Many managers with degrees in business might remember taking a course that dealt with the subject of infrastructure. Certain terms have probably remained in their memories, terms such as span of control, centralization, organizational charts, departmentalization, and specialization. In fact, managers use these concepts frequently in the everyday running of the firm; they are not just textbook terms.

Infrastructure is a nebulous term. We will describe it in more concrete terms and show how it has roots in the bureaucratic (or administrative) management background, introduced in Chapter 4. Its lack of definite form has prevented it from being developed sufficiently by many businesses, as contrasted with technology. This chapter will show why it is important and what kinds of changes are necessary for an infrastructure to be effective.

The term "infrastructure" has various meetings to different people. Some may think of roads, bridges, and airports, or a transportation infrastructure. Others may envision the computer networks that join a firm's headquarters and field offices together, or the IT infrastructure. In this chapter, we will use it to describe the components of a business that tie it together into an integrated unit.

So what are these components of infrastructure? There is no set list as that would vary from business to business. However, as a starting point, the following components could be considered part of most business infrastructures:

- Strategies
- The four classical management functions
- Organization structure
- Knowledge management

We will look at each of these elements in more detail.

Strategies

Where do you start when looking at the infrastructure of an organization? Probably the best place is to begin with its strategies. A company's strategies are what drives it toward its goals. When the strategies are in place, the other components of the infrastructure can be put together. Of course, when we talk about strategies, we are talking about a number of different items, so we will clarify what the different types of strategies are.

Corporate Strategy

This strategy looks at the types of industries in which the business wants to compete. Some companies, such as McDonald's, pretty much stay in one line of business, which is fast food. Other companies, such as The Walt Disney Company, choose to compete in several different industries, such as parks and resorts, studio and media entertainment, and consumer goods products. Every business must decide how extensive it wishes to grow in terms of breadth of industries. Sometimes a company will decide it is competing in too many industries. When that occurs, it may decide to retrench a bit and get back to basics. One of the authors once worked for what originally started out as a vending food company. Eventually it expanded into a number of different industries including trucking, contract food service, uniforms and linen, and even day care. Although it later retrenched from the trucking and day care industries, that company, ARAMARK, remains one of the premier service contract companies in the areas of facilities management, food service, and uniform/career apparel.

Corporate strategy also looks at the issue of growth for the company. In general, there are only three ways to go — stay the same size, grow bigger, or get smaller. Some industries, such as those in new technologies, offer plenty of growth opportunities so that companies competing in these industries can choose to grow larger if they wish. Staying the same size may be an option for other companies that do not wish to complicate operations any more than necessary. These companies may feel that expanding too much could sacrifice the quality of their product or service. Furthermore, if the company is in a slow growth industry, trying to expand may turn out to be costly.[1] Retrenchment, or growing smaller, is also an option if a company has encountered financial problems. Some big name companies have cut back a bit over the years, including IBM, Proctor & Gamble, Heinz, Nokia, and Union Carbide.[2]

Business Strategy

Once a company knows what industry (or industries) it wants to compete in, it needs to decide how to best compete in those industries. This statement may sound obvious, but actually, many companies do not survive because they choose the wrong business strategies. What is especially noteworthy is that a firm can be successful for many years, but then eventually fail because it does not update its business strategies. For example, a number of formerly large, successful companies have fallen by the wayside due to changes in their environment and the selection of poor business strategies. Some of these include Eastern Airlines, Howard Johnson's restaurants, and F. W. Woolworth's.

Michael Porter, a Harvard business professor and leader of the Institute for Strategy and Competitiveness, has led the thinking on what constitutes effective business strategies. His original book in 1980, *Competitive Strategy,* has been acclaimed as the premier guide to business strategy. According to Porter, a company can follow one of three strategies: low cost, differentiation, or focus.[3] A company following a low-cost strategy (sometimes referred to as a cost leadership strategy) seeks to be efficient in the production and offering of its goods and services. Cost savings in every aspect of the business operation is sought after. Most college textbooks will mention Wal-Mart as an example of a firm following a low-cost strategy. In fact, Wal-Mart takes extreme measures in cutting its costs, including requiring its suppliers to reduce their prices over the length of the contract.[4] Usually, companies that follow a low-cost strategy will also pass those savings on to their customers. Again, Wal-Mart serves as the premier example, as they are absolutely committed to providing products at the lowest costs possible.

Suppose you were to buy your next set of jeans from Wal-Mart. Chances are, you would find jeans that were a good value (low price), and of generic quality (in other words, adequate quality, but not necessarily high quality). For buyers who shop at low-cost firms, the expectation is always low prices, not quality, as the first priority. It is understood that appliances will wear out sooner, clothes will not be the top name brands, and foods may be packed in large containers, offered by generic suppliers. For the consumer who is interested in more quality, shopping at a firm that follows a differentiation strategy is a better option.

A differentiation strategy puts the emphasis on better quality, not low prices. Companies following this strategy seek to provide products and services that are unique and different from their competitors. Following this strategy means that more resources can be expended on adding value to the product. As a result, the goal is a quality product, not a cheap product at a low cost. Now suppose you decide to buy that next set of jeans at Nordstrom's, a high-end retailer. The product line would offer better name brands, better quality, and of course, higher prices.

Sometimes a firm seeks to compete in just a narrow market segment, or niche. Instead of offering clothes for all ages, men and women, and all sizes, they choose to focus on a select market. For example, Tall and Big Men clothing stores focus on large sizes of clothing for men. Some stores offer only maternity wear. Others specialize in clothes for babies or kids. These companies are following the third strategy discussed by Porter, the focus strategy. The focus strategy looks at a narrow market and zeroes in on those customers only. These companies realize they do not want to be all things to all people, so they focus on that part of the market they can serve best. Another example is the quick oil change companies that change your oil and check fluids in less than 30 minutes. They are focus companies because they do not perform all auto repairs, just those related to oil and fluids.

There is actually a fourth strategy that some companies may follow, although it is not recommended. This business strategy occurs when a company tries to pursue a low-cost and differentiation strategy simultaneously. Porter calls this strategy "stuck in the middle." Simple logic tells us that this strategy is hard to pursue because it suggests that quality and low cost can exist in the same product. Take the example of Ruth's Chris Steak House (a differentiator) and McDonald's (a low-cost provider). Both companies sell beef, but Ruth's Chris Steakhouse cannot sell its products at low prices like McDonald's, that is simply not the goal. Of course, this example is a bit extreme, but it does illustrate what stuck in the middle means, two strategies going divergent directions cannot be used at the same time. It is better for a company to pick one or the other.

The Four Classical Management Functions

Pick up any college-level Principles of Management textbook and you will find it outlined according to the four functions of planning, organizing, directing, and controlling. Indeed, this describes what managers do, although these functions are not necessarily sequential and segregated from each other, as many new students might believe. Instead, the functions are tightly interwoven together, often overlapping and sometimes hard to distinguish from each other.

The four classical management functions can be traced back to the thinking of the French engineer, Henry Fayol (1841–1925). As a young manager of a large company, Fayol began to see that technical competence and managerial competence were two separate skill sets. In fact, he felt that a good manager with mediocre technical skills was more beneficial to an organization than a mediocre manager with good technical skills.[5] The teaching of management skills was not a practice during Fayol's time because the discipline of management as a field of study was virtually unheard of. Fayol gets the credit for getting the ball rolling in that direction though. He felt a need for a theory of management to be developed, and as a result, spent a great deal of time writing and presenting papers on what were the fundamental principles of management at that time. Today, his fundamentals are the backbone to much of the field of modern-day management.

So we will look again at those four functions originally proposed by Fayol.[6] To illustrate the functions, we will envision a fictitious pizza restaurant chain[7] that is about to go through some major changes. For many years, the company offered one style of pizza dough, a very thin, yet tasty crust that launched the company into a regional chain, well known particularly in the Midwest. Of course, a variety of toppings could be ordered with each pizza, but the crust was the same for all of them: thin, very thin. And that is where we will begin with the four classical management functions.

Planning

This part of management is involved with setting goals and deciding how to achieve those goals. At some point in the company's history, it realized that if it were going to ever grow and be a nationwide chain, it would need to make a dramatic change in its menu. Consequently, the decision was made to add a new style of pizza dough to its menu, a dough that was thick, in other words, one that was the opposite of the current thin dough. The plan was to have two styles of dough for the customer to choose from, thin or thick.

Obviously, this was going to take a lot of planning. Goals and subgoals would need to be set. Timetables and budgets would have to be managed. In fact, if the company grew as hoped, the whole growth process would have to be planned as well. In short, this was a big undertaking, but one that offered a huge payoff if successful.

Organizing

So how do you organize such a big task? Well, with a lot of planning, of course! Planning and organizing are very much intertwined. However, there is enough difference in the two processes that we can separate them out for a moment. Organizing deals more with gathering the resources needed and allocating them in a controlled manner. Basically, there were at least four major resources that needed to be organized: time, money, people, and materials. Within these four broad categories, there were various departments represented in the company that would be impacted. Some of these departments included research & development (to invent the new crust), finance (to arrange the extra funds needed to develop the product), purchasing (to buy the new equipment needed to make the crust in the stores), human resources (to arrange training on how to make the new crust), marketing (to advertise this new product), regional managers (to coordinate the implementation of the new product into the restaurants), and unit managers (to actually make and sell the product). A look back at the Input–Transformation–Output Model discussed in Chapter 3 will remind you that the resources and departments just described are found in the major components of the model (Figure 8-1).

Directing

Leading the company through the development of this new product will be a series of upper-level executives and mid- to lower-level managers. Directing then can be thought of as the leadership that must accompany the desired changes in the organization. Upper management is responsible for seeing the need for a new change in the menu philosophy. They must also sell the idea to the rest of the organization, which may think that a new pizza crust is totally unnecessary. You can almost hear the older managers murmuring among themselves: "We've been selling this same style of pizza for many years now, why change it at this point?" Indeed, much of what directing involves at the higher levels of the organization is selling the skeptics at the lower levels of the organization.

Somewhere in the mid-management area of the company, the hard work of developing and implementation will need to be conducted. Research & development (R&D) will need to plan how it will organize its efforts to develop the new crust. Regional managers will need to prepare their store managers for the big

Figure 8-1   ITO Model at a pizza chain developing a new type of crust.

changes ahead. Marketing managers will work on promoting the new product while finance managers figure out how to pay for it all.

The directing will continue at the lowest levels of the organization as store managers teach their hourly cooks how to make the new product. Servers will be taught how to use suggestive selling, enticing customers to try the product. The directing process is always occurring, at all levels of the company. Without successful leadership, the effort to promote the new pizza dough may fail.

Controlling

Throughout the entire process of envisioning, developing, and selling this new product, a separate process will need to take place that monitors the efficient use of resources. That function is called controlling. This process seeks to answer two major questions: (1) are we reaching our goals, and (2) are we being wise in our use of resources? Above all, the goal of controlling is to introduce the new product, in a timely fashion, at a cost that does not bankrupt the company.

The resources mentioned earlier in the transformation process will need to be monitored. For example, time is a limited resource. Most likely, the company will have certain deadlines that must be met. Above all, the product will need to be introduced to the market in a reasonable time period. That time period must be long enough to ensure adequate developing and testing of the product so that it is of the highest quality possible. However, it must be short enough so that the competition does not step in and introduce the same product before the company does. Delays in product introduction can also be expensive, as each day lost past the introduction date becomes an inefficient use of cash. The use of budgets and timetables are important aspects of the controlling process. Without them, there is no way of knowing if the company is hitting its goals in a timely and efficient manner.

Organization Structure

Most managers will think of the company organizational chart when you mention organizational structure to them. This is, in fact, true to an extent. A visual representation of an organization's structure can be seen through its organizational chart. However, it needs to be remembered that different types of structures are good for different types of company goals. There is no one-size-fits-all formula for organizational structure. Basically, structures fall into four categories: functional, product, geographic, and matrix. We will look at each one in more detail.

Functional

Many organizations are set up exclusively by functional departments. There is an accounting department, finance department, marketing department, and so on. These departments serve the entire organization. Those organizations look similar to Figure 8-2.

Functional department organizations are good at grouping expertise into one department, thus creating opportunities for more new ideas to be developed relating to innovation and efficiency. However, on the downside, functional organizations are not set up according to product lines; therefore, it is difficult to isolate costs according to a particular product.[8] In addition, functional departments are notorious for infighting among departments, of which the classic example cited in many articles is what can occur between marketing and production. Typically, marketing wants new products with lots of features, regardless of the cost. Production desires standard production items with cost efficiencies as a top goal. Needless to say, these two have been going at it for years.

Product

Some organizations are set up by product line. Functional departments also exist, but they are set up under each product line. The goal is to view the business as a set of different businesses with each mini-business representing a product line. An organization set up on this basis may look like Figure 8-3.

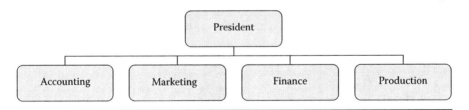

Figure 8-2   A functional organization.

Figure 8-3   A functional organization.

An organization set up by product lines offers several advantages. First, there is more financial accountability for the success or failure of that product line because it can be set up as a profit center.[8] A second advantage is that managers can learn the entire scope of a business by being involved in the running of a product line,[9] something that cannot be accomplished if they work in a functional area most of their careers. On the downside, organizations set up on this basis are generally not as efficient in terms of operating costs. This is because functional departments need to be created for each product line.

Geographic

Some companies will set their structure up according to a geographical framework. For example, there may be an area and regional division, with several areas comprising a single region. If the company is located in several countries, there may be a domestic and an international division. There are numerous ways companies can be set up by geographic area. One illustration appears in Figure 8-4.

Organizations set up by geographic region offer the advantage of being able to adapt better to the local regions they are serving without formal approvals from company headquarters. This is common in the food service industry, where local tastes can be accommodated in different regions. Imagine serving grits in the Northeast? Or scrapple in the South? In geographic arrangements, these kind of mistakes can be avoided. The geographic organization does have the disadvantage of being somewhat costly to administer, and there is also the problem that one region of the company may try to upstage another region.[10] However, for large organizations that cover a wide geographic area, this arrangement is suitable.

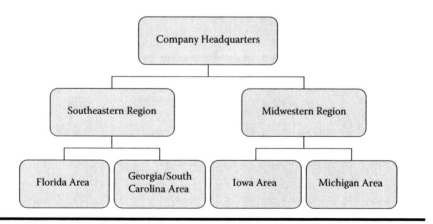

Figure 8-4 An organization set up by geographic region.

It should be noted that combinations of the above organizational structures can occur as well; in fact, they are quite common. Take a large hotel chain, for example. The company may be set up in a geographic configuration, with regions of hotels in Europe, Asia, and North America. However, each hotel may be set up on a functional level, with housekeeping, front desk, food service, and security all operating as separate departments.

Alternate Organizational Structures

The vanishing boundary between manufacturing and services has blurred some of the characteristics of the traditional organizational forms described above. Newer organizational forms have arisen that have the characteristic of being able to adapt better, both to changes in the company and to changes in the business environment.

The Matrix Organization

Imagine combining a functional and divisional structure together into one format; the result would be a structure that looks like a matrix. A few companies actually follow this arrangement in some areas of their infrastructure. Temporary projects and even some permanent product lines may be set up in a matrix organization. Figure 8-5 shows what a matrix format might look like.

Companies that are extensively involved in projects or are highly geared toward technological innovation may be well suited to using a matrix format. This struc-

Figure 8-5   A matrix organization.

ture offers the advantage of putting specialists on specific projects, allowing managers to be freed up to work on other duties. On the downside, there is the problem of a single employee reporting to two supervisors, a situation that can create confusion and divided loyalties.

The Horizontal Organization

Organizations can be tall or flat. Tall organizations have many levels of management tiers. Tall organizations were typical of many businesses prior to the 1980s. But then, a wave of downsizings began, which eliminated many middle managers from the company. More noble terms were chosen for this type of downsizing, including the very popular "re-engineering." The goal of re-engineering was to reorganize the company so it would function more efficiently and, at the same time, better serve customers.[11] No doubt, re-engineering has accomplished this feat in many companies, as many were heavy with too many managers and inefficient in responding quickly to customer needs. Countless stories are written in business textbooks about companies that cut their number of employees and responded more quickly to customer demands. Some of these cuts in staff occurred among hourly employees. Other cuts were more far-reaching as entire levels of the company, usually in middle management, were eliminated.

The impact of re-engineering has also led to the creation of a new organizational structure, the horizontal organization. "Unlike traditional vertically oriented organizations with functional units such as production, marketing, and finance, horizontal organizations are flat and built around core processes aimed at satisfying customers."[12] A horizontal organization may look something like that shown in Figure 8-6.

Replacing the traditional functional departments are cross-functional process teams.[13] These teams are designed to address the specific needs of the customer better as opposed to running an internal process of the organization. Hopefully, the result is an expedited process transaction with the customer. For example, instead of

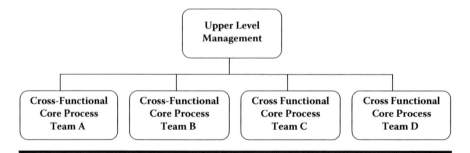

Figure 8-6   A horizontal organization.

taking weeks to process an insurance policy or a mortgage, the goal is to complete the transaction in a matter of days. Streamlining the transaction process requires a careful analysis of the job, a process human resource management specialists like to call job analysis.

Of course, a lesser form of the horizontal structure is simply a flatter organization. De-layering the organization compresses the hierarchy and widens the organization chart. The result is a flatter organization that is more flexible in responding to change. Translated, management is able to make decisions faster because there are not as many levels of approval needed. Because fewer personnel are on the payroll, particularly in middle management, a savings in labor cost is also expected.

The Virtual Organization

There appear to be at least two notions of what the virtual organization is.[14] The first notion is a business operating with a small core of employees, while outsourcing much of the other activities involved in its infrastructure. For example, a business may perform the tasks that it does best, sometimes called its core competencies, while contracting other businesses to perform those tasks that are not in its expertise. One Smooth Stone, an event-planning company, uses this approach as it plans out major business meetings for its clients. Their core competency is in the planning end, a feat accomplished with just 17 employees. All other functions are contracted out.[15] This type of virtual organization has been given a fancy name by some, the "inter-organizational virtual organization."[16] One depiction of what this type of virtual organization may look like appears in Figure 8-7.

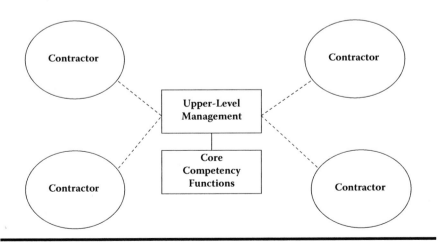

Figure 8-7 A simple depiction of an inter-organizational virtual organization.

In this figure, the dashed lines indicate a contractual relationship between the contractor and upper management. The solid line between upper management and the core competency functions indicates that both groups are in the same organization. Of course, the number of contractors can increase or decrease, depending on the number of projects the company is managing at the moment. Thus, virtual organizations are constantly in a state of flux and change.

Another viewpoint holds that the virtual organization is a business that performs most of its own functions, but is linked over a wide geographic area via computer networks. The organization may still hold a traditional structure, but in terms of communications and performing various processes, it is linked through smaller virtual teams. The accepted term for this arrangement is the intra-organizational virtual organization.[16] These organizations may have virtual teams working on advertising campaigns, product development, or a number of other common business functions. Instead of working in a department in one building though, virtual teams are linked via computer networks over a geographic area.

Several advantages are available using virtual teams. First, the expertise that is closest to the point of action can be tapped into. Field personnel can work closely with the home office on a real-time or close to real-time basis. This advantage is especially important when developing new products or advertising campaigns. A second advantage is the best talent in the company can be utilized for specific projects, regardless of where those personnel are located.[17] Larger companies that cover a vast geographic span are especially able to utilize these types of virtual teams.

Depicting a visual chart of a "typical" intra-organizational virtual organization is not possible because many organizations of any size already have elements of virtual communications in place. However, one commonality that may exist for specific project management is the matrix format. This is because virtual teams are often composed of temporary members who work on a specific project for a set length of time. When the project is completed, that virtual team is disbanded.

Trends in Organizational Structures

In addition to the alternate forms discussed above, there are a number of trends that have been occurring in organizational structures that should be noted. We will look at these trends next.

Moving from Centralization to Decentralization

Centralization deals with where the major decision making takes place in the running of the organization. Highly centralized companies have major decisions made at the upper levels of management, or the home office. Most restaurant chains are highly centralized, as major decisions such as menus, advertising campaigns, logos, store design, and prices are made at corporate headquarters. Decentralized companies push the major decisions to the lowest levels of the organization possible. Consider a major food service contract company like Sodexo or ARAMARK. These companies are responsible for running corporate, hospital, and college cafeterias all over the country. Major decisions concerning menus and how to set up the local facility are made by the food service managers at the location where the account is administered.

In general, businesses must learn to adapt in changing times, which means being flexible and responsive to the customer. As a result, there appears to be a trend toward decentralization.[18] The blurring of the service–manufacturing boundary is showing us that managers at the lowest levels of the organization are closest to the action, and are often the best equipped to make decisions involving adapting to customer needs. This is especially important to note in larger companies that have tended to be more centralized in their decision making. One of the authors remembers hearing a student lament the fact that fashion decisions were set by her company (Sears) in Chicago, and she was a buyer in Memphis, Tennessee, where fashions could be quite different. The top-down, centralized approach of the company did not work well in her opinion because local fashions trends were not being accounted for.

Moving from Vertical Structures to Horizontal Structures

Tall, vertical organizational structures are typical of highly centralized companies. It is no surprise then that as decentralization becomes more popular, horizontal structures, which work better in decentralized environments, are becoming more popular.

A related topic concerns span of control, the number of people that report to a manager. Simple logic tells us that as organizations become more horizontal, a single manager will take on more employees to supervise, thus increasing the span of control. Increasing span of control has good and bad points. On the positive side, the company saves money because less labor cost is needed because fewer managers are employed. Also, employees must be given more authority to make decisions, a process we like to call empowerment in the management field. On the downside, too wide a span of control can cause lapses in supervision, because a manager can only keep up with so many employees. Nonetheless, spans of control are increasing in today's economy, and so is employee empowerment, discussed next.

Moving from Autocratic Managers to More Empowered Employees

Although not directly an organizational structure concept, a discussion on the movement toward employee empowerment is appropriate here. Autocratic management is a style of leadership that can be effective in some environments. As management researchers Hersey and Blanchard have noted, the style of leadership needed may vary according to how ready the employees are to perform the task. Employees who are unwilling and have low abilities will need a more directive style of supervision.[19] Most readers of this book have probably also made this same observation: some employees seem to need more direction than others.

Sometimes, a particular business or organization will adopt an autocratic style because it fits with the culture of the industry and seems to work well. Branches of the military have traditionally adopted this approach. Nonetheless, the trend in many organizations is to move towards more employee empowerment and away from autocratic styles of leadership. From a structural perspective, this trend seems logical. We have mentioned that managerial spans of control are increasing and organizations are getting flatter. The result is that managers will expect their employees to perform more of the "thinking" aspects of their jobs, roles traditionally taken care of by managers. Autocratic and micro-managing styles of leadership are thus replaced by participative styles of leadership, where employees have more say in the decision making of the operation.

One example of a successful service operation that has adopted this approach is the Ritz-Carlton Hotel company. "The Ritz," as it is often called, goes to extreme measures to ensure that guests are well satisfied. After all, a five-star hotel is expected to deliver only the best in service and food. Employee empowerment is one way the hotel accomplishes this goal. Each employee is empowered to spend up to $2,000 without managerial approval to fix a customer problem.[20] That may sound extreme, but one thing service industries have taught us is that it is more cost efficient to keep a current customer than it is to find a new one.

Moving from Job Specialization to Higher Skill Variety

Thanks to the work of Frederick Taylor (1856–1915), we have a great deal of job specialization that has permeated our working landscape. Taylor, if you remember, was the engineer who founded the principles of scientific management. His own identification with the working class was impressive, but misunderstood in his later years, as critics from labor unions thought of him as an exploiter of manual labor.[21] His contribution to the modern economy is the work philosophy of finding the one best way to perform a task. Usually, this involves breaking a job down into its component parts and analyzing the most efficient manner to perform that job. Selecting the best worker for the job, training the worker in the exact methods, and giving the worker the right equipment were important for the highest productivity. Taylor's hope was that both the worker and management would prosper under this approach.[22]

Today, we see evidence of scientific management in both manufacturing and services. Fast food restaurants have rigid methods on "flipping burgers," assembling sandwiches, frying food items, and serving the customer. The rigidity of job specialization has resulted in faster production and a standardized product that can be made to taste the same across a wide geographic area. Indeed, the fast food industry owes much of its success to Frederick Taylor. But job specialization does have its drawbacks. For one, the work, over a period of time, can become boring. Boring work creates its own problems, as workers can become apathetic, which can result in sloppy work, accidents, and employee turnover. As a result, many restaurant managers will attest that much of their work involves training new employees.

Job specialization may be eroding in the future, at least somewhat. We will look at how organizational structure enters into the picture. First, there is the simple math consideration. As companies reduce their numbers of employees, the remaining employees will perform more work. In some ways, this is a negative occurrence as it can result in work overload. However, in some jobs, there could be a surprising positive benefit: the work will become more interesting. This is because the skill variety to perform the job may increase. Two management researchers, Richard Hackman and Greg Oldham, discovered this fact back in the 1970s. Simply put, when the skill variety of a job goes up, worker motivation will go up also.[23]

Another reason skill variety may be increasing is due to the flattening of organizational structures. As fewer managers are in the hierarchy, more demands will be put on those remaining managers. Again, skill variety may increase, which can increase the meaningfulness of that managerial job.[23] So, reducing the number of employees and flattening out the structure of the business can have a positive side effect besides saving money: it can be motivating to the worker. In some companies, it should be noted that management has purposely decided to broaden jobs and reduce work specialization. Avery-Dennison, Hallmark, American Express, and Ford Australia have all made decisions in this direction.[24]

One caveat is worth mentioning at this point. Broadening the scope of a job, but overworking the employee can be a de-motivator. Managers at all levels need to make sure they are not overworking the managers or employees below them. Otherwise, the gains that could be made by changing the structure of the organization could be lost due to worker apathy and turnover. Also, not all jobs are suitable for enlarging. Fast food jobs are a case in point as broadening these jobs could significantly slow down service to the customer.

Moving from Line Managers to Self-Directed Work Teams

The factory floor setting has traditionally been set up using line managers to direct a group of employees. This format is effective and is still in use in a number of work settings today. The underpinnings for this approach have some tie-in with Taylor's scientific management principles. The line manager (sometimes called a supervisor) directs a smaller group of employees using precise work methods that have been carefully created to accomplish the job. When workers deviate from these precise work methods, the supervisor is always there to correct the situation by putting the worker back in line with the required method of doing the job. Supervisors are also responsible for scheduling the work, ordering the supplies, and arranging nonwork items, such as vacation time for the employees.

The line manager arrangement is also popular in the service industries, particularly among fast food restaurants. In these environments, supervisors often lead younger employees, many of whom are working for the first time, into job tasks that require precise methods and timing. For example, the cooking of a food item involves a step-by-step procedure, with critical timing events within the process. In addition, stringent safety and sanitation standards must be maintained. As a result, a strong supervisor who leads a group of inexperienced workers must guide and direct those employees on a close basis.

There are some work situations where a different format is used. In these cases, the work group does not have a line manager one hierarchal level above them who directs their work. Instead, the group itself does all of its own directing. These groups, or self-managed work teams, perform a complete work process while managing themselves and arranging nonwork related items, such as sick days and vacation time. Self-managed teams require workers to move away from performing only a specialized task to becoming proficient at a number of skills, a concept called multi-skilling.[25] This concept is consistent with the trend discussed above, where skill variety among workers is being increased. Self-managed teams are found primarily in manufacturing environments and can be seen at well-known companies such as Corning, Boeing, Pepsi, and Hewlett-Packard.[26]

Moving from Specialized Departments to Cross-Functional Teams

Specialized departments (sometimes referred to as "silos") are among the more traditional ways to organize employees. The functional organization discussed above is an illustration of a specialized department at work. In contrast, cross-functional teams are composed of employees from different departments in the same organization. The matrix organization, also discussed above, illustrates the use of cross-functional teams.

Cross-functional teams have enjoyed a fair amount of popularity in contemporary organizations and their use appears to be growing. They are not departments in themselves, but groups of employees set up on a temporary basis and assigned to various projects. Some readers might remember the best selling book, *In Search of Excellence* that came out in the 1980s. In their account of excellent companies, Tom Peters and Robert Waterman, Jr. actually documented an early example of a cross-functional team, called a "skunk works."[27] In this somewhat amusing example, a member of a company skunk works was called in to analyze a piece of machinery that was overheating. Engineers within the company had proposed several expensive solutions, but a skunk works member solved the problem by aiming an inexpensive fan at the machine. From skunk works to cross-functional teams, the concept of taking a "multi-disciplinary" approach to solving business problems has become a popular practice.

Moving from Top-Down to Multi-Directional Communications

The traditional model of communications in organizations is what has been referred to as the "top-down" approach. In this arrangement, orders and directions always originate from the manager to the employee. Employees are not expected to question their manager or participate in decision making. The basic mentality is to take the order and run with it. This arrangement works fine in some companies and industries, but cannot be expected to be effective in all situations. Top-down communication formats are common in tall, vertical organizations, with more specialized jobs.

Many organizations have found that using multi-directional communication formats is more to their advantage. Multi-directional formats use top-down, upward, and lateral communication pathways. Top-down communications, just described, are useful when giving directions, communicating goals, and indoctrinating employees.[28] Upward communications occur when lower-level employees are actively communicating with their immediate supervisor. One of the goals of upward communication is to foster an environment of participative management. Participative management encourages lower-level employees to work with their managers to solve problems related to the unit they work in. Companies have found that using participative management practices can result in better employee motivation, improved product quality, less stress for the employee, higher job satisfaction, and ultimately, higher profits.[29] The Ford Motor Company was one of the first pioneers in the participative management arena. This feat was a tall task, given that the United Auto Workers union was initially against the move.[30] However, although the company still has its problems competing, it does well in working with its union.

Multi-directional communication also emphasizes lateral communication among peers in the organization. The trend is to encourage this type of communication more than in the past. Cross-functional teams described earlier are an example of lateral communication. The advent of E-mail has made lateral communication faster and more effective. Lateral communication is especially useful when seeking advice from other departments or coordinating projects across departments.[31]

From Rigid Policies and Procedures to More Flexibility

There is no doubt that businesses need some degree of formalization. As companies grow, it is typical to develop policies and procedures to handle situations that seem to occur over and over again. In doing so, life is simplified because the employee does not have to stop and think to himself, "What is it I need to do in this situation?" This need for formalization also allows large companies, particularly in the service arena, to offer a consistent product over a wide geographic area. That is why fast food giants such as McDonald's and Wendy's maintain their consistency, through rigid policies and procedures.

We are also learning that sometimes a rigid procedure can backfire and cause a dissatisfied customer not to shop with us again. Consider this example: a retail store has the policy that all returns must have one manager sign off on the return in the presence of the employee and the customer. This, in itself, is not a bad policy. But suppose that at one point during the day, all managers are briefly off the premises at a meeting with a prospective supplier. A customer arrives with an item that is defective and wants a refund. Under this rigid policy, the customer is told they must return several hours later when the right manager is on duty. Not a good situation. Of course, this example is basic and hypothetical, but it does illustrate a myriad of other possibilities where customers are sometimes the victims of a rigid store policy.

Because there are situations like the one described above where some leeway may be appropriate, there is a trend in many companies to extend some discretion to employees in enforcing policies and procedures. Ritz Carlton Hotels described previously is an example of one of many companies that offer flexibility in decision making to their service employees. The movement away from rigid procedures and rules does not mean an abandonment of these items. Certainly, the advantage of formalization is the consistency in service and quality that results. However, when that rigidity works against the best interests of the company, then relaxing those rules a bit and letting employees make "informed" decisions is often a better practice.

Moving from Mechanistic Structures to Organic Structures

By now, it may seem like all of these facets of organizational structure change in concert with each other. In other words, as one item changes, so does another. For example, as we mentioned earlier, as organizations become flatter, the span of control of the manager automatically increases. The fact is, the parts of organizational structure all do work together and this was acknowledged over 40 years ago when management researchers Tom Burns and G.M. Stalker outlined the concepts of mechanistic and organic organizations.[32] Although the terms sound fancy, the concepts are easily described in Table 8-1 and summarize our discussion on this topic so far.

Table 8-1 can actually be viewed as two end points on a continuum, with mechanistic characteristics on one end and organic characteristics on the other. In practice, most companies operate somewhere along the continuum. Within an organization, it is possible to see that some departments may be mechanistic and others are more organic. This is not uncommon, but rather, makes sense given the mission of different departments. For example, the production department (or

Table 8-1 Mechanistic versus Organic Organizations

Organizational Structure Element Being Discussed	Mechanistic Organizations Look More Like This	Organic Organizations Look More Like This
Degree of centralization	Highly centralized. Main decisions are made by top management	Highly decentralized; main decisions are made at the unit level
Number of management levels	Tall organizational structure with many managerial levels	Flat organizational structure with fewer managerial levels
Power source	In the hands of management	In the hands of employees and managers
Job design	Jobs are more specialized	Jobs are broad in scope and contain more skill variety
Control source	Control is in the hands of line managers	Control is in the hands of self-directed work teams
Coordination of work	Coordinated primarily by functional departments	Coordination is still accomplished by departments and also by cross-functional teams
Communication	Communication is top-down, meaning orders come from the manager down to the employee	Communication is multi-directional
Policies and procedures	Rigid policies and procedures exist	More flexibility allowed in procedures

any department, service, or manufacturer that actually assembles the product) is likely to be more mechanistic. That would make sense given that order, precision, and quality are necessary to produce a product such as a food item in a restaurant. Another department in the same organization, such as research & development, would function better under an organic framework. This, too, would make sense because the creative development of a product idea is not a rigid process.

The Role of the Internet in Changing Organizational Structure

A discussion on organizational structure would not be complete without acknowledging the impact the Internet has had on businesses. Indeed, the Internet has changed many aspects of running a business as well as being an informed consumer.

To understand these changes at a more fundamental level, it would be good for the reader to understand two concepts, one related to economics and the other related to consumer information.[33] The first is a type of expense called a "transaction cost." The second concept we will discuss is that of "information symmetry" and "information asymmetry."

Transaction costs are the costs of doing business that do not actually add value to a product or service, but must be paid by the business nonetheless. For example, paying brokerage fees is a transaction cost. Driving to the store to buy an item involves transaction costs when you consider the gas needed to drive your car to the store. The further away the store is, the higher the transaction costs. Jokes have been made about frugal consumers who drive endless miles to multiple stores, just to save several cents on an item. The problem is that they do not consider the transaction costs of driving around, nor the price on their time, which is also important. For a business, a transaction cost could be the startup cost paid to use a certain vendor, or the one-time costs needed to begin selling your product to another business. In a sense, transaction costs may be viewed as the number of "touches" a product or service passes through on its way to the customer.

So where does the Internet factor into all of this? In short, it has helped to reduce transaction costs. Consider the example of driving around town comparing prices on a new outdoor grill. With the Internet, driving around is no longer necessary because prices can be compared by researching online. Hence, the "driving-around transaction cost" has been eliminated. Now consider what happens when all businesses and consumers use the Internet for researching and shopping — the net effect is a reduction in transaction costs, which is a good thing for consumers, but at the same time, it intensifies competition in an industry. When competition intensifies, businesses need to evaluate all of their expenditures carefully. One area is the old "make or buy" decision. Should our company make the product, or buy it? Assuming that quality is the same whether we make or buy (economists like to make those kinds of assumptions), the final decision will most likely be made on how much it costs to make the item. If it costs more to make the item, then we outsource it. If it costs less to make the item, then we make it.

It is not difficult to see what the trend is in our business environment today — to outsource the making of that product whenever it results in a cost savings. From an organizational structure perspective, this means companies are getting smaller by laying off production employees so the same product they used to make can be manufactured somewhere else (usually overseas), where costs are lower, usually due to cheaper labor. To summarize, the Internet lowers transaction costs, which accelerates competition, which causes the production jobs to go overseas. Of course, this is a very simplistic (and yet realistic) scenario, and it should be qualified by saying the pressure on low prices is also influenced by global competition.

Now, on to the information symmetry/asymmetry discussion. Again, these terms sound daunting, but the concepts are really quite simple. The basic idea revolves around the availability of information to the parties in a business transaction. When all parties have access to the same information about the details of a transaction, then information symmetry exists. We can continue with the previous example discussed above. Suppose that while you are shopping for an outdoor grill, you construct a chart with the features and prices of each grill you are interested in. You also list the various retailers that offer these grills to minimize your driving time when you make your final decision. As we mentioned above, the Internet makes this analysis easier by offering the information online (an example of information symmetry) and by reducing the number of miles you have to drive to get this information (an example of lowering your transaction costs). Before we had the Internet, analyzing this cost information was more difficult. It was difficult for prospective buyers to compare retailers and models, and it was difficult for retailers to see what the competition was offering and charging for their products. Because information was not readily available to all of the potential parties in the transaction, a situation termed information asymmetry existed. The impact of the Internet has been to move toward information symmetry, which is advantageous for consumers. Again, as in the reduction of transaction costs, moving toward symmetry intensifies competition that can lead to more outsourcing. Organizational structures ultimately change as a result of these processes.

The Integration of Knowledge Management into Organizational Structure

A discussion on new movements in organizational structure also involves noting a recent trend — knowledge management. Quite simply, knowledge management seeks to use the intellectual resources that already reside in the organization, to a more effective degree. To put it lightly, it is the concept of working smarter by working smarter. Of course, this definition is not sufficient, so we will begin by walking through several knowledge management terms, with which you may already be familiar.

Data, Information, Knowledge

You already know that data is simply facts and figures. Suppose that your company is experiencing a problem with employee turnover. You are concerned because the employees that are leaving seem to be good workers, many of whom are younger managers with college degrees. This is perplexing to you because it appears that although your company is able to attract good candidates, it is not able to keep those "up and coming" managers that every company needs. You decide to do some investigating into the problem and begin by noting, by month, the number of employees who are submitting their resignations. After some reading through employee files, you construct the chart shown as Table 8-2.

Table 8-2 Data on Employee Turnover for Managers

Name	Month Left	Age	Years with Company	Reason for Leaving	Performance Level
John	January	23	1	Went to work for another company	Excellent
Shirley	January	25	2	Went to work for another company	Excellent
George	February	62	32	Retired	Good
James	March	28	4	Unknown	Good
Mary	April	29	5	Went to work for another company	Good
Rick	May	33	8	Went back to graduate school	Excellent
Mitchell	June	28	3	Went to work for another company	Excellent
Justin	August	30	4	Unknown	Good
Samantha	September	29	2	Moved out of the area	Good
Thomas	November	24	0.5	Went to work for another company	Still on probation

At this point in your analysis, you have data in front of you. It contains some quantitative and qualitative features. The quantitative features relate to numerical observations, such as how long the manager worked for you and what their age was when they left. The qualitative information focuses on why they left and how satisfactory their job performance was. At this point, the data is not very useful to you. It needs some interpretation to understand the patterns that exist in your company concerning managerial turnover. What is needed next is called information.

Information attaches meaning to the data. Suppose now that as you look at your newly constructed chart, you note the following trends:

- Six of the ten managers or 60 percent have purposely left your organization to go to work for another company or to pursue graduate school.
- Of these six managers, the average age was only 28 age years old.
- These same six managers worked an average of 3.25 years for your company before leaving.
- All managers who left had performance ratings of good to excellent.

The data that existed before in the table has now been interpreted to produce statements that reflect information. Information, then, is data with relevance and purpose.[34] What appears to be a problem for your business is the inability to keep good managers. But the obvious next question is, why? Now suppose you meet with other executives in your company, including the director of human resources. During your meeting, you discuss the possible causes of and remedies to the turnover problem. One obvious suggestion is that the compensation may be too low. But the human resources director mentions that a salary survey indicates your company is competitive with other companies in your industry. Another suggestion is that perhaps the culture of the organization does not support younger managers. In other words, maybe the "old guard" mentality is so entrenched in the company that it is difficult, if not impossible, for young, up and coming managers to express their ideas about change. After a lengthy discussion on this cause of turnover, it is decided that the six managers should be contacted to see why they left the company.

Some readers may comment at this point that an exit survey should exist as standard practice when a manager leaves. However, this is not always the case in every company. Businesses committed to learning may conduct such surveys, but an organization not committed to learning from the past may overlook such obvious measures. We will suppose for now that the six former managers are successfully contacted and an exit interview is conducted. The exit interview seeks to find the underlying reasons why employees leave the organization, or voluntarily turnover as we like to say in management speak. Sure enough, we find out that, overall, the former managers were satisfied with their pay and compensation, but felt frustrated in that they were not able to grow professionally at the company. Furthermore, they expressed concern over their inability to make changes in their departments without a lot of approval from upper management. Five of these employees found companies that were more "forward thinking" and open to change, so they jumped ship, so to speak. The one employee who left for graduate school also stated the problem with the inability to change the organization, but overall, felt a calling to go into higher education as opposed to staying in the business world.

At this point, your company has now acquired bits of knowledge that can be useful as upper management seeks to staff the company with young new managers. Therefore, knowledge is information that has been subjected to human reasoning and in-depth analysis. We will review the knowledge that has been acquired.

- To keep good managers, it is not necessary to increase the compensation suddenly. In fact, doing so will raise the cost of the problem, and ultimately, not solve the problem.
- To keep good managers, the company must become more user friendly to its younger managers. Becoming more user friendly means letting these managers make changes in their departments without requiring a lot of red tape to make those changes.
- Developing a culture that values younger managers and more senior managers is necessary in your company.

So far, it sounds straightforward and within the grasp of management to begin implementing changes using the new "knowledge" that has been acquired. Nevertheless, the problem at this point is only a few executives within the company actually "know" this knowledge. Getting the word out to the rest of the company is important. In short, the company needs to become a learning organization. Knowledge management is the practice of helping companies become learning organizations.

Why Knowledge Is Not Transferred

Although it would seem plausible to "get the word out" to the rest of the organization on the matter of managerial turnover and why it may exist, the fact is, the word often does not get out for various reasons. There are three reasons why knowledge management may be impeded:

1. The organizational structure makes it hard to transfer knowledge. This chapter is about organizational structure factors, and it should be noted that structure can hurt the transfer of knowledge under some circumstances. Management researchers have told us that companies with rigid functional structures, that is, departments that may actually compete against each other, are at a disadvantage when it comes to sharing knowledge within the organization.[35] It should come as no surprise, then, that flatter structures, virtual organizations, and companies that use cross-functional teams described above do a better job of knowledge management.
2. The right technology is not in place to share knowledge. Moving knowledge throughout the organization should be a systematic process, not a random series of events. Achieving this feat will involve a new culture of openness among management and staff. It will also help if the right technology is used to move knowledge through the company. Intranets are good vehicles for doing this, especially in larger companies where employees are geographically separated from each other. There are also knowledge management software programs that can aid in this process.

    Knowledge management may have already saved your life, or at least kept you from becoming very sick. Pharmacists and doctors use knowledge management to make sure certain prescription drugs are not taken together, lest the patient have a serious reaction, or even die. Hackensack University Medical Center uses such a system to make sure patients do not receive dangerous combinations of medication. The medical center also utilizes a robot to help doctors make rounds of their patients while at home. The device, called Mr. Rounder, can be operated from the doctor's laptop computer from home. Mr. Rounder can enter a hospital room and use a two-way video to talk to the patient about his or her condition. The robot even wears a white lab coat and stethoscope. Hackensack University Medical Center may be slightly ahead of the curve when it comes to using knowledge management techniques, and the

results have been good. According to a recent *Business Week* article, patient mortality rates are down, while productivity and quality of care is up.[36]
3. Some managers and executives fear that sharing knowledge will erode their power base. On a personal level, there is the fear among some managers and executives that sharing too much of their expertise may somehow compromise their power base. This mindset could actually be true in some businesses, especially ones not committed to knowledge management. In fact, the willingness to share, along with a supportive organizational culture, is viewed by some as being the main factor needed for knowledge management to flourish.[37]

One final note should be made about this thing we call knowledge. It is not limited to special problems such as the investigation of employee turnover. It can be anything that is of value to someone in the business. It might be tips on how to perform a task better. Or, it could be a series of workshops on how to prevent certain types of operational crises, such as machine failures or safety mishaps. At Ernst & Young, best practices are documented and then shared through a computer application called COIN (community of interest). Other companies actively involved in knowledge management include General Electric, Toyota, Hewlett-Packard, and Buckman Laboratories.[38]

Does Your Business Need a Change in Its Infrastructure?

In this chapter, we have provided an overview of the key elements in an organization's infrastructure. It is normal for businesses to make infrastructure changes as the company grows or as the competition in the industry becomes more intense. Infrastructure changes also occur when manufacturing and services become more integrated within a single business. As we have mentioned in this book earlier, manufacturing companies are adding more services, which means changes in infrastructure are not only desirable, but also inevitable.

So what does it take to make changes in your infrastructure, and thus ease the transition of integrating your manufacturing and services functions?

- Awareness of the need to change. The first step is to realize that all businesses are, to a degree, in a state of transformation. This means that management should periodically assess whether improvements are needed in the infrastructure.
- Top management endorsement and continuing support. Change can occur anywhere in the business, whether it be in the manufacturing shop, the front line contacts with the customer, or the internal financial operations at the home office. It is important that top management be visible in its support of change. Change is difficult in most organizations, so when enthusiasm for

making changes wanes, employees will look to top management and gauge their degree of wanting to change based on what they see in their superiors.
- A plan to make the change, including the project plan. Change takes time, and to be successful it should be organized into smaller steps. Making changes should also involve the input of your lower-level managers and employees. Ultimately, they will be affected by the change, so they should have a say on how to make the proposed changes work.
- The resources to complete the change. Resources are another word for monetary support. It takes money to make the required changes. Money buys new equipment, new employees, and additional training. There was a saying going around where one of the authors had worked that went something like this: "You can change anything you want, just as long as it doesn't cost anything." Needless to say, that particular organization was not very innovative or open to change.
- The energy to carry out the plan (probably over a multi-year period). Some changes can be carried out fairly quickly, but most changes in infrastructure will involve several years. Be prepared for a longer time span to work with and do not get discouraged if progress is slow. It is better to make slower progress and be moving than to be standing still and ready to topple over.

Notes

1. Parnell, J., *Strategic Management: Theory and Practice* (Second Edition), Atomic Dog Publishing, Cincinnati, 2006.
2. Robbins, S. and Coulter, M., *Management* (Ninth Edition), Pearson-Prentice Hall, Upper Saddle River, NJ, 2007.
3. Porter, M., *Competitive Strategy: Techniques for Analyzing Industries and Competitors*, Free Press, New York, 1980.
4. Fishman, C., *The Wal-Mart Effect: How the World's Most Powerful Company Really Works — And How It's Transforming the American Economy,* The Penguin Press, 2006.
5. Wren, D., *The Evolution of Management Thought* (Third Edition), John Wiley & Sons, New York, 1987.
6. Fayol, H., *Industrial and General Administration*, Dunrod, Paris, 1916.
7. Actually, this example will be "loosely" based on the Pizza Hut restaurant chain during the early 1970s. However, the example is meant to be more universal in that it illustrates the four classical management functions within the context of introducing a new product.
8. Lester, D. and Parnell, J., *Organizational Theory: A Strategic Perspective,* Atomic Dog Publishing, Cincinnati, 2007.
9. Wright, P., Kroll, M., and Parnell, J., *Strategic Management: Concepts,* Prentice Hall, Upper Saddle River, NJ, 1998.
10. Lester, D., and Parnell, J., *Organizational Theory: A Strategic Perspective,* Atomic Dog Publishing, Mason, OH, 2007.

11. Hammer, M. and Champy, J., *Reengineering the Corporation: A Manifesto for Business Revolution*, Harper Collins, New York, 1993.
12. Kreitner, R. and Kinicki, A., *Organizational Behavior*, McGraw-Hill Irwin, New York, 2007, quote from p. 551.
13. Ostroff, F., *The Horizontal Organization*, Oxford University Press, New York, 1999.
14. Kasper-Fuehrer, E. and Ashkanasy, N., The interorganizational virtual organization: Defining a Weberian ideal, *International Studies of Management & Organization*, 33(4), 34–64, 2004.
15. Book publisher McGraw-Hill Irwin provides an excellent video series to accompany its management textbooks. *One Smooth Stone* was recently featured in one of these videos.
16. Kasper-Fuehrer, E. and Ashkanasy, N., The interorganizational virtual organization: Defining a Weberian ideal, *International Studies of Management & Organization*, 33(4), 34–64, 2004.
17. Kerber, K. and Buono, A., Leadership challenges in global virtual teams: Lessons from the field, *S.A.M. Advanced Management Journal*, 69(4), 4–10, 2004.
18. Robbins, S. and Coulter, M., *Management* (Ninth Edition), Pearson-Prentice Hall, Upper Saddle River, NJ, 2007.
19. Hersey, P. and Blanchard, K., *Management of Organizational Behavior* (Fifth Edition), Prentice Hall, Englewood Cliffs, NJ, 1988.
20. Mosely, T., Are you being served? The route to good customer care, *Consumer Policy Review*, 12(6), 223–228, 2002.
21. Wren, D., *The Evolution of Management Thought* (Third Edition), John Wiley & Sons, New York, 1987.
22. Taylor, F., *Principles of Scientific Management*, Harper, New York, 1911.
23. Hackman, R. and Oldham, G., Motivation through the design of work: Test of a theory, *Organizational Behavior and Human Performance*, 16(2), 250–279, 1976.
24. Robbins, S. and Coulter, M., *Management* (Ninth Edition), Pearson-Prentice Hall, Upper Saddle River, NJ, 2007.
25. Newstrom, J., *Organizational Behavior: Human Behavior at Work*, McGraw-Hill Irwin, New York, 2007.
26. Robbins, S. and Coulter, M., *Management* (Ninth Edition), Pearson-Prentice Hall, Upper Saddle River, NJ, 2007.
27. Peters, T. and Waterman, R., Jr., *In Search of Excellence: Lessons from America's Best-Run Companies*, Warner Books, New York, 1988.
28. Jennings, D., *Effective Supervision: Frontline Management for the 90s*, West, St. Paul, MN, 1993.
29. See Chandler, P. and Schraeder, M., Will the TEAM work for employees and managers? *The Journal for Quality & Participation*, 26(3), 31–37, 2003; Crandall, W. and Parnell, J., On the relationship between propensity for participative management and intentions to leave: Re-opening the case for participation, *Mid-Atlantic Journal of Business*, 30(2), 197–209, 1994; Gonring, M., Communication makes employee involvement work, *Public Relations Journal*, November, 39–40, 1991.
30. Carrell, M., Jennings, D., and Heavrin, C., *Organizational Behavior*, Atomic Dog Publishing, Cincinnati, 2006.
31. Jennings, D., *Effective Supervision: Frontline Management for the 90s*, West, St. Paul, MN, 1993.

32. Burns, T. and Stalker, G.M., *The Management of Innovation,* Tavistock Publications, London, 1961.
33. See Lester, D. and Parnell, J., *Organizational Theory: A Strategic Perspective,* Atomic Dog, Cincinnati, 2007, for an expanded explanation of these concepts within the framework of organizational structure and strategic management.
34. Lee, J., Sr., Knowledge management: The intellectual revolution, *IIE Solutions,* 32(10), 34–37, 2000.
35. Mohamed, M., Stankosky, M., and Murray, A., Applying knowledge management principles to enhance cross-functional team performance, *Journal of Knowledge Management,* 8(3), 127–142, 2004.
36. Mullaney, T. and Weintraub, A., The digital hospital, *Business Week,* March 28, 2005, p. 77. See also Kreitner, R. and Kinicki, A., *Organizational Behavior,* McGraw-Hill Irwin, New York, 2007, p. 379, for an excellent summary on Hackensack University Medical Center and its use of knowledge management techniques.
37. Lee, J., Sr., Knowledge management: The intellectual revolution, *IIE Solutions,* 32(10), 34–37, 2000.
38. Robbins, S. and Coulter, M., *Management* (Ninth Edition), Pearson-Prentice Hall, Upper Saddle River, NJ, 2007.

Chapter 9

Understanding Organizational Culture — the Elusive Key to Change

John noticed that one of the park attendants was looking anxiously in his direction. Was he aware that John was a newcomer to the park? Did John need a ticket to get in? What was the attendant's concern? John decided to face the issue directly. He walked up to the attendant and said, "Looks like you've got your hands full keeping up with all of the activity." The attendant sighed, "You're right. It's not like the old days when I first started working here. Then the parents would bring their children into the park and they would walk around enjoying the flowers and the water fountains. Sometimes they would sail their toy boats in the lake or even play catch in one of the ball fields. They were so orderly that they were never a problem. The worst problem we had to deal with was a skinned knee or a bruised elbow. I do remember one child that broke his arm, but that was a real exception. Today we never know what to expect. The children are so unruly. They walk on the grass, play ball where they shouldn't, and some climb trees that are off limits. Even the parents are always asking us to change things. Why don't we have lattés at the hot dog stand? Why can't we have a train ride, or pony rides for the children? We try to explain that they should be happy just to have a park, and warn them that if they aren't careful, they may not even have it that much longer if they continue to destroy it."

The attendant was visibly distressed. "Why are you so opposed to change?" John asked. "Why, we do change," replied the attendant. "Just last year we changed the slides to make them safer, and the accident rate has gone down. This year, we plan to put in barbeque pits so families can cook their food. I really hate to see that; it's going to be a mess, I know. Next year, and I hope to be retired by then, we are going to change the music on the merry-go-round. We change, but we change slowly. We're the culture and we are supposed to decide how the park is to be run. Sometimes we don't understand how to make the changes because we don't understand how the change will help. It may just make our jobs harder."

"I noticed the signs are all worded in the negative," remarked John to the attendant. "It is as if you are saying 'don't do this' and 'don't do that' around every corner." The attendant replied defensively, "That's the way we have always done it around here. The signs are repainted every three years by the way."

John noticed the schedule of events on the sign coming into the park. "I see you have events in the spring and fall only, but nothing during the summer or winter. Did you ever consider something for these seasons? It seems like a lot of people would want to enjoy some park events then." The attendant remarked, "We tried something five summers ago. We had so much traffic in the park that it was nothing but a mess. Too much work for our staff to handle."

John wondered if they had ever thought of hiring additional staff during these events, but at this point, he thought it best not to ask another question. Whatever culture was in place at this park definitely was not in favor of change.

Another pause to reflect.

John woke up again, this time in a cold sweat. That last dream was too realistic to be much fun. Could this be another message from his subconscious trying to tell him something about how to improve his business? Here he was, in the middle of the night, thinking about how to apply what he had seen in his dream.

John began to think about his family and two kids. Technology was like children — active, rambunctious, irreverent at times, wanting to please, but not always sure how. It may not always be exactly what a business would like, but given time, technology can become a tremendous asset. It *must be nurtured carefully, and yet, it must sometimes be allowed to introduce change into the business.*

The parents were like the infrastructure of a business — the organization structure, the plans, the policies, and all of the other guidelines that businesses use to conduct their normal operations. They know that the business needs to have a framework in which the employees operate. Just as parents change, a business can change its framework to accommodate new conditions in their business environment.

The attendants at the park were like the culture of the business. They have deeply entrenched beliefs and practices, often extending back before the present management of the business. John remembered that his father's influence still permeated his present business. Cultures change, but ever so slowly. If you try to change too quickly or without adequate explanation, the employees will rebel, just as the park attendant admonished you if you broke the rules.

John thought he now understood. In most cases, technology is going to be the driver of change. It can be enabled by changing the infrastructure to fit the technology or adjusting the technology to fit the infrastructure. Finally, culture will resist change initially and will be willing to adjust only after the employees see that there is a need to change and that the change will be beneficial. Blending the three elements into a successful improvement will take a great deal of management skill and perseverance.

John was eager to get started, but after a quick look at the clock, decided he still had time to get a good night's sleep. Turning the pillow over to get to the cool side, he went back to sleep.

Introduction

Every organization has a culture. Think about that a moment. The authors, like you, have been to many different types of businesses and government agencies. One trip to the state drivers' license agency was especially memorable.[1] The line was long; the employees were not particularly friendly; and the facility was small and in need of modernization. Getting serviced at that place was a frustrating experience. What a different experience when we later ate at a nice restaurant. The service was friendly, the food arrived fast, and the facility was very modern. In short, the organizational culture in those two places was very different.

Of course, you might be thinking at this point, "Wait a second, how can you compare a state agency with a modern restaurant?" Not a problem, remember the Input–Transformation–Output (ITO) Model? In both organizations, this model was at work, taking resources and other inputs, transforming them into products and services, and then distributing them back to the customer. However, there are differences. State agencies are funded differently from restaurants. Agencies depend on tax revenue to run their divisions, and restaurants depend on private investors and the revenue from customers to run their businesses. State agencies must be efficient; after all, offering extra frills to the customer means that tax rates will have to go up. Restaurants must also be efficient, but to differentiate one restaurant from another, owners or managers spend extra money to enhance décor, food, and entertainment options.

The missions of the state agency and the restaurant are also different. State agencies serve the general public by ensuring that all motorists are properly licensed and qualified to operate a vehicle. A restaurant, on the other hand, serves a small segment of society by providing food, and perhaps an entertaining experience. The processes between the two organizations are also different. State agencies provide a service; a restaurant is both a manufacturer of a product and a service provider. As a result, their external facilities will look different. State agencies rely more on information systems to maintain the correct data for its citizens. A restaurant relies more on production equipment because food must be stored, prepped, cooked, and then served in an appetizing manner. One final difference should be acknowledged between the two organizations, and perhaps this one is the most obvious — the profit motive. The restaurant must make a profit to survive; the state agency's existence is more certain.

To summarize, it appears the state agency and the restaurant share one thing in common, the ITO Model, and that is about it. Everything else is different, including one item we have not mentioned yet: the organizational culture. Both have a culture, but they are very different from each other. Organizational culture is an intriguing thing because it is easy to see, yet we seldom stop to think about it. This chapter is about organizational culture. We will look at what it is, and why it is so important. To help understand it better, we will examine the different components of culture. You will already be familiar with some of these components, just by reviewing your everyday shopping experiences. We will also look at different types of culture and the role ethics plays in shaping an organization's culture. Finally, we will discuss how you can change your company's culture. Changing cultures is often an essential part of implementing improvement programs. First, we will look at what organizational culture is.

What Is Organizational Culture?

Organizational culture is the belief system that members of an organization share. For example, members of a church, synagogue, mosque, or other religious organization share a common belief system. Religious organizations are an obvious place to begin discussing culture because their beliefs separate them from each other. The belief systems of these organizations are written in sacred documents and taught to their members. Each religious group has important people (some called prophets) that played a key role in the founding of that particular religion. The belief system includes a code of conduct for members. Such guidelines are important when members face difficult situations in their lives, or to help them know how to act when facing a new situation. In addition, each religious organization has a set of rituals that help members act out and celebrate their belief system in a more formal manner.

Businesses and nonprofit organizations can have their own culture, just like religious organizations. For example, Southwest Airlines is one of the most talked-about companies in the business media. One reason is the airline's unique culture that is based on having fun at work and incorporating a sense of humor into the workplace. The culture stems from the founder, Herb Kelleher, whose zany and outgoing personality has dazzled admirers from the business world for years. The culture is reflected through the employees, who like to make flying fun for their passengers. Gag announcements are common, such as the now-famous one where passengers who wanted a smoke were encouraged to visit the lounge, located on the wing of the aircraft, where they could also enjoy the movie, *Gone with the Wind*.[2]

One of the fundamental beliefs at Southwest is that employees, not customers or shareholders, are the number-one priority. Herb Kelleher explains, "The employees come first. If they're happy, satisfied, dedicated, and energetic, they'll take real good care of the customers. When the customers are happy, they come back. And that makes the shareholders happy."[3] Because that is a fundamental belief of the company, the treatment of employees reflects that. Many companies boast that employees are their number-one asset, but not all follow through with the rhetoric. For Southwest Airlines, there is much documentation about the positive treatment of its employees. In fact, the company consistently made *Fortune Magazine*'s "Best Companies to Work for" list from 1997 to 2000, although it has chosen not to participate in this ranking since 2000. *Fortune* also ranked Southwest as the fifth most admired company in 2004.[4]

Based on what we have just said, it sounds like organizational culture is important; but then again, we have only talked about one company. Before the chapter is over, we will include discussions on a number of other companies and how culture is important to their success. We will also look at how culture is an elusive element of organizational change, one that many managers forget about when they are implementing reforms in their companies. First, we will look at why organizational culture is so important.

Why Is Organizational Culture so Important?

Organizational culture is important, yet it is also one of those "touchy-feely" kinds of topics that make some executives feel uncomfortable. However, one of the early writers in the area, Linda Smircich, has identified four down-to-earth reasons why organizational culture is important.[5] She prefers to call them functions of culture, so we will describe each of them in some detail. The functions of culture give the company an identity, help an employee make sense of things, enable employees to be committed to the company, and add stability to the organization.

Organizational Culture Gives the Company an Identity

First, culture gives the company an identity. Many companies demonstrate a unique culture. In the 1980s, IBM was the epitome of a conservative, suit-oriented culture, while its rival, Apple, and especially the Macintosh Division, were considered the bad guys of the computer industry because of the outlandish "insanely great" slogan[6] that permeated the company culture and its darling new product, the Macintosh. Members of organizations with unique cultures take pride in being part of that culture. In essence, it is similar to how we felt when we were part of the "in group" in high school or college. This simple behavioral quirk is important for management to remember because employees want to be part of the groups with which they derive an identity. Consequently, it is important that employees take pride in their company's work culture.

A number of companies are identified with their cultures. We will begin with an easy one, Microsoft. If we were to sum the culture in two words, we might say "really smart." Indeed, that would be true as prospective employees go through grueling interviews that include such questions as "Why are manhole covers round?" or "Given a gold bar that can be cut exactly twice and a contractor who must be paid one seventh of a gold bar a day for seven days, what do you do?"[7] Of course, the people at Microsoft are smart, and they know they are smart, and one of the cultural beliefs, according to author Mark Gimein, is that they know they are different from other people. Hard work, long hours, and innovation are corporate values. One of the biggest values is the push to ship a product to the market successfully.[8] Every team person that gets his or her product to market earns the SHIP-IT plaque. The plaque is also an example of a cultural artifact, which will be discussed later in the chapter. With Microsoft, it should be easy to see how the culture helps employees derive an identity.

In a manner similar to Microsoft, two younger companies are also taking in an abundance of high-tech talent, Google and Yahoo! Software engineers traditionally flocked to places like Microsoft or Sun Microsystems in the past. Both Google and Yahoo! have done well recruiting lately; in fact, some of the recruits have come from Microsoft.[9] Like Microsoft, the talent that goes to the Internet search companies is very smart, and interested in working on cutting-edge technical problems. Companies like these are not just a place to work, but a chance to be with the best minds in the business and work on real problems. Culture drives these companies to seek employees who want to make a difference, particularly with problems that impact the Internet.[10]

Another company with a unique culture is REI, which stands for Recreation Equipment. The Sumner, Washington-based retailer sells high-quality outdoor recreation equipment for the serious backpacker. It recruits many of its employees from its customers, who value the mission of the company and its products. Love of the outdoors and a concern for the environment are two fundamental values of the corporate culture. Quite conveniently, they are also part of their customers' personal culture as well, which makes them good recruits. As a result, about half of the sales force consists of customers.[11] At REI, culture is both a recruiter and an identity for employees.

One other company that links its identity through its employees is Hot Topic. The central values that come through this culture are music and young fashion. Customers are young and mostly female, stores are in malls, and the merchandise is clothing identified with music and entertainment promotions. Executives must live the corporate culture by attending concerts and movies, and by staying tuned to MTV. Comments Cindy Levitt, Hot Topic VP, "For us, Hot Topic is a corporate culture. We must live the lifestyle."[12]

Organizational Culture Helps Employees Make Sense of Things

We all want our worlds to make sense to us. When something good or bad happens, it helps if we understand a reason for it. Religions have pondered this dilemma for centuries. Likewise, companies also offer their spin on the meaning of life, particularly business life, through their organizational cultures. In turn, employees can understand why their company does things the way it does.[13]

While in college, one of the authors[14] worked for a company that sold books door to door. The work was very difficult, and rejection was a common occurrence from customers who were not interested in the product. However, the culture of the company was to treat disappointments and rejections as a way to become a better sales person, and to grow on a personal level as well. Student sales persons were reminded that each rejection meant they were one step closer to a sale. In addition, hard work and perseverance would eventually pay off in the end. These values permeated the company and helped college students make sense of the disappointments that awaited them in the "the real world."

The Boston Consulting Group (BCG) is a think tank committed to making sense of the business environment. The organizational culture can be seen in its mission statement: "BCG aims to help the world's best organizations make decisive improvements in their direction and performance by sparking breakthrough business ideas." Three value statements solidify this mission to make meaning of the complicated business world:

- We continually strive to generate deep insight into what drives value creation and competitive advantage in our clients' businesses and the economy as a whole.
- We work collaboratively with clients to convert insight into strategy that will have a substantial positive impact on performance.
- Consistently delivering impact earns the trust that is the foundation of lasting relationships. These relationships serve as a platform for still-deeper insight and more significant impact.[15]

This example illustrates a common thread in organizational culture; the mission statement of the organization serves as a basis for the culture. In fact, if the two are not harmonious with each other, then something is wrong. Either management is not communicating the mission well, or the employees have rejected it altogether. The result is going to be a culture with an identity complex. At BCG, this is not a problem; mission and culture do go well together. In fact, *Fortune Magazine* ranked it 11th in its "100 Best Places to Work for 2006" issue. According to their report, "Knowledge really is power. If you begin work at this management consulting firm with a B.A. degree, they will send you to a top institution for an M.B.A., pick up the tuition bill, and double your salary if you agree to stay on."[16]

If culture is to help employees understand why things happen the way they do, then it must also enable employees to understand why the company does things the way it does. This is a key ingredient to change; in fact, the elusive ingredient to change, as we like to refer to it in this chapter. We will look at a basic relationship in a business to illustrate this point — the hourly employee and the unit manager. Frequently, hourly employees do not understand why managers sit in their office and talk on the phone all day. This is because they do not understand the manager's job. The results are complaints about the manager, and even the company in general. It may be too simplistic to say that the employees are just unmotivated, or like to complain. However, knowledge of organizational culture helps us understand that the employees are not learning the culture that the company wants them to know. The authors want to acknowledge that there are unmotivated employees, but what we are trying to point out here is that an otherwise decent employee can become disenchanted if he or she does not understand management's vision and culture for the organization.

Management writers Robert Kreitner and Angelo Kinicki discuss the need for all employees to understand the company's mission and culture. They discuss how, in 1971, Southwest Airlines originally had a vision to compete with ground transportation. As a result, the strategy of the company consisted of offering low fares, short flights, high frequency of flights, and point-to-point service (a fancy way of saying, no hubs, like the big airlines use). Translated, this means a plane must be in the air more than it is on the ground. To accomplish this, flights that land must be quickly serviced and put back into the air, sometimes in as little as 20 minutes.[17] This means that all employees, sometimes even the pilots, must help clean the plane so it can be put back in service. Imagine the sneers this might get from employees who work at other airlines. Usually, a separate cleaning crew does this kind of work. The Southwest Airlines culture is to communicate to all employees the importance of fast turnaround. They understand they are competing against ground transportation companies as well as a potential customer's personal vehicle. To the employees, the work they do makes sense, because they understand the company's mission and culture.

Organizational Culture Enables Employees to Be Committed to the Company

Culture helps an employee stay committed to the employer. When employees feel like they can identify with the values of their employer, they are more inclined to want to work for that employer. Usually, the values the company puts forth as being important include acknowledging the dignity and worth of the work force. Thus, a company with a good organizational culture values its employees.

Let's look at the Container Store. The company sells items that help get your home or office organized. If you visit the Container Store's Website, you will eventually come across the cliché, "employees are our greatest asset." So far, nothing really new there, all companies say that. Then, the corporate culture suddenly leaps out in these statements: "One of the Container Store's core business philosophies is that one great person equals three good people in terms of business productivity. So, why not hire only great people? With that said, when it comes to selecting who will join the team, we go to great lengths to look for that one great person to fill a position. Most employees at the Container Store are college educated and most were customers first."[18]

The statement, "one great person equals three good people" puts a quantitative dimension to a qualitative subject. Obviously, the company has thought this through and realizes one of the key elements of organizational culture — hire only those applicants who will perpetuate the culture you want in your company. Furthermore, customers make good prospective employees because they are already interested in what the company sells — organization. Note how the Container Store and REI share a similar idea on recruitment by looking at their customers as potential employees. From an organizational culture standpoint, it makes perfect sense.

Does the Container Store actually have committed employees? The answer appears to be yes. The company has appeared consistently in the Fortune 100 Best Companies to Work for, and in 2007, they finished an impressive fourth.[19] Furthermore, voluntary turnover, a key indicator of the success of the culture was at a low 19 percent. According to this year's *Fortune* report, the "retailer pays sales employees 50 to 100 percent above industry average. Nearly one tenth of all employees take advantage of the 'family friendly' shift, from 9 a.m. to 2 p.m., allowing for school drop offs and pickups."[20]

As we have seen with the Container Store, if a company proclaims that employees are a great asset, then that should translate into excellent human resource practices. This, in turn, creates a positive organizational culture because employees feel as if they really are important (as opposed to just hearing that they are important). One company that many recognize as being one of the best in its industry area (software design) and in the area of treating its employees well is SAS out of Cary, North Carolina. The culture created by the company has made it one of the most talked about places to work, and to show that, it boasts a 97 percent retention rate of its employees.[21]

When management writers talk about the SAS culture, they inevitably expound on the great benefits the company offers, and these are important to mention:

- On-site (or in SAS talk, on campus) medical facilities
- On-campus swimming pool, exercise room, and basketball courts
- On-campus massages, dry cleaning, auto detailing, and even haircuts
- On-campus Montessori day care center[22]

However, to stop there would be missing an important part of the culture — the ability to manage highly creative employees. Managerial Comment 9-1 explains the philosophy behind fostering creativity at SAS.

Managerial Comment 9-1. Fostering Creativity at SAS

"Creative people ... crave the feeling of accomplishment that comes from cracking a riddle, be it technological, artistic, social, or logistical. Though all people chafe under what they see as bureaucratic obstructionism, creative people actively hate it, viewing it ... as the enemy of good work."[23]

The application of this approach to creativity is to enable employees by keeping them intellectually engaged and removing such distractions as unnecessary meetings. It also means removing distractions that can occur off the job. That is why SAS has an on-site medical facility, so employees can return to work faster, rather than trekking across town and waiting for hours to see a doctor. The basketball courts, gym, and swimming pool also simplify the employee's life of distractions because they do not have to travel off campus to take an exercise break.

One final note about the SAS culture. Much has been written about the company's 35-hour work week. In truth, the creative software developers probably work a little more than that, but they certainly are not encouraged to work the 70-hour weeks known throughout Silicon Valley. A company proverb states, "After eight hours, you're probably just adding bugs."[24]

Commitment can also be found in the health care industry. The Mayo Clinic offers a glimpse of how a positive culture can be cultivated in the industry. Perhaps the most revealing aspect of the culture is the team approach to patient care. Only the top physicians are recruited to go to work at the Mayo Clinic, which should be remembered is not just one facility, but over 42,000 employees in various clinics and hospitals across the United States. However, doctors who want to be stars, work independently, lack interpersonal skills, or want to maximize their income, need not apply. What most hospitals desire is a team player who derives satisfaction from seeing a complicated medical case successfully treated because a number of professionals were involved.[25]

To fit into this type of culture, hospitals seek a certain type of individual, one who can perpetuate the culture, not fight it. That is why the selection process at the Mayo Clinic, and other organizations we talk about in this chapter, must be stringent. Once the right person is on board, that employee can work within that culture because that is what they are best suited for. As a result, commitment to the organization usually follows. In the case of the Mayo Clinic, *Fortune Magazine* reports that 89 percent of the employees say they are proud to tell others they work at this hospital.[26]

Organizational Culture Helps Add Stability to the Company

Culture adds traditions to the organization, which can add stability to the workplace. As a result, a positive culture can help buffet the company as it goes through rough times. Some cultures have allowed companies to endure for many years. The William Wrigley Jr. Company makes candy and gum, and has been around since 1891. Four generations of the Wrigley family have led the company, with the latest being CEO Bill Wrigley, a great-grandson of the founder. Currently, Wrigley serves as executive chair while Bill Perez serves as the company's first outsider CEO. Although the company has always been committed to innovation, it relies on its culture of excellence and quality in all its products. Traditions remain strong at the Wrigley Company, as the Gum Lady still distributes free gum at the company headquarters in Chicago. The traditions and culture seem to appeal to the employees as well, as one third of the employees have been with the company more than 15 years.[27]

Nordstrom's, founded in 1901 as a partnership, is another company that endured over the years by providing excellent customer service, a foundation of their organizational culture. The Seattle-based retail chain translates its culture of superior customer service into high commissions for its store sales personnel and promotion from within. This type of culture also means that employees have some leeway in how to meet a customer's needs. This is an important consideration when formal rules dictate much of an employee's behavior. In the area of excellent customer service, sometimes an absence of stiff rules is more appropriate.[28] The cultural norm is to "use your good judgment in all situations."[29] As a result, it is the sales person who makes the decisions on exchanges and refunds, not the "supervisor." The formula works at least at Nordstrom's as this company has withstood a century of retail turbulence. Today, sales associates can earn commissions over $100,000 a year, which in a retail store, is almost unheard of. Culture then, adds stability to the company, and income to employees.

Some work cultures are destructive for employees because they can reduce morale and lead to poor productivity, absenteeism, turnover, and even lawsuits. Many discrimination lawsuits result from a culture that tolerated racism, sexual harassment, or a hostile environment of some type. Health care researchers Stephen Crow and Sandra Hartman of the University of New Orleans have reported how "rogue doctors" can contribute to a dysfunctional culture in healthcare.[30] They describe a rogue doctor as one who exhibits severe ethical and performance problems that can expose a patient or someone close by to certain risks. Such doctors can be disrespectful toward others, particularly nurses and aides. Some health care institutions look the other way at such behavior, but the result is a culture that tolerates the rogue physician. As the culture solidifies, the result can be regular intimidation of health care staff that is not addressed by management. In such environments, a "zero tolerance" approach to identifying and removing these doctors is being encouraged.[31]

In summary, organizational culture can be an asset or a liability. Learning to identify the elements of culture, both good and bad, is the focus of the next section.

What are the Components of Organizational Culture?

The Components of Culture

As we have seen, organizational culture emanates from a belief system. This belief system manifests itself in a number of ways. First, and most probably, it will show in the company's mission statement. The CEO expresses the mission in speeches and written statements. Culture can also be expressed through more indirect means, labeled as cultural artifacts by management researchers who study these types of things. In this section, we will look at the components of culture beginning with values. Next, we will look at cultural artifacts in more detail.

Values

The values of a company often come from the founder. These values then become part of the culture. Certainly, we can look at well-known companies and see their founder's personality and values expressed through that company. Herb Kelleher's personality and values are obviously engrained at Southwest. Ray Kroc, although technically not the founder of McDonald's (two brothers actually owned the restaurant; Kroc just built the empire), was instrumental in its success and growth as a chain by instilling the values of quality, service, cleanliness, and value.[32]

With all of the hype about Southwest Airlines' culture, it makes one wonder if there is a company that bases its values and culture on Southwest? Actually, there is. HomeBanc Mortgage based in Atlanta, Georgia, borrows some of its culture from Southwest Airlines as well as the United States Marine Corps, and the Walt Disney Company. Like Southwest, employees come first, not customers. The value here is that satisfied employees will take good care of the customers. Another value is the emphasis placed on character. "The director of human resources is a minister, and internal meetings are often opened with a prayer. No experience is necessary, but good character is."[33]

Values appear in a number of ways. At the Midwest-based accounting firm of Plante and Moran, values translate into a lower than usual turnover rate. The atmosphere, which has been informally labeled "jerk-free," thrives on discouraging bad supervisory styles and instead puts an emphasis on the Golden Rule.[34] As a result, the turnover rate (8 to 15 percent over the previous decade) is about half that in other similar sized accounting firms.[35]

A quick way to discern an organization's values is to look at its mission statement. If values appear anywhere, it should be there. The mission should express the organization's reason for existence, as well as a list of key values that are important.

Artifacts: The Display of Organizational Culture

When the culture is visible, it translates into "how things are done around here."[36] Looking at the physical manifestations of culture or the artifacts is important to understanding what the culture is about. In other words, artifacts help reveal what the values are of the organization. Organizational culture researchers have identified various types of artifacts.[37] These include stories, language, symbols, ceremonies, rituals, identifiable value systems and behavioral norms, the physical surroundings characterizing the particular culture, and organizational rewards and reward systems.

Stories

These are also referred to as myths and sagas. The stories reflect the ideals that the company wishes to convey. Stories are often told during orientation and training sessions to let newcomers know what is valued in the organization. Innovation is often a value reflected in such stories. For many years, one of the most popular stories told in business textbooks was the invention of those sticky notes that we take for granted today. The author remembers the most common version of the story going something like this: a 3M inventor was looking for a way to mark his hymnal in church with pieces of paper that would stick to the page, but not stick so well that it would tear the page. The invention of what later became Post-Its was a huge success and apparently occurred on "bootlegged" time, time that engineers could spend on pet projects that were not related to their current company-sanctioned projects. Although the above story is true, credit should be given where credit is due. Actually, Post-It Notes were the result of two inventors in the 3M Company. Dr. Spencer Silver discovered an adhesive in 1968 that was very thin, but did not stick very well. For several years, he tried to find a use for the discovery, but to no avail. Another 3M inventor, Art Fry, needed a way to mark his hymnal in church with something that would stick, but not stick so well as to tear the delicate pages of the hymnal. Fry worked on the project and eventually came up with the sticky notes that we are familiar with today.[38] As for that term "bootlegging" used earlier, that is 3M's informal term for letting inventor/scientists spend 15 percent of their work time on any product of their choosing.[39]

Stories serve several purposes in perpetuating the culture. First, they involve the listener's emotions as most everybody likes a good story and can relate to some degree to the events going on. Second, they show listeners how the company dealt with past mistakes, and yet managed to overcome those mistakes and forge on. Third, stories communicate the traditions of the company. Finally, stories help listeners become cohesive with their organization by showing them how they can change to become like the organization.[40] From this discussion, it should be apparent why stories are often included as part of a company's orientation and training; it is a way to show new employees the values the organization cherishes.

Language

Management researchers like to refer more formally to this area as language systems and metaphors.[41] What is apparent in many organizations is that vocabulary and expressions used among employees can convey a sense of the culture. We will consider vocabulary first. There is something inside each of us that enjoys using a secret language that is indiscernible to an outside audience. When we use that language, there is a sense of pride that comes with the type of work we do and, perhaps, the organization we work with. Sometimes, that language uses words or expressions that mean something very different from what the average person thinks when using that same word or expression. For example, in the restaurant industry, the following words or phrases take on different meanings:

- "86" This number is understood to mean the chef (or the restaurant) is out of that menu item. For example, "86 the crab cakes on the menu tonight."
- "In the weeds" An expression used among staff, particularly in the kitchen, to mean that they are very busy. For example, "we can't talk right now, we're in the weeds."
- "Slammed" Similar to "in the weeds," but conveys more of a sense that things got busy all of a sudden. For example, the tour bus came in, the tickets started jamming up in the window, and "we got slammed."
- "Give me an echo on that" Repeat the order, please.
- "Run it through the garden" Load that sandwich with everything: lettuce, tomatoes, onions, etc.
- "Give me a 31 over easy with bacon and a 21 scrambled" This is breakfast talk among cooks and staff. Translated, the first order gets three eggs over easy with bacon. The second order gets two scrambled eggs, no meat.
- "I need a 62 scrambled, a 21 poached with ham, and a Western" A variation of the previous example. In this case, two of the customers want exactly the same thing, three scrambled eggs with no meat. The second customer wants two eggs poached with ham. The third customer wants a Western omelet.[42]

No doubt, whatever industry the reader is working in, a vocabulary has emerged that may sound foreign to those outside the company. With this vocabulary comes a sense of pride and ownership in one's line of work. This in itself is an aspect of culture, that there is pride in one's work.

Language can also take another route in expressing culture through the use of phrases, mottos, and slogans. A favorite expression can go a long way in perpetuating the values sought. Managerial Comment 9-2 lists some favorite company slogans that communicate the organization's values and culture.

Managerial Comment 9-2.
Innovation, Slogans, and Culture

The following list shows various company slogans, some past, some present, which have been used. These particular examples have a common thread in that the companies strive to have a culture (and hence, a strategy) of innovation and staying on the cutting edge of their chosen industry.

- Apple: "Think Different." This expression comes from the same company that referred to its Macintosh PC as "insanely great."
- Avis: "We Try Harder." The original slogan was preceded by the phrase "we're number 2," and as a result, "we try harder."
- Bayer USA: "Science for a Better Life"
- Cingular Wireless: "Raising the Bar"
- FedEx: "Absolutely, Positively Overnight" was the former company slogan that remained popular for many years. To convey the same sense of innovation and reliability, the current slogan is "Relax, It's FedEx."
- Ford: "Quality Is Job One." This was a former slogan with obvious implications. The present slogan is "Built for the Road Ahead."
- General Electric: "Imagination at Work"
- Nike: "Just Do It"
- Orbital Sciences Corporation: "The Leader in Small Space and Rocket Systems"
- Rolls-Royce: "Trusted to Deliver Excellence"
- Toyota: "We Are Driven"
- 3M: "Innovation"

Symbols, Ceremonies, and Rituals

This category of artifacts is comprised of two subcategories, with symbols focusing more on physical items, and ceremonies and rituals focusing more on processes. We will begin with symbols. Some companies have found ways to use a material item as a symbol. WorldNow is an Internet technology company with the company value of "drilling down to solve problems" for their customers. To help support this culture, a used drill was purchased (for two dollars) at a local thrift shop. The drill is awarded for one month to an employee who displays the quality of working hard (drilling down) to solve a problem. After the month is over, another employee who has also performed great work receives the drill. The drill must be personalized in some way by the employee before presenting it to the next employee. One employee was innovative and put a Bart Simpson trigger on the drill. Another employee put an antenna on it to show it was "wireless."[43] Another symbol was used by a dean at a major university who wanted faculty to have an open door policy. To illustrate his point, he provided each faculty member with a rubber doorstop. The doorstops became a powerful symbol which exemplified the cultural value of working hard to meet the needs of students.[44]

Ceremonies and rituals are also an important aspect of organizational culture. These events often celebrate an accomplishment of a key employee or division of the company. Perhaps the best-known examples are annual sales meetings, held at many companies, where top achievers are recognized and rewarded for their work. In education, the commencement exercise celebrates the graduation of students. When an employee retires, there is usually some recognition for his or her work in the form of a party and material presents. Some of these ceremonies also include personal comments of friends and colleagues. These types of events serve several purposes: (1) they acknowledge the worth of the employee, (2) they celebrate the hard work of the employee, and (3) they encourage younger employees to identify with the accomplishments of the recognized employee. Thus, the values of the company, its culture, continue through these ceremonies.

Some ceremonies serve a different purpose: they seek to unite the members of the organization into a more cohesive unit. These types of events seek to bring employees together and help them to feel good about themselves and each other. In this way, employees build commitment to the company. Hamburger giant Wendy's holds an annual hamburger cooking contest at its annual meeting. Teams are put together to compete in making the best-quality hamburger. This event seeks to build on a cultural value of Wendy's, the making of a quality product.[45]

Identifiable Value Systems and Behavioral Norms

These can actually be fun to observe as you visit various organizations. The dress code is one behavioral norm that is tied directly back to the company's value system. Conservative dress codes reflect a value in the organization that one should look professional at all times. More lenient dress codes, such as those followed in a university setting, emphasize the importance of being comfortable and less formal, particularly when among students. Although no one dress code is superior to another, what is important to remember is that the code should match the particular organization. A casually dressed executive among a conservative board of directors would not be appropriate. Neither would a professor in a three-piece suit teaching an art history class unless, of course, that is the culture the university wants to convey.

Another behavioral norm that could also fall under the category of language is the use, or absence of the use, of profanity in the workplace. The use of profanity is a reflection of the professional atmosphere of the workplace.[46] To put it simply, the atmosphere of the workplace is perceived to be less professional if profanity is allowed. Some organizations have actually taken steps to clean up the language used in the workplace, and the first place they look is to make sure management is setting a good example. Other companies have realized that profanity may end up being a form of harassment, which could result in lawsuits. As a result, written policies about the use of inappropriate language have been encouraged.[47] Regardless of what a company's policy is on profanity, inferences can be made about the culture of that company based on the kind of words that are allowed or not allowed in conversations.

The Physical Surroundings Characterizing the Particular Culture

The buildings, offices, furniture, and even the landscaping help convey the organizational culture. Most readers should be able to relate to this idea because the majority of us own cars and homes. The automobiles we choose to drive say volumes about who we are. Exotic, expensive, and fast cars convey a much different mind set than more conservative cars that are fuel efficient and less expensive to own and operate. Indeed, the owner of a Jaguar and the owner of a Prius are probably very different types of people. The way people decorate their homes and the type of house they live in convey an impression of those who live in that house. A large home with all of the latest technological gadgets shows what is important to the person who owns that home, and hence a glimpse of who that person is. On a different note, most of us have witnessed people who seem to save everything and end up with a cluttered house. This observation tells us that one of the values of that person is to be frugal and not throw anything away. Indeed, people who have grown up during the Great Depression may illustrate this type of home owner. The point of this discussion is not to be critical of the person with the cluttered house, but to see the values that are within that person conveyed through how he or she manages the home.

Now, we will apply this discussion to the way an organization sets up its buildings and offices. Some organizations keep their home offices simple, inexpensive looking, and plain in atmosphere. Wal-Mart keeps its corporate headquarters far away from being elaborate or expensive looking. Instead, the surroundings are frugal compared to corporate offices at other companies. Wal-Mart's retail stores also convey this simple, inexpensive approach to business — a large box, simple store structure, and an eye for efficiency. Indeed, these are the same values that permeate the company. In other words, the physical structures illustrate the company values.

Other companies of course are more flamboyant in how they set up their physical surroundings. Oakley, maker of products for athletes and probably most well known for its sunglasses, illustrates this approach. The company headquarters is located in Foothill Ranch, California, and resembles a spaceship, a bunker, and a futuristic fortress, all in one. The building is intentionally designed to reflect the values stated in the company's innovation philosophy located on the company Website. Managerial Comment 9-3 states these values.

Managerial Comment 9-3.
Innovation at Oakley

1. Redefine products by redefining what is physically possible. Nothing is impossible — we know that from experience. But new ideas demand new technology, and some are insistent. CAD/CAM engineering lets us experiment and explore.
2. Reject conventional ideas, and when necessary reinvent from scratch. That means sending out hunter-gatherers to roam the world for inspirations, and locking ourselves in a design bunker until even the raw materials get the science treatment.
3. Erase the line between form and function by elevating physics to the level of art. We think they should fuel each other until they reach a critical mass called innovation.
4. Transcend performance innovations into solutions for everyday applications. By serving the demands of professional athletes, we create innovations that serve all.
5. Deliver the unexpected. That's what invention is all about. Science and art come together, producing a new formula of performance and style. And the occasional nasty explosion.

Source: Oakley corporate Website.[48]

Some physical surroundings are a bit offbeat, yet sophisticated and innovative. At Google, employees can race their remote-controlled blimps through their large offices. Digital toilets are also available, with adjustable seat temperatures that can be controlled via a remote.[49] Such surroundings help convey a culture of creativity, high technology, and fun.

The arrangement of offices is also an indicator of the values the company wants to put forth. Cubicles and open office arrangements do save money, but they also promote better communication among employees. This value is important when cross-functional teams are part of the organizational structure. On the other hand, individual offices limit the type of face-to-face communication that can occur. For some, such offices are status symbols; for others, a necessary retreat when quiet and concentration are required. Some companies incorporate the benefits of both arrangements by having open office areas and individual offices for those who need time away from the noise.

Organizational Rewards and Reward Systems

In general, rewards tend to motivate employees. To perpetuate a desired culture, an organization will reward the types of behavior it is seeking from its members.[50] For example, Timberland, an outdoor boot and gear retailer, takes the environment seriously and wants its employees to do the same. Employees who buy a hybrid car can get a $3,000 credit.[51] The company also promotes a culture of community volunteerism, and pays up to 40 hours each year if employees do volunteer work in their communities.[52]

Most companies want to perpetuate a culture of excellent performance from their employees. As a result, most companies also have pay for performance compensation programs to encourage this behavior. Across-the-board pay raises should be a thing of the past for any organization seeking to motivate employees. Such programs, which still hang on in some organizations, only reward seniority, not performance. Consequently, these programs reward the mediocre employees, who are glad to have a job, and they drive off the excellent employees, who feel offended by such out-of-date pay programs.

It is important to select carefully the performance criteria; otherwise, the wrong performance, and hence, cultural values may be encouraged. One of the authors remembers traveling to a conference recently and receiving a form letter and survey from the hotel manager a day before the stay was over. The survey was worded so that a "1" was a poor indicator and a "7" was an excellent indicator, with the numbers in between indicating a variation of each. Each question in the survey addressed an operational aspect of the hotel. The letter thanked the guest(s) for the stay and asked that they complete the survey. Within the letter, there were several appeals to contact the manager or designee if there was a reason a score of "7" could not be entered with each question. In other words, that manager wanted to see "7s" because another party was going to be reviewing those surveys, perhaps his or her supervisor. It did not seem to be a genuine method of capturing data on where improvement areas were needed in the hotel because the tone was to contact the manager before scoring a number other than "7" on the survey.[53] This practice violates common survey research design. Thus, the performance criteria and culture were encouraging perfect surveys, not excellent customer service.

Management Review writer Dean Spitzer once researched disastrous reward systems and found the following examples:

- A pizza delivery company was rewarding its drivers for speedy, on-time deliveries. Unfortunately, the drivers were also causing numerous accidents.

- An insurance agency was rewarding its sales agents for the number of telephone calls actually made (as opposed to the number of sales), and found out that all they were getting from this reward system was a high telephone bill and few sales.
- A manufacturing company was rewarding its maintenance mechanics based on the amount of time they were actually spending making the physical repairs. Time spent on problem solving and preventative maintenance was not rewarded. The result, there was little problem solving and preventative maintenance being performed.
- A freight company came up with the idea of basing rewards on the number of containers shipped. The amount in the container did not matter, just that the container was shipped. Unfortunately, a check on the capacity of the containers shipped revealed that only 45 percent of them were full when shipped. This example encouraged a culture of waste and high shipment numbers, a practice that eroded profits for the firm.[54]

In summary, the monetary reward system must match the desired cultural behaviors.

Of course, other rewards can be dispensed that are not monetary in nature. These types of rewards also serve to perpetuate the organizational culture being sought. One such reward is recognition. That is why many companies have annual sales meetings so they can identify and formally recognize their top sales people. No doubt, there are some monetary rewards given out as well, but the formal recognition serves not only to honor the exceptional employee, but also to encourage the "rookies" or "up and comers" that there is a place for them if they perform well.

What Types of Organizational Culture are There?

Some business researchers have been able to identify various types of organizational cultures. Certainly, we can observe that some cultures are conservative; others are more innovative and free flowing. Some cultures seem to thrive on providing excellent customer service, and others seem not to care whether the customer is treated well. Usually, this divide is seen more when comparing for-profit organizations, such as a private business, with a government agency that must provide a service, regardless of the number of customers.

Types of Cultures

One of the best-known classifications of organizational culture focuses on four distinct groups: adhocracy, market, clan, and hierarchy cultures.[55]

- Adhocracy. The main characteristics of this culture are the striving for a spirit of entrepreneurship, innovation, and risk taking. These businesses are always developing new products. There is an emphasis on growing the company and acquiring new resources. As a result, these cultures are externally focused. The type of person that does well in this culture is one who is flexible, creative, and adaptable.
- Market. Like the adhocracy, this culture is externally focused in that it is seeking to grow and be competitive in the market. The main characteristics of market cultures are their enthusiasm for competitiveness and wanting to lead the industry in which they operate. As a result, this culture is also very goal oriented. Individuals who do well in this culture are achievement oriented.
- Clan. As the name implies, this culture is more like an extended family. The culture emphasizes harmonious relationships among members, solid mentoring of newer employees, and working in groups or teams. Unlike the adhocracy and the market culture, the clan culture is internally focused in that it seeks to run well as a cohesive unit. The type of person who does well in this culture desires to be part of the team and seeks the good of the company. Individual stars who are performance-driven fit better in a market culture than a clan culture.
- Hierarchy. The hierarchy culture thrives on rules, policies, procedures, and order. Hierarchy cultures are internally focused in that the goal is to make the operation predictable and smooth running. These cultures do best in slow-changing environments. As a result, innovation is not one of the strengths of the hierarchy. Individuals who function best in this type of culture like order, efficiency, and predictability.

Aligning Culture and Strategy

Robert Quinn and John Rohrbaugh felt that all four cultures were useful in the right environment. As a result, each culture had its own unique strengths and weaknesses. However, when the culture and the business environment were mismatched, problems could occur that would hinder the long-term survival of the business.[56] The problem for management, then, was what to do when the culture did not match the environment and, hence, the business strategy.

Another consideration when aligning cultures with strategy is determining which culture produces the best business performance. Using the measures of profitability, size, growth rate, and market share, Rohit Deshpandé and his colleagues found that market cultures perform best, followed by adhocracy, clan, and hierarchy cultures.[57] Following this line of thinking, it comes as no surprise that many government agencies exhibit hierarchy cultures, an indicator that efficiency and predictable service to the public is needed over growth and innovation strategies. Likewise, highly competitive, for-profit companies often exhibit market and adhocracy cultures, a necessity in the types of business environments in which they operate.

What happens when a hierarchy culture exists in a company that is seeking to become more competitive? Or a clan culture exists when a hierarchy culture is needed? Mismatches in culture are common, and can often be detected when new technologies are introduced into the workplace. The following section discusses what companies can do to change their organizational culture.

Changing Organizational Culture

To make changes and improvements in an organization, three elements need to be addressed: technology, structure, and organizational culture. This chapter describes the power of culture as a key to change. Many companies do well at changing the technology when needed; they may be less adept at changing the structure. Changing the culture is often overlooked. It is that elusive element often forgotten in the change process.

The good news is that culture can be changed, but it is often a slower process than changing the technology or structure of the organization. Changing organizational culture involves addressing a number of different focus areas. Figure 9-1 overviews the different change focus areas that will be discussed in the next section.

State What You Want the Culture to Be

Organizational leaders that want a specific culture state what they want that culture to be. Such statements will appear in several formats. Because culture is a belief system, followed by actions throughout the organization, it makes sense to state those beliefs in the company mission and values statements. Putting these statements on the company Website is, of course, standard practice now. However, there are other creative places where mission statements can appear: on business cards, company letterhead, lobby headquarters, even on coffee mugs. For these statements to be truly effective, employees should be able to recite the statements word for word (if the statement is short). Many organizations have a short mission statement, which can be recited, followed by a longer list of values, which can be referred to.

The values of the desired culture should also be verbally repeated by key company spokespersons, such as the CEO, president, or vice presidents. These values can be stated in company literature as well as advertising. At the University of North Carolina at Pembroke, there is a key value statement that we[58] are very proud of: "where learning gets personal." Classes are purposely kept small so students and professors can have more contact with each other. This one value appears in university advertising, and is frequently mentioned by the chancellor of the university.

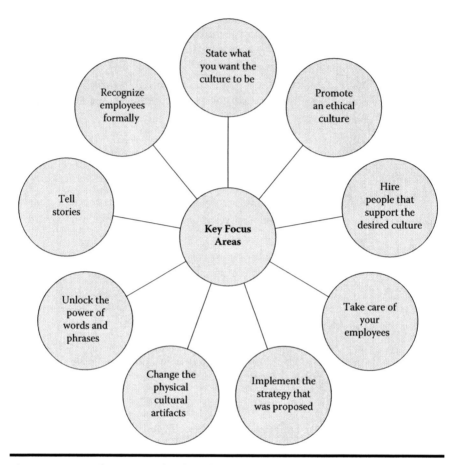

Figure 9-1 Key focus areas in changing organizational culture.

Although the university holds other values, this particular value is the one that is promoted the most and, hence, permeates the culture of the institution.

Promote an Ethical Culture

Although it may seem obvious in this day and time that a company would want an ethical culture, it should be remembered that wanting and promoting such a culture are two different things. Promoting an ethical culture involves extra work on the part of management in that ethical behavior must be communicated, nurtured, encouraged, and expected. Not promoting such a culture leaves management only hoping that the company will be ethical.[59]

Management writers Stephen Robbins and Mary Coulter offer five excellent guidelines in promoting an ethical culture.[60] First, managers and supervisors need to be visible role models to the rest of the employees in the company. Think about that! If employees are not sure how to act in a particular situation, to whom do they look for guidance? If managers act in an ethical manner, there is a better chance the employees will, too. However, unethical behavior on the part of managers and executives will also be imitated. Management textbooks are full of cases where unethical behavior permeated a company like cancer — the process metastasizing when lower-level employees imitate the behavior of their supervisors.

The second guideline is to communicate the ethical expectations to be part of the culture. Again, these guidelines can be expressed in a number of ways, through written means, such as company documents as well as its Website, and through verbal statements by management. Codes of ethics or conduct are also used by many organizations. These codes are usually a series of statements that address ethical areas including complying with all laws and regulations, avoiding conflicts of interest, maintaining the security of company information, avoiding gifts of a certain monetary amount or lavish entertainment from clients and suppliers, and general behavior while in a public setting.[61]

The third guideline is to provide ethics training to employees. This practice is usually offered within the format of brief company workshops and training sessions. Likely subjects covered in training include those mentioned above from the code of conduct. In addition, sexual harassment has been a frequent topic of ethical training in more recent years.

The fourth guideline is to reward ethical acts and punish unethical ones. The practice of punishing unethical acts is one that is frequently overlooked in organizations. A simple example might help illustrate this point. When a supervisor makes a derogatory or humiliating statement to an employee, the practice may continue if another manager or another employee does not challenge it. The danger then is that such statements become an established part of the organizational culture, no matter how uncomfortable the remarks may sound to whomever they are directed. Almost any undesirable behavior on the part of an employee can be perpetuated and become an undesirable part of the company culture, if left unchecked.

The fifth guideline is to provide mechanisms whereby employees can discuss and report ethical dilemmas. Some companies have an ethics officer that oversees this function. Utilizing ethics hotlines within the company may also be implemented. Ethics committees can also make recommendations to management on how to handle specific ethical situations.[62]

Table 9-1 Potential Scenarios when Hiring a Professor

Strength of the Potential Professor	What the College or University Wants		
	Research Oriented[a]	Teaching Oriented[b]	Both Research and Teaching Oriented[c]
Research oriented	Ideal fit	Not a good fit	A somewhat good fit
Teaching oriented	Not a good fit	Ideal fit	A somewhat good fit
A combination of research and teaching oriented	A somewhat good fit	A somewhat good fit	Ideal fit

[a] Large universities have as part of their mission the goal of generating primary research to add to the body of knowledge of an academic discipline.
[b] Small colleges typically focus on teaching undergraduates as their primary mission.
[c] Medium-sized colleges and universities stress both; however, research is not usually as intensive as at the large university.

Hire the Kind of People You Want to See Perpetuate Your Desired Culture

As we noted from the examples of the Mayo Clinic and REI, it makes sense to hire not just good potential employees, but ones who share the culture and vision of the organization as well. It is necessary to make this distinction because even an excellent new employee in the wrong culture can be a problem to the company. One of three scenarios is likely: employees will become discontent and leave, they will try to influence the culture in a different direction than what the organization wants, or they will do a combination of these scenarios.

Colleges and universities see this type of occurrence often when they hire new professors. Table 9-1 illustrates the potential scenarios that can occur. The table overviews how well a potential professor will "fit" into a particular college or university. By fit, we mean the ability to perform well on the job and to collaborate effectively with other personnel inside the institution. The table indicates (in the second row, left side) that a research-oriented professor will be an ideal fit with a research-oriented university. These types of professors enjoy conducting research, attending professional meetings to present their research, and publishing their research in journals and books. Because research is their passion, they may see teaching as just another part of their job, but not a part they are particularly enthusiastic about. As a result, research universities usually allow these professors more time to research and less time to teach, a practice that fits both the potential professor and the mission of the university quite well.

As we move across the row, the next scenario depicts a research-oriented professor in a teaching-oriented college. This situation is not a good fit. The culture of the college supports teaching, not research, but the strengths of the professor are in research, not in teaching full time. The college will usually assign professors a heavy teaching load, leaving little time for research. The research-oriented professor will be frustrated in this environment and can either leave the institution or stay and try to change the culture so it is more accepting of research. Leaving is usually the better alternative because it will be difficult, if not impossible to change the orientation of the college to a research agenda. Such teaching-oriented colleges will do best if they make sure they are hiring professors who are passionate about teaching. This is not to say the potential professor cannot also be a good researcher, but between teaching and research, the best potential fit will be a professor who favors teaching over research.

Moving across to the far right of the table, an interesting, yet flexible institution emerges, the university that stresses some degree of teaching and research. Teaching- and research-oriented professors can be a suitable fit in most of these universities, because the cultures of these schools support both. However, the intensity of research will not be as great as at the larger research university discussed earlier. Still, a research-oriented professor will be at home in this environment. What is required, however, is that professors be fairly competent at both teaching and research, not just one area over the other. As a result, the best professors in these environments find they need to balance the time demands of research with the student-centered demands of teaching. This can be a difficult task for some and can result in some professors leaving the university. For example, teaching-oriented professors who do no research will find they are not eligible for promotion or tenure. Likewise, research-oriented professors who do not perform well in the classroom may also find that the university does not want to keep them.

The rest of the table depicts the other combinations of possible scenarios. It shows that the fit of the potential employee to the organization is an important factor in encouraging the desired culture. Teaching-oriented professors will find they do best at teaching-oriented colleges. Research-oriented professors will prefer the larger universities. Those who choose universities with a hybrid of both teaching and research cultures will have to become proficient at both if they want to continue in this type of school.

Take Care of Your Employees, and They Will Take Care of Your Customers

The examples of Southwest Airlines and top-end retailer Nordstrom's illustrate how taking excellent care of your employees can result in excellent care to the customers. A sobering reality is evident, though. When employees are unhappy with their employer, they may reveal this discontentment to the customer. Most of us have experienced a service worker who "grumbled" about the boss or their company. Typically, such comments may sound similar to "They need to hire more people around here," or "These procedures they (the company) require are useless."

Many times, these types of comments come from employees who either do not feel valued by their company or feel the company does not care about their work situation. Why else would they complain about not having enough help? Inside, they too may be frustrated by the level of service they are providing because the company only gives them so much in the way of resources to work with. Front-line employees, those who deal directly with the consumer, are generally observant of the feelings of their customers. Most want to serve the customer in a way that makes them happy; otherwise, if the customer is upset with the service, the employee's job will not be pleasant either.

An organizational culture that puts employees first will go a long way in helping the attitude of the front-line employee. When their attitude and feelings are good, they will perform better in their roles. Resources will be available (translated: more help will be provided when business gets busy) and needless restrictions will be lifted. As we learned in Chapter 8 with the Ritz-Carlton Hotel chain, front-line employees need the freedom to remedy customer problems without undue restrictions.

Implement the Strategy Proposed

This statement might sound obvious, but unfortunately, many proposed changes by management only get partially implemented, which ultimately means they are not fully implemented at all. Furthermore, it is impossible to formulate a distinct culture if the business strategies of the firm are not implemented. Managerial Comment 9-4 elaborates on this point.

> ### Managerial Comment 9-4. The Strategy–Culture Linkage
>
> "An effective culture is closely related to business strategy. Indeed, a culture cannot be crafted until an organization has first developed its business strategy. The first criterion for using culture as a leadership tool is that it must be strategically relevant."[63]

Most readers should be able to relate to the following scenario. Your organization has proposed a significant change in operations. To accomplish that change, a major shift in how employees think about their work will be necessary. If management fully expects this change to take place, then actions will follow. Management will follow up with announcements, actions, and implementations, causing the change to take place, and ultimately, we will suppose the change is successful. It will follow that the employees involved will also have to change their mindset about their work and support the change.

To see that this scenario is not just an abstract example, we will now suppose that management only announced that a change would take place; they intend to implement an improvement program to be more competitive and ensure the company's continued prosperity. Employees, at this point, may be skeptical of the usefulness of the change. Some of the "old-timers" (translated: more senior employees who do not like change in general and want things to stay the same because it makes their job easier) may voice their opposition to the proposed change. Suppose that a small part of the implementation process begins, but then is halted because of a shortage of employees in a certain sector of the unit. Perhaps there is sickness, absenteeism, or imminent vacations. At any rate, this stops the momentum of the change. Management hopes that in several weeks things will get back to "normal" and change efforts can be renewed. Unfortunately, another demand develops that takes management off its focus on the improvement program. Perhaps this time, it is the preparation for a key inspection or audit. Or maybe a series of critical reports are due. Whatever the reason, momentum shifts away from the changes, those changes are not implemented, and the culture stays the same. The old-timers comment that "we knew it wasn't really going to happen. Management always talks about making changes, but it never happens."

Most readers can probably remember when proposals were made by management for change (maybe it was you and your management team that proposed the change), but because the implementation was weak and incomplete, those changes never occurred. When this occurs, the organizational culture does not have a chance to change either, and the status quo becomes even more engrained in that culture. To remedy this situation, it should be obvious that management must mean what they say when changes are proposed. Make sure the strategy proposed is enacted. Otherwise, the organizational culture will expect to "hear" of changes that will never be implemented.

Change the Physical Cultural Artifacts

"Physical cultural artifacts" is a fancy term that encompasses a number of areas, including the layout and décor of the offices, the design of the building, symbols and mascots, and for that matter, almost any item that takes up space. Although this description may sound a bit facetious, it means to convey the degree of creativity that is available to management.

Several examples may be helpful to illustrate the power of changing artifacts. Consider the impact a face-lift in décor has on the motivation of personnel. Redesigning offices, changing colors, providing new furniture are all visible indicators that a change is in the works. Of course, change for the sake of change does have its merits, but changing items in an office should also lead to a higher level of functionality and efficiency. The Crummer Graduate School of Business at Rollins College made a number of changes of these types, including providing breakout rooms for small group discussions (to enhance face-to-face interaction and creativity), upgrading the building for wireless Internet service, designing tiered classrooms, providing practice presentation rooms for student projects, and installing plasma monitors in group study rooms.[64]

Many organizations change their physical surroundings to blur the level of hierarchy that exists among management and hourly employees. One such change includes providing "rankless" parking lots, where all employees park on a first-come, best-place allotment, regardless of whether they are management or in the hourly ranks. A variation on this theme is to provide a common food-service area for all employees, leaving the executive dining room for special occasions only.

A physical artifact can also be a highly symbolic item, such as the old drill used at WorldNow to convey the depth that employees will go to meet customer needs. Physical artifacts can also be used during speeches. A favorite tactic by some speakers who crusade against excessive bureaucracy is to show physically the amount of paperwork required to perform certain tasks. President Reagan sometimes produced stacks of paper during his speeches to illustrate the endless requirements that the federal government required of its constituents. Other speakers have thrown papers and forms in the air and let them fall to the floor to show that such paperwork is not necessary. All kinds of props can be produced during a speech. The point is to use the prop (artifact) to convey some value that is of importance to the speaker. Such displays during a speech can be very powerful in communicating a message to its listeners.

Unlock the Power of Language and Phrases

Introducing phrases can be a powerful solidifier of the desired culture that you want to convey. Phrases can be used in speeches by the CEO or upper management, in company advertising, in the corporate logo, as part of the company mission statement, and on physical artifacts, such as coffee mugs and letterhead stationery.

Continental Airlines used this approach to help change the corporate culture of the company. In 1994, the airline was in dire straits. The previous ten years had seen two bankruptcies and ten CEOs. The company was the object of jokes by personnel from competing airlines. Some employees even cut the Continental name and logo off their uniforms so other people would not know the identity of their employer.[65] The new CEO, Gordon Bethune, and COO, Greg Brenneman, came on board and worked at making the airline competitive again by addressing aspects of organizational culture. One area they focused on was the use of slogans. In fact,

Table 9-2 The Interplay of Slogans, Culture, and Strategy

Slogan	Strategy Area	Purpose
"Fly to Win"	Marketing	The new marketing plan emphasized different approaches to seeking out markets.
"Fund the Future"	Finance	The emphasis was on raising cash and becoming more liquid.
"Make Reliability a Reality"	Product/service package	From a customer's perspective, the term "reliability" is important — the strategy was to improve flight reliability.
"Work Together"	Employees and teamwork	This strategy focused on getting employees to accept the plan and work harmoniously to implement the turnaround.
"From Worst to First"	Employee motivation	In 1994, Continental employees knew the airline performed poorly; this slogan was meant to motivate performance, yet not deny that the past had occurred.

Source: Higgins, J. and McAllaster, C., If you want strategic change, don't forget to change your cultural artifacts, *Journal of Change Management*, 4(1), 63–73, 2004.[65]

the new strategy for turning the airline around was called the "Go Forward Plan." Other components of the strategy included slogans to help employees remember that this was also going to be a cultural change as well. Table 9-2 lists the other slogans used in the corporate turnaround.

Tell Stories

One of the key areas to communicate cultural values is in the orientation and training of new employees. This is an excellent time to relate personal stories about former or existing organizational members that exemplify the qualities of an outstanding employee. Orientation sessions should cover a history of the company, including the personality characteristics of the founders. These types of stories often show how ordinary people overcame great obstacles to establish their company.

At the Crummer Graduate School of Business, students are told stories about previous students who achieved extraordinary goals. A sampling of these stories[66] appears next:

- Students in Free Enterprise (SIFE) is an organization that pits student teams against each other in regional and national competitions. SIFE teams at the Crummer School have won several regional competitions and one national competition. Stories about SIFE are important because they perpetuate a culture of competition and excellence.
- Students are told stories about how consulting projects have resulted in profitable changes at the project companies. These types of stories are useful to stress a culture of problem solving.
- Students learn that international projects provide career offers for many graduates. Many colleges and universities try to promote a culture of international awareness. Stories like these help to motivate students to travel overseas, many for the first time.

Recognize Employees Formally

Reward systems should include recognizing employees formally for the type of achievements that are desired in the culture. This is particularly true when the desired behavior has not been reinforced previously through the existing reward system. When management says it wants "something" (e.g., a desired performance level), but does not recognize when that behavior has been achieved, employees will wonder if it is truly important.

Many colleges and universities have found themselves in this type of dilemma. The scenario goes something like this. Faculty spend most of their time teaching, yet are also held to one quantifiable goal — writing a certain number of research articles in a specified time period. The rationale for this goal is actually a good one: faculty who are inquisitive and want to research outside of the classroom will ultimately find that they will be better teachers. Research helps professors not to become "stale" in the classroom. Perhaps the reader remembers a class where it was obvious that the professor had not changed the lecture notes in quite some time. The result is usually a bored professor and a bored class. Conducting research helps alleviate this boredom and sharpens the mind of the professor, who by the nature of the job, should always be looking for new things to teach.

However, a dynamic tension usually exists in the university setting. As we illustrated in Table 9-1, many professors enjoy a combination of teaching and research. Now, we will address the culture shift that occurs at many colleges and universities. Some schools emphasized only teaching from their faculty, but are now moving to requiring more research. This type of movement is an example of a major culture shift. Such a shift is understandably difficult for faculty who are accustomed to teaching. To make the transition to research, the dean needs to emphasize continually the importance of research. These statements seek to address the values and beliefs held by the professors. To accompany these statements, formal recognition is needed for faculty members who publish research articles. Many schools recognize these professors each time they publish an article. Some who publish regularly receive research awards. Many schools set up a display case in the lobby and show the articles of their publishing faculty. The point is that recognition is made in a formal manner. Only when this occurs consistently will employees begin to understand that this level of behavior is actually desired by the organization.

Continental Airlines also acknowledged formal recognition practices as the company was seeking to change its culture. Each time the company placed in the top five of its airline rankings, employees were rewarded with a $65 bonus. Furthermore, the amount was not just included as part of the regular pay check, but issued as a separate check for more emphasis and visibility.[67]

Conclusion

In this chapter, we have looked at organizational culture as a key component in addressing change in the company. We have referred to it as the "elusive key to change" because of its more subtle and hidden nature. Unlike technology and infrastructure, which are more externally oriented, organizational culture resides "beneath the surface" in the form of embedded values and beliefs held by organizational members. Fortunately, culture can change, although it is a slower process than changing technology and infrastructure. Implementing major improvement programs almost always requires a change in the organization culture.

The key to changing organizational culture is to address the values and beliefs of the employees first. Fundamental values, held dearly by many, may have to be changed or at least challenged. Usually, the source of this value change will come from top management, most likely the president or CEO. In addition to changing beliefs, more external cultural elements, called artifacts, will also need to be addressed. New slogans may need to implemented, the physical structure may need to be improved, stories will have to be told of employees who achieve outstanding goals, and reward systems will need to acknowledge excellent performance by organizational members formally.

In short, organizational culture is an elusive key to change, but only if it is not addressed. When culture is considered part of the equation for making organizational change, the rest of the process has a better chance at being successful.

Notes

1. The particular state in this example will remain undisclosed, because overall, the author liked living there. Also, the author wants you to know that he has also been to drivers' license agencies where the employees were very friendly — even though the lines were long.

2. Carrell, M., Jennings, D., and Heavrin, C., *Organizational Behavior,* Atomic Dog, Cincinnati, 2006.
3. Godsey, K., Slow climb to new heights, *Success,* October, 1996, p. 21.
4. Kreitner, R. and Kinicki, A., *Organizational Behavior* (Seventh Edition), McGraw-Hill Irwin, New York, 2007.
5. Smircich, L., Concepts of culture and organizational analysis, *Administrative Science Quarterly,* 28, 3, 339, 1983.
6. This phrase was made famous by Apple cofounder, Steve Jobs, who was fond of describing the next generation of PCs that would emerge from the company.
7. Gimein, M., Smart is not enough, *Fortune,* 143, 1, 124, 2001.
8. See Gimein, M., Smart is not enough, *Fortune,* 143, 1, 124, 2001, for the answer to these questions and others.
9. Elgin, B., Hof, R., and Greene, J., Revenge of the nerds — again, *Business Week,* August 8, 2005, pp. 28–31.
10. Elgin, B., Hof, R., and Greene, J., Revenge of the nerds — again, *Business Week,* August 8, 2005, pp. 28–31.
11. Arnold, J., Customers as employees, *HR Magazine,* 52, 4, 76, 2007.
12. Cooney, 2003 Retailer of the Year, *License!* January, 20–30, 2004.
13. Kreitner, R. and Kinicki, A., *Organizational Behavior* (Seventh Edition), McGraw-Hill Irwin, New York, 2007.
14. On a much too trivial and personal note, between his junior and senior years at college William "Rick" Crandall traveled the small towns of southern Texas during the summer of 1977, peddling books door to door. Unfortunately, the younger Crandall was perhaps one of the worst sales persons the company had ever seen, and when the summer was over, actually owed money to his employer, which at that time, was paid by his father, Richard E. Crandall. Nonetheless, the lessons the company taught are still appreciated to this day.
15. Boston Consulting Group Website. Retrieved 7/5/2007, from http://www.bcg.com/
16. 100 Best Places to Work for 2006, *Fortune Magazine,* 153, 1, 71, 2006.
17. Kreitner, R. and Kinicki, A., *Organizational Behavior* (Seventh Edition), McGraw-Hill Irwin, New York, 2007.
18. Excerpt from the Container Store website. Retrieved 7/5/2007, from http://www.containerstore.com/
19. 100 Best Companies to Work for 2007. Retrieved 7/6/2007, from http://money.cnn.com/magazines/fortune/bestcompanies/2007/snapshots/4.html
20. http://money.cnn.com/magazines/fortune/bestcompanies/2007/snapshots/4.html
21. Rawe, J., A homey cubicle helps a little, *Time,* 161, 21, 50, 2003.
22. Florida, R. and Goodnight, J., Managing for creativity, *Harvard Business Review,* 83, 7/8, 124, 2005.
23. Florida, R. and Goodnight, J., Managing for creativity, *Harvard Business Review,* 83, 7/8, 124, 2005.
24. Florida, R. and Goodnight, J., Managing for creativity, *Harvard Business Review,* 83, 7/8, 124, 2005.
25. Berry, L., The collaborative organization: Leadership lessons from the Mayo Clinic, *Organizational Dynamics,* August, 2004, pp. 228–241.

26. Levering, R., Moskowitz, M., Levenson, E., Mero, J., Tkaczyk, C., and Boyle, M., *And the winners are...*, *Fortune*, 153, 1, 63, 2006.
27. Levering, R., Moskowitz, M., Levenson, E., Mero, J., Tkaczyk, C., and Boyle, M., *And the winners are...*, *Fortune*, 153, 1, 63, 2006.
28. Chatman, J. and Eunyoung, C., Leading by leveraging culture, *California Management Review*, 45, 4, 20, 2003.
29. Spector, R. and McCarthy, P., *The Nordstrom Way*, John Wiley & Sons, New York, 1995.
30. Crow, S. and Hartman, S., Organizational culture: Its impact on employee relations and discipline in health care organizations, *Health Care Manager*, 21, 2, 22, 2002.
31. Crow, S. and Hartman, S., A prescription for the rogue doctor, *Clinical Orthopaedics and Related Research*, 411, 334, 2003.
32. Pepin, J., Ray Kroc, *Time*, 152, 23, 176, 1998.
33. McGarity, M., Happy at HomeBanc, *Mortgage Banking*, 170, 234, 26, 2005.
34. *Fortune Magazine*, 100 Best Places to Work for 2006, 153, 1, 71, 2006.
35. Bufe, B. and Murphy, L., *How to keep them once you've got them*, *Journal of Accountancy*, 198, 6, 57, 2004.
36. Triandis, H., *Culture and Social Behavior*, McGraw-Hill, New York, 1994.
37. Higgins, J., McAllaster, C., Certo, S., and Gilbert, J., Using cultural artifacts to change and perpetuate strategy, *Journal of Change Management*, 6, 4, 397, 2006.
38. Anonymous, 1978 — Post-It Notes — 3M, *Manufacturing Engineer*, 83, 5, 36, 2004.
39. Useem, J., *Jim McNerney thinks he can turn 3M from a good company into a great one — With a little help from his former employer, General Electric*, Fortune, *146, 3, 127, 2002.*
40. Newstrom, J., *Organizational Behavior: Human Behavior at Work* (Twelfth Edition), McGraw-Hill Irwin, New York, 2007.
41. Higgins, J., McAllaster, C., Certo, S., and Gilbert, J., Using cultural artifacts to change and perpetuate strategy, *Journal of Change Management*, 6, 4, 397, 2006.
42. Just a few of the many expressions the second author used, and heard, in his former career in the food service industry.
43. Canabou, C., Here's the DRILL, *Fast Company*, 43 (February), 58, 2001.
44. Higgins, J., McAllaster, C., Certo, S., and Gilbert, J., Using cultural artifacts to change and perpetuate strategy, *Journal of Change Management*, 6, 4, 397, 2006.
45. Carrell, M., Jennings, D., and Heavrin, C., *Organizational Behavior*, Atomic Dog, Cincinnati, 2006.
46. Anonymous, What to do about swearing in the workplace, *Leadership for the Front Lines*, 413 (November 1), 1, 2001.
47. Wah, L., Profanity in the workplace, *Management Review*, 88, 6, 8, 1999.
48. http://oakley.com/innovation/design. The reader is also encouraged to view pictures of the company's corporate headquarters, also available on the Website.
49. Elgin, B., Hof, R., and Greene, J., Revenge of the nerds — again, *Business Week*, August 8, 2005, pp. 28–31.
50. Kerr, J. and Slocum, J., Managing corporate culture through reward systems, *Academy of Management Executive*, 19, 4, 130, 2005.
51. Carlson, L., Companies chip in for workers' hybrid cars, *Employee Benefit News*, 19, 12, 73, 2005.
52. *Fortune Magazine*, 100 Best Places to Work for 2006, 153, 1, 71, 2006.

53. The author did not fill out the survey, but used it as a classroom example on how not to collect this type of data.
54. Spitzer, D., Power rewards: Rewards that really motivate, *Management Review*, 85, 5, 45, 1996.
55. The original framework was developed by Quinn, R. and Rohrbaugh, J., A competing values approach to organizational effectiveness, *Public Productivity Review*, 5, 122, 1981. We also draw from the later work of Deshpandé, R., Farley, J., and Webster, F., Jr., Corporate culture, customer orientation, and innovativeness in Japanese firms: A quadrant analysis, *Journal of Marketing*, 57, 1, 23, 1993; and Petrock, F., A quadrant of organizational cultures, *Management Decision*, 34, 5, 37, 1996.
56. Quinn, R. and Rohrbaugh, J., A competing values approach to organizational effectiveness, *Public Productivity Review*, 5, 122, 1981.
57. Deshpandé, R., Farley, J., and Webster, F., Jr., Corporate culture, customer orientation, and innovativeness in Japanese firms: A quadrant analysis, *Journal of Marketing*, 57, 1, 23, 1993.
58. William "Rick" Crandall is Professor of Management within the School of Business at the University of North Carolina at Pembroke. He wants the reader to know that he is very proud of this value the university promotes, "where learning gets personal." The phrase is not just rhetoric; the university truly lives up to this value.
59. Egan, M., Infusing ethics into your organization, *Insurance Advocate*, 117, 25, 14, 2006.
60. The discussion that follows is based on Robbins, S. and Coulter, M., *Management* (Ninth Edition), Pearson Prentice-Hall, Upper Saddle River, NJ, 2007, p. 69.
61. Rotta, C., Rules of behavior, *Internal Auditor*, 64, 3, 33, 2007.
62. Post, J., Lawrence, A., and Weber, J., *Business and Society: Corporate Strategy, Public Policy, Ethics*, McGraw-Hill Irwin, New York, 2002.
63. Chatman, J. and Eunyoung, D., Leading by leveraging culture, *California Management Review*, 45, 4, 20, 2003. Quote is from p. 21.
64. Higgins, J., McAllaster, C., Certo, S., and Gilbert, J., Using cultural artifacts to change and perpetuate strategy, *Journal of Change Management*, 6, 4, 397, 2006.
65 Higgins, J. and McAllaster, C., If you want strategic change, don't forget to change your cultural artifacts, *Journal of Change Management*, 4, 1, 63, 2004.
66. Higgins, J., McAllaster, C., Certo, S., and Gilbert, J., Using cultural artifacts to change and perpetuate strategy, *Journal of Change Management*, 6, 4, 397, 2006.
67. Higgins, J. and McAllaster, C., If you want strategic change, don't forget to change your cultural artifacts, *Journal of Change Management*, 4, 1, 63, 2004.

Chapter 10

Integrated Supply Chains from Dream to Reality

John walked to the next stage. He thought he recognized some of the same people he had seen at one of the earlier stages, the ones who were using ITO modules to build a supply chain. Sure enough, they were the same people, but what they were working on now didn't look anything like the pontoon bridges he had seen earlier.

John approached the same consultant he had talked with before. "Hi," he said, and was greeted with a "Hello, how have you been?" John said, "This doesn't look like what you were doing before. Didn't your pontoon bridge idea work?" "Oh, it worked very well," the consultant replied. "We were able to get the goods over to the other side much faster than before and we made a lot of friends on the islands on the way over. In fact, it worked so well that we came up with this new idea. We are going to build a high-speed train to carry the goods to the other side. Of course, we have to build the train and the track and get all of the different groups to coordinate their activities, but the result will benefit everyone involved.

"We are going to call this agile and integrated supply chain the CI train, for continuous improvement. It will have a lot of Cs and a lot of Is in its structure. This pamphlet entitled 'Agile and Integrated Supply Chains' will tell you all about it. We are going to continue to operate the pontoon supply chain and gradually replace it with the new supply chain as we are able

to complete each section. In the end, however, it will be a completely integrated supply chain with very little delay in the movement of goods from here to the other side. We have used our combined thinking, including ideas from our customers, and we think we have a winner."

John was intrigued. "How long do you think this new supply chain will last?" The consultant pondered carefully. "The pontoon bridge was good for several years and we think this new one will be good for a while, too. But you never know what the next generation will come up with. We can deliver the goods, but some customers are already complaining that we need to add some services to go along with those goods."

Introduction

Supply chain management (SCM) is widely recognized as important. Increased competition is forcing businesses to consider how best to gain a competitive advantage or at least to hold their own against global competitors. One way is to work with selected customers and suppliers to build lean and agile supply chains.

Present Status of Supply Chains

The benefits of successful supply chains are direct and measurable — lower product costs, lower inventories, higher quality, faster response times, fewer stockouts, and higher on-time deliveries. Other benefits that are more difficult to measure and verify include increased market share, improved customer satisfaction, increased customer retention, and a more competitive position overall. In speaking of global strategies, Edward Davis and Robert Spekman suggest that "the extended enterprise is really about creating a defensible long-term competitive position through strong supply chain integration, collaborative behaviors, and the deployment of enabling information technology."[1]

If the benefits are so attractive, why are so few companies building supply chains that satisfy not only the critics, but also the companies who want to participate in a successful supply chain? An annual survey conducted jointly since 2003 by Computer Sciences Corporation (CSC) and *Supply Chain Management Review* acknowledges progress, but observes that the majority of companies are still struggling to implement true partner collaboration practices.[2,3] Davis and Spekman summarize the situation as follows: "We advocate close ties among supply chain partners. Our observation is that most firms are not even close to developing the requisite mind set; they lack the skills and competencies needed and cannot implement the processes that lie at the heart of the extended enterprise."[4]

Table 10-1 Status of Lean Supply Chain Implementation

Survey Variable	Companies Reporting in Each Area (%)				
	Internally Focused	Little SC Focus	Some SC Focus	Significant SC Focus	Supply Chain Focused
Managing the demand signal	14	39	14	19	15
State of demand collaboration	13	34	30	19	5
Waste reduction effort	21	31	29	16	4
Value-added activities	13	28	34	22	4
Planning and production process standardization	7	33	41	13	6
Company product standard	12	30	19	28	12
Industry product standard	17	14	43	14	13
Data standard	21	42	19	15	3
People (expendable or valuable asset?)	9	22	27	26	15
Continuous improvement/ change culture	3	20	32	27	17
View of team	10	46	19	21	4
Average (%)	13	31	28	20	9

Source: Adapted from Manrodt, K.B., Abott, J., and Vitasek, K., Understanding the Lean Supply Chain: Beginning the Journey, 2005 Report on Lean Practices in the Supply Chain, APICS.[5]

Another study evaluated progress toward lean supply chains and found that the majority of companies were still more internally oriented than externally oriented on supply chain matters. The researchers defined the lean supply chain as "a set of organizations directly linked by upstream and downstream flows of products, services, finances, and information that collaboratively work to reduce cost and waste by efficiently and effectively pulling what is needed to meet the needs of the individual customer."[5] Tables 10-1 and 10-2 summarizes the results of this study.

Why is it so difficult to build lean and agile supply chains? In Chapter 3, we described some of the obstacles in trying to link companies with their customers and suppliers in a meaningful way. In this chapter, we will describe in more detail the steps required and why it is difficult. At the same time, we hope to encourage businesses to realize the opportunities that a lean and agile supply chain can provide.

Before we begin our discussion, we would like to call your attention to one of the most comprehensive descriptions of how a company can build a supply chain. In his book, *Supply Chain Architecture, A Blueprint for Networking the Flow of Material, Information, and Cash*, William Walker calls on his 30+ years of experience with Hewlett-Packard and its spin-off, Agilent Technologies, to provide the reader with a detailed description of the individual components and how they fit together to form a world-class supply chain.[6]

Background of Supply Chains

There have always been supply chains. The early traders bringing silks and incense from the Far East to Europe are some of the best known. The ships that sailed from the colonies to England represented another supply chain. It has only been in recent years that the supply chain has become a popular topic in the management literature. The concept of supply chains has evolved from several different research areas. First, the term "supply chain" suggests a movement of goods or services from an origin to a destination. Second, the supply chain builds on research from facets of organization theory. Finally, systems theory has also contributed to our understanding of supply chains.

Material Flows

Figure 10-1 shows how this fits with the movement of materials from suppliers to customers along the horizontal axis. The concept of integrating entities is illustrated along the vertical axis. Beginning at the bottom, we show separate functional areas that move product in discrete steps. As companies progress in an upward direction, they begin to link internal functions together and, even more progressively, link with their customers and suppliers. At the most advanced stage (the top level), they exist within an integrated supply chain.

Organization Hierarchy

Supply chains have also evolved as an organizational concept. Historically, organizations have tended toward a vertical hierarchy, with levels in the organization indicating levels of authority and responsibilities. These organizations also tended

Figure 10-1 Evolution of the supply chain. As companies extend their supply chain, both forward to the customers and back to the suppliers: (1) complexity increases, making communications more critical; (2) variability in demand decreases, making reduced inventories possible; (3) interdependence increases, making delivery reliability more important; (4) supplier dependability increases, making fewer suppliers necessary; and (5) material flow smooths, making lead (response) times shorter.

toward specialization, or the separation of functional areas into departments. The division of labor concept to increase productivity was first applied in production operations and later extended to staff functions, no doubt helped along by Taylor's "functional foreman" recommendations.[7] Because of this specialization, functions within the organization tended to be somewhat isolated from one another. The current term, "functional silos," describes this condition. As a result, it is more difficult for information to flow horizontally from department to department.

To remedy this situation, companies have moved to cross-functional teams as a means of stimulating a freer flow of information and faster resolution of problems. This move has led to closer collaboration of functions within a company and to extending collaborative efforts along the supply chain to customers and suppliers. The result has the effect of reducing the rigidity of vertical organizations and facilitating the horizontal flow of information, much as in the flow of goods and services.

Table 10-2 Progress in Developing Lean Supply Chains

Status of Program to Develop a Lean Supply Chain	%	Obstacles to Developing a Lean Supply Chain	%
Just beginning	12.7	Lack of resources	27.2
Some sporadic implementation	47.3	Lack of training	25.7
Formal integrated approach ready	25.3	Lack of top management commitment	21.2
Entire product line has flow	9.0	Other activities (Six Sigma, TQM)	11.0
Lean is a standard procedure	5.7	Hard to apply in my industry	8.5
		No clear benefit of lean	6.3

Source: Adapted from Manrodt, K.B., Abott, J., and Vitasek, K., Understanding the Lean Supply Chain: Beginning the Journey, 2005 Report on Lean Practices in the Supply Chain, APICS.

Systems Theory

Systems theory has also played a role in the evolution of the supply chain. A system includes individual entities that link together in a progressive way to achieve a final result. It may be a system to circulate blood in a human being or a supply chain that moves products and services from suppliers to customers. We described this linking in Chapter 3 as the extension of the ITO (Input–Transformation–Output) Model. Systems theory is especially useful in describing the role of information technology (IT) in supply chains. Without the advances in IT, especially in the area of interorganizational systems (IOS), the modern supply chain would not be possible.

Role of the Supply Chain

Modern supply chains perform three primary functions. They deliver products and services to customers. They also collect and assimilate information about customer wants and needs. Finally, they facilitate the flow and equitable distribution of money among the various participants in the supply chain.

In this chapter, we will assemble the management modules we described in Chapter 2 through Chapter 9 into a strategically planned and integrated supply chain. We will also show how continuous improvement programs fit into this progression from a traditional batch flow operation to a lean and agile organization of the future. First, we will explore the implications of moving to an agile and integrated supply chain.

Setting the Stage

Is your supply chain all you expected it to be? On the other hand, are you, like many other businesses, finding that the reality is far different from the dream? What did you expect? The *APICS Dictionary* offers the following definitions:

- Supply chain: The global network used to deliver products and services from raw materials to end customers through an engineered flow of information, physical distribution, and cash.
- Supply chain management (SCM): The design, planning, execution, control, and monitoring of supply chain activities with the objective of creating net value, building a competitive infrastructure, leveraging worldwide logistics, synchronizing supply with demand, and measuring performance globally.[8]

Apparently, supply chain management is more than finding a few good vendors who will deliver the right product at the right time in the right quantity and with good quality.

What about demand? Is the customer part of the supply chain? In case there is any doubt, another definition clears up that point. The lean supply chain is "a set of organizations directly linked by upstream and downstream flows of products, services, finances, and information that collaboratively work to reduce cost and waste by efficiently and effectively pulling what is needed to meet the needs of the individual customer."[9]

Does the supply chain have to be lean, too? Yes, and according to some authorities, the supply chain should be agile, adaptable, and aligned.[10] It should also be integrated, innovative, globally optimized, involve inter-organizational collaborative relationships, and share the benefits equitably among the supply chain members.[11] Herbert Kotzab and Andreas Otto describe SCM as "a strategic, co-operation oriented, business process management concept, cutting across organizational boundaries (i.e., integration oriented), which leads to increased results for all supply chain members. SCM demands integration and co-operation beyond the logistics dimension. SCM processes are customer and/or end-user driven."[12]

Does your competitor have a dream supply chain? Probably not. That is, unless your competitor is among the handful of companies that is continuously cited in the literature as having one of the best supply chains around (Dell, Nokia, Hewlett-Packard, and Wal-Mart, to name a few). The facts are that the reality of supply chains is quite different from the dream. Here is what a few researchers have found out:

- SCM can deliver powerful results — reducing costs, boosting revenues, and increasing customer satisfaction and brand equity by improving on-time delivery and product or service quality. Yet overwhelmingly, senior executives at large companies worldwide believe SCM has failed to live up to the promise during its first two decades.[13]
- Approximately 75 percent of the firms in the United States and the United Kingdom are small- and medium-sized enterprises. The authors have concerns that these firms have "grave" problems in their supply chains, both because of legacy systems (hardware and software) and legacy mental models (how they manage their businesses).[14]
- In a comprehensive study of over 600 companies, in which the researchers used 11 variables to assess the status of supply chain implementation, less than one third of the participants indicated that their efforts had "significant" or full supply chain focus. The remaining two thirds indicated their program was internally focused or had little supply chain focus.[15]
- In another global study, Computer Sciences Corporation used a five-level scale to place companies' progress toward "full network connectivity," or level 5. Levels 1 and 2 were in the early stages of supply chain evolution. They found that approximately 50 percent of the companies were at levels 1 or 2, with 35 percent at level 3, leaving only 15 percent at levels 4 or 5.[16]
- Supply chains of the future will include reverse logistics, or green supply chains. Businesses will design processes to handle returns, repairs, remanufacturing, and redistribution of resalable products, and recycling of disposable materials. This is already occurring, but will grow in importance.[17]

There is much more. For example, most supply chains focused on cost reduction, even though some studies suggest that improved customer service and increased revenues will have an even bigger payoff.

Does the gap between reality and dream mean that supply chains are not a good idea? Of course not! It just means that building a really good supply chain is not easy, and takes time. What makes it so difficult?

- Supply chains are complex. How many suppliers do you have? Are they all alike? How many products do you have? Are they all alike? How many customers do you have? Are they all alike? How many combinations of supplier/product/customer do you have? Are they all alike? Even in a small company, the complexity is daunting; in a large company, it is overwhelming. See the study by Deloitte Touche Tohmatsu that discusses the complexity paradoxes of optimization, customer collaboration, innovation, flexibility, and risk.[18]
- If the inherent complexity described above were not enough, most companies are making the relationships even more complex by outsourcing both goods and services. Outsourcing is here to stay, and assimilating it into everyday practice is a challenge for even the most experienced global participant.

Thomas Friedman calls outsourcing one of the ten "flatteners" in his best-selling book *The World Is Flat* (a worthwhile read).[19]
- Competition is increasing; this means that every business has to improve itself continually. The supply chain that was innovative just a few years ago is barely adequate today. The order winners of yesterday are only order qualifiers today.
- Variability is rampant along the supply chain. Not only are the relationships complex, they are constantly changing, and change breeds variability. It is difficult to build interpersonal relationships between individuals, but how do you build inter-organizational relationships between companies? It is difficult and takes time, longer than most of us have the patience to devote.
- Another major element of change is the introduction of E-commerce in the equation. That is another whole subject, but the impact on supply chains is indisputable. You probably cannot use the identical supply chain for your E-commerce efforts that you do for your regular products.
- Supply chains, as a management concept, are getting more comprehensive. What started as a transactional process is rapidly becoming a strategic process. What started in purchasing or logistics is becoming a cross-functional, horizontal flow process that includes all of the functions of a business as well as top management. Concepts like sales and operations planning are definitely in vogue, after years of languishing in the shadows.

Is there any hope to improving supply chain successes? Are businesses doomed to be forever mired in the early stages of supply chain development? The results of surveys indicate slow progress, not lack of positive results. However, the gap is getting wider between those who are getting it right and those who are struggling.[20] Whether or not to have an effective and efficient supply chain is no longer an option; it is an imperative!

Several sources provide direction and encouragement. The consulting group PRTM describes the transformation required to move from a functionally focused supply chain to cross-enterprise collaboration: (1) functional focus, (2) internal integration, (3) external integration, and (4) cross-enterprise collaboration.[21] Cigolini and colleagues offer a similar progression from traditional logistics, with a focus on reducing inventory, to modern logistics, with a shift from cost reduction to improving customer service, to integrated process redesign, which provides a systemic vision of the supply chain, to an industrial organization, which focuses on the strategic alliances among the various members of the same supply chain.[22] On a similar note, Togar Simatupang and colleagues describe the "recursive interplay" of logistics synchronization, information sharing, incentive alignment, and collective learning with an integrated supply chain. There are plenty of answers; it is just that none of them is easy.[23]

Some see a brighter future if managers want to work for it. As Kempainen and Vepsalainen emphasized, "information sharing and coordination are often considered the preconditions for successful supply chains. Our analysis of supply chain practices in industrial supply chains shows that visibility is still limited. The companies have a realistic view on the advantages and risks of information sharing, and so planning information is shared only selectively. Coordination efforts focus on inter-functional operations and relations with selected partners. The companies hesitate to expand their coordination efforts beyond order processing and operational scheduling within the dyadic supplier–buyer relationships. Some mechanisms are expected to emerge for supply chain and networkwide coordination by 2010."[24]

Vijay Kannan and Keah Tan found, in a survey of APICS members, that information sharing was not among the top criteria used by companies in selecting suppliers; yet, information sharing was one of the major positive influences on firm performance, in such areas as market share, return on assets, product quality, and competitive position. They suggest increased attention to effective information sharing.[25]

In the interest of promoting improved coordination, Chee Wong and colleagues describe the advantages of loosely coupled supply chains. They categorize tightly coupled supply chains where partners closely connect in a one-to-one relationship, such as through traditional EDI. In contrast, "a loosely coupled supply chain is an extensively integrated but loosely connected supply chain network. These concepts have emerged from the Web services, Web portal and Extended Mark-up Language (XML) technologies." They suggest that a loosely coupled supply chain has a specific niche for short- and medium-term buyer–seller relationships, small- and medium-size enterprises, secondary partners, and for a large number of partners.[26]

Chuck Poirier and Frank Quinn also offer cautious optimism. In reporting the results of a survey, they comment, "The findings further suggest that those running the business today continue to think of supply chain management as a short-term cost-reduction effort, not worthy of strategic importance. Only 37 percent of the companies indicated that their supply chain initiatives were 'mostly' or 'fully' aligned with corporate strategy. The survey results indicate that collaboration across an extended enterprise is still more theory than practice. Collaboration was the single most pressing need — both internal collaboration and external collaboration with suppliers and customers. Perhaps the most important insight from our survey is that the real business benefit of advanced supply chain management remains largely untapped."[27]

Consider this scenario. Companies report that developing reliable demand forecasts is one of their major challenges. To make better forecasts, a company needs more information from downstream members of the supply chain and the company needs to collaborate with downstream customers and upstream suppliers, to develop a mutually beneficial demand forecast. To obtain better information, partners need effective inter-organizational communications. Improved communications involves more than technology. Not only do you need to have a better way to communicate, such as through electronic inter-organizational systems (IOS), but you need to agree on what information to communicate. Here it gets tricky.

Collaboration requires sharing information. What information are you willing to share, and what information are the other members of the supply chain willing to share with you? This is where trust enters the picture, and trust has been largely lacking in most supply chains. Most of those who have studied trust, and the list goes well beyond business researchers to include those in organizational science, industrial psychology, economics, and operations research, have found it a difficult ingredient to integrate into working relationships.[28] Much of the research boils down to the fact that, to trust someone or some entity, you have to work with them to verify that the benefits of trust outweigh the risks. At the same time that you are evaluating those parties in the supply chain, they are evaluating you. At some point, you have to make a judgment decision to trigger the upward spiral from trust to collaboration, to better demand forecasts, to improved execution, to increased profits along the supply chain.

So there it is. Communicate, collaborate, trust, and share the benefits equitably with the members of your supply chain; everything will be all right. Just as long as you do the hundreds of other things necessary to make your multiple supply chains become the reality, not the dream.[29]

At the same time, care must be taken that investments in supply chains are monitored closely and overspending does not occur. Suvankar Ghosh and colleagues warn that some companies have invested two to three times the level that would be desirable from a stakeholder perspective.[30]

Supply Chain Models

A number of researchers have proposed models or narratives to describe the evolution to an effective supply chain. We describe several of them below and then introduce the model that we believe extends the existing models to a more complete state.

Evolution of Supply Chain Models

One approach is to classify SCM into different schools of research. In a comprehensive review of the literature, Cigolini, Cozzi, and Perona developed the following classifications that indicate the progression of supply chain thinking over the past 20 years:

- Supply chain awareness: Recognize that there is a continuous chain of functional areas through which materials flow and that extends from suppliers to final distributors.

- Traditional logistics: Main objective is to improve supply chain efficiency by reducing inventory levels, whereas little emphasis is given to supply chain effectiveness.
- Modern logistics: The focus shifts from mere cost reduction to include also service and quality improvement.
- Integrated process redesign: Through quantitative models applied to a systemic vision of the supply chain studies show how to redesign the entire supply system to obtain more efficient and effective flows of materials and information.
- Industrial organization: Focuses on the strategic alliances between the various actors of the same supply chain.[31]

The following eight supply chain management processes are included in the Global Supply Chain Forum (GSCF) framework:

1. Customer relationship management provides the structure for how relationships with customers are developed and maintained.
2. Customer service management provides the firm's face to the customer, a single source of customer information, and the key point of contact for administering the product service agreements.
3. Demand management provides the structure for balancing the customers' requirements with supply chain capabilities, including reducing demand variability and increasing supply chain flexibility.
4. Order fulfillment includes all activities necessary to define customer requirements, design a network, and enable the firm to meet customer requests while minimizing the total delivered costs.
5. Manufacturing flow management includes all activities necessary to obtain, implement, and manage manufacturing flexibility and move products through the phases of the supply chain.
6. Supplier relationship management provides the structure for how relationships with suppliers are developed and maintained.
7. Product development and commercialization provides the structure for developing and bringing to market new products jointly with customers and suppliers.
8. Returns management includes all activities related to returns, reverse logistics, gatekeeping, and avoidance.[32]

The objectives of the Supply Chain Council (SCOR) framework are

- Plan: Balances aggregate demand and supply to develop a course of action that best meets sourcing, production, and delivery requirements.
- Source: Includes activities related to procuring goods and services to meet planned and actual demand.
- Make: Includes activities related to transforming products into a finished state to meet planned or actual demand.

- Deliver: Provides finished goods and services to meet planned or actual demand, typically including order management, transportation management, and distribution management.
- Return: Deals with returning or receiving returned products for any reason and extends into post-delivery customer support.[33]

Four conceptual perspectives on purchasing versus SCM have been proposed:

1. Traditionalist: SCM is a strategic aspect of purchasing, with emphasis on supplier development, and partnerships with first- and second-tier suppliers.
2. Relabeling: Simply change the name of purchasing to SCM. The scope of SCM equals purchasing. Other fields have also been relabeled as SCM, such as logistics.
3. Unionist: Purchasing is a part of SCM, as SCM completely subsumes purchasing, as well as logistics, marketing, and operations management.
4. Intersectionist: SCM is not the union of logistics, operations, and purchasing. Rather, it includes elements from all of these disciplines as well as marketing, organizational behavior, and strategic management.[34]

The PRTM Group describes the transformation from a functionally-focused supply chain to a cross-enterprise collaboration arrangement. The stages are

- Stage 1: Functional focus. Functional departments within an organization focus on improving their own process steps and use of resources. Managers typically focus on their individual department's costs and functional performance. Processes that span across multiple functions or divisions are not well understood, resulting in limited effectiveness of complex supply chain processes.
- Stage 2: Internal integration. Division- or companywide processes are now defined, allowing individual functions to understand their roles in complex supply chain processes. Cross-functional performance measures are clearly defined, and individual functions are held accountable for their contributions to overall operational performance. A well-defined demand/supply balancing process that combines forecasting and planning with sourcing and manufacturing is evident at this stage.
- Stage 3: External integration. Stage 2 practices are now extended to the points of interface with customers and suppliers. The company has identified strategic customers and suppliers, as well as the key information it needs from them to support its business processes. Joint service agreements and scorecard practices are used, and corrective actions are taken when performance falls below expectations.
- Stage 4: Cross-enterprise collaboration. Customers and suppliers work to define a mutually beneficial strategy and set real-time performance targets. Information technology now automates the integration of the business

processes across these enterprises in support of an explicit supply chain strategy.[35]

An alternative approach is to describe the evolution of a supply chain along the following steps.

1. Enterprise integration (functional/process)
2. Corporate excellence (intra-enterprise)
3. Partner collaboration (inter-enterprise)
4. Value chain collaboration (external)
5. Full network connectivity (total business system)

Overall, approximately 50 percent are at levels 1 or 2; approximately 35 percent at level 3; and the remainder at levels 4 or 5.[36]

A Comprehensive Supply Chain Model

We have developed a model to describe the journey from batch flow to an integrated, agile supply chain. Figure 10-2 is a diagram of this model. The general structure of the model is as follows: the horizontal axis, reading from left to right, shows the progression, over time, from batch flow to an integrated and agile supply chain. Although the progression occurs over time, the axis does not represent a specific time scale. The journey varies for each supply chain but generally the time is in years, not months or weeks.

The vertical axis is composed of three major levels. The top level shows the stages that a company goes through: (1) present batch flow, (2) make the strategic decision to change, (3) develop a strategic direction for their program, (4) effect the internal changes needed, (5) establish linkages with key external entities, (6) attain an effective level of supply chain integration, (7) enhance the supply chain with agility, and (8) enter the virtual environment state. The change agents needed to move from stage to stage are interspersed between the stages.

The middle area of the model shows the key change components, or "Cs," needed to make the change successful. In addition to the Cs, the diagram shows the initial starting conditions of functional departments, legacy systems, and intermittent process flow. Further to the right, it shows the changes to a responsive organization, integrated information flow and synchronous product flow. Even further to the right, it shows the beginnings of a supply chain and subsequent progression to integrated and agile supply chains.

At the bottom of the diagram, we show the action steps necessary, or the "Is" of the program. These are the activities necessary to bring about the Cs described above.

Integrated Supply Chains ■ 321

Figure 10-2 Transition from batch flow to an integrated and agile supply chain.

As an overview, the left third may be viewed as "starting the journey," the middle portion may be seen as "achieving internal synchronization," and the right third as "building the supply chain."

We believe the model shown as Figure 10-2 illustrates several salient points about the development of effective supply chains:

- Developing a supply chain is an evolutionary process. The movement from batch to a lean enterprise precedes the progression to a supply chain, then to an integrated supply chain, and finally, to a lean and agile supply chain.
- The progression along the model shows movement from discrete, even fragmented, functions of an entity to the integration of multiple entities into a seamless vibrant entity.
- The movement to integration requires the blending of inter-disciplinary functions within a company and between companies.
- As the evolution progresses, there is a shift in emphasis from transactions to strategies. Although attention to detail persists throughout, there is also a need to develop a holistic perspective of the program.
- Just as there is a need for inter-disciplinary activity, there is a requirement for interactive movement from connectivity to coordination to cooperation to collaboration.

Each element of the model will be explained more fully in the following sections with examples from reported case histories. First, we will look at why many authorities advocate the shift from batch flow to lean production.

The World of Lean Production

In Chapter 5, we used lean production as an example of one of the major improvement programs that has become popular in the past few years. Although it arose in the manufacturing sector, many are now advocating its use in service operations. Lean thinking has largely replaced Just-in-Time (JIT) in the number of articles written. The terms "lean production," "lean manufacturing," and "lean thinking" have become almost synonymous in the literature. The key values are overviewed below:

- Create value for the customer. Lean production creates value for the customer by reducing the amount of waste within companies and along the supply chain.
- Create a smooth flow of goods. One of lean production's objectives is to smooth the flow of goods along the supply chain by eliminating problems and other delays that interrupt flow.

- Provide a fast response. By eliminating waste and creating a smooth flow, lean production reduces the response times in product and service deliveries.
- Detect quality problems quickly. Lean production continues an emphasis on maintaining quality of products and services. The need for fast, even flow makes it possible to detect quality problems sooner.
- Develop cross-trained employees. Cross-trained employees are necessary to maintain the even flows and high quality required by lean production.
- Illustrates a contemporary outlook. Lean production reflects some of the latest thinking in operations management and fits nicely with the supply chain concepts.

Although lean production provides a number of benefits, it is not without some limitations or constraints in its implementation:

- It takes time to implement lean production. Companies that implement lean production should not view it as a quick fix. It takes time to implement it effectively. We will describe some of these steps in the following section.
- Lean production requires significant change. To implement a complete version of lean production, a company will probably have to make substantial changes in the way it operates. Changing the physical layout of a plant and moving from job specialization to job enlargement and enrichment are two areas that involve the greatest adjustment.
- It is difficult to achieve and maintain. The road to lean production is long and difficult. For those companies that achieve a level of competency, consistent effort is needed to maintain that level of competency. Just as it is difficult for world-class athletes to maintain their physical conditioning, it is difficult for companies to stay lean. To continue with that metaphor, they can become "lax and fat."
- It is difficult to extend throughout the supply chain. Attaining a lean condition in one company is difficult; achieving that condition throughout the supply chain is even more difficult because there may be trade-offs that benefit one member of the supply chain while adversely affecting another. For example, reducing inventory at the retailer level may impose greater inventory levels upstream in the supply chain.

The potential benefits of a lean manufacturing world sound wonderful. So how does a company get there? To oversimplify, a company needs to reduce variation in its products or services, its purchasing process, its production process, and in its delivery process. These steps relate to the internal management of a business.

Product

Many companies are attempting to increase revenues by expanding the variety of products or services they offer. Although this approach may generate sales in the short term or until a competitor offers the same products or services, it also increases complexity and variation in the product lines. This complexity inhibits the desire to go lean. To resolve this dilemma, a company needs to develop product or service modules that can be configured to suit the variety required. An example is kitchen cabinets that can be customized by assembling modules to fit the available space. A service example is a will prepared by a lawyer who adapts modules of features to prepare a customized will for specific clients.

Purchasing Process

The traditional purchasing process has involved placing a variety of potential suppliers in a competitive bidding situation to obtain the lowest cost. This introduced variation in the suppliers used, quality levels, response times, and confidence levels in the suppliers. To reduce these variances, companies need to develop long-term relationships with carefully selected suppliers. Contemporary lean purchasing practices dwell on the total cost of ownership and consider both tangible and intangible factors in their decision making.

Production Process

In production, variations can take the form of uneven work flow, smaller lot sizes, extra setups and changeovers, employee work assignments, more rush orders, and a host of other problems that makes the production of goods and services a demanding task. To reduce the adverse effects of variation, the production process must become more flexible with shorter setup times, more versatile employees, and general-purpose equipment — just to name a few. In addition, the planning horizon must be extended, with increased feedback from downstream customers, so that there is an increased opportunity to cope with the variation.

Delivery Process

Increased variation means more frequent but smaller deliveries. The increased frequency of deliveries has increased the scheduling difficulties, especially in large distribution centers. It is no longer possible for suppliers to deliver at their convenience; they must deliver at the time specified by the customer. If the supplier is early, they wait. If they are late, they pay both financial penalties and perhaps face the loss of future business. The answer is a focused approach to their delivery process that is dedicated to eliminating potential problems before they occur and in designing an integrated delivery system that considers customer needs, delivery equipment, employees, external factors, and potential acts of nature, such as hurricanes or snowstorms.

Demand Variation

Companies also need to reduce the variation in demand. There may be greater variation in demand than in any of the internal areas described above. Reducing demand variation may sound counterintuitive at first. Who wants to tell the customer when to place an order or request a service? However, there may be situations where customers benefit if they change their ordering habits to obtain faster service or avoid long waits. Collaboration with customers can sometimes pay dividends in reducing demand variation.

How are companies doing in their quest for a lean supply chain? Tables 10-1 and 10-2, introduced earlier in the chapter, shows results from a survey of several hundred companies.

Why is it so difficult to achieve the lean status? The literature suggests a gap in understanding the transition and change process. Managers understand the need, but they do not always understand how to make the necessary changes. Even if they understand, it is difficult to anticipate the problems and challenges in making the implementation. They need an integrative model of the common drivers (barriers and facilitators) of the change process. In the following section, we will describe why achieving a lean and integrated supply chain is difficult.

Steps to Achieve a Lean and Agile Supply Chain

It is easy to be enthusiastic about the possibilities of change, but it is difficult to reap the benefits of those changes. We believe there are several important phases a company must go through to achieve success.

Commitment

The first requirement is a commitment to change. A decision to change should come only after a careful investigation reveals the necessity to change. A commitment to change requires top management support. Although this support does not necessarily require day-to-day participation in the details of the program, it does mean that top executives support the program and understand its importance in the company's future. Management should provide adequate funding for the project and commit to future allocations as needed. Top management should remember its role as the drivers of change in the structure of the organization and its practices. They monitor the progress of the program and recognize the achievement of milestones in the program's progress.

As part of the commitment to change, the entire organization should recognize and accept a role in the program. This realization means that employees must understand why the program is important and what their role is in its implementation. They should have a role in the design of the program to ensure their "buy-in" as a stakeholder in the process.

The commitment phase should involve enough of a plan to make sure that the initial presentations about the objectives, scope, and potential consequences of the program are accurately and objectively portrayed to employees at all levels of the organization. The intentions of the company should also be shared with other key stakeholders, whether it is customers, suppliers, financial institutions, or major investors. Planning should mean that the company is thinking several steps ahead and not just responding to random events.

Researchers have found that top management support is important, as evidenced by the following comments:

> Supply chain management continues to fall short of its great promise, according to a new assessment. Surveying nearly 200 senior executives from manufacturing and industrial companies in North America, Europe, Asia, and Latin America, a consulting firm cites reasons for the disciplines' failure, including (1) top management's narrow view of SCM and superficial involvement in the design and guidance of the function, and (2) an unrealistic view of what technology can achieve.[37]

> Supply chain management requires strong top management leadership. Our survey found that in companies where responsibility for SCM resides below senior management, annual savings in the cost to serve customers are just 55 percent of what they are when SCM is a component of the overall business strategy.[38]

Once a company has made the commitment to change, it has to act. However, action without a plan is premature. At this point, planning should focus on developing the concept that will carry the program through to a successful conclusion.

Concept

The next step is to develop a concept of the program — an understanding of what to do, what to accomplish, who will be responsible, and when it will be completed. The concept should be a "big picture" view of the program that provides a direction and scope for those who will participate in the design and implementation process.

Now is the time to involve the entire organization. This type of program needs the participation of the management team and nonmanagerial employees. Program managers should anticipate where they will find enthusiastic acceptance of the change plan and where they may encounter some resistance. Although it is desirable to gain acceptance throughout the organization, it is essential to have support during the initial change stages.

It is also important to adapt the program to fit the company. Although there are many elements of the program that may be generic to most companies, there are unique characteristics in most companies that will be positive factors if correctly incorporated into the improvement program. On the other hand, these same characteristics may become negative factors if ignored or disturbed.

As part of the program plan, project managers should decide what changes will be required in the structure of the company — the organization, policies, processes, and especially the effect on job contents and employee responsibilities and relationships. They can then decide how best to present these required changes to the rest of the organization.

As the planning progresses, it will become obvious that the program to adopt a lean and agile supply chain consists of a variety of activities and events. In other words, it is a project, and needs a project planning approach. As with most improvement programs, the implementation efforts run parallel with the regular sustaining operations of the business. Unless there is a way of monitoring progress, the program will likely falter or even lose its identify as an improvement effort.

Configuration

This phase involves the physical changes necessary in the marketing, purchasing, manufacturing, distribution, finance, and top management processes. Although these changes may be relatively straightforward, they can cost significant amounts of money and often take substantial amounts of time. Many of the changes involve physical or process changes; however, in all cases, the employees involved have to change their operating paradigms. We will discuss this aspect later in the Culture section. In the sections below, we list a number of the changes for each of the major functions of a business.

Marketing

Marketing will assume greater responsibility for retaining customers. The current thinking, represented by customer relationship management (CRM) programs, is that it is more beneficial to a company to cultivate life-long relationships with customers than it is to keep trying to find new customers. They will have to change from a "make and sell" paradigm to a "sense and respond" paradigm.[39] The supply chain basics for the marketing function include the following:

- Build databases to capture more data about existing customers.
- Analyze existing customers to determine why they buy the products and services they do.
- Develop programs designed to retain the customer.
- Obtain feedback from customers for use in the design of products and services (sense and respond).
- Participate in cross-functional teams to develop customer service programs.

To accomplish these and other changes, the marketing department will need to develop a structure that is flexible and operates with employees that endorse and participate in this new way of thinking.

Purchasing

Purchased goods and services represent a major portion of a company's expenditures. This cost category can range from 50 to 70 percent of sales in manufacturing industries and even higher in distribution and retailing industries. Consequently, the purchasing function will assume greater responsibilities in the future. The role of purchasing is changing from processing quotations from a wide variety of vendors to find the lowest price, to building relationships with suppliers. The supply chain basics for the purchasing function are listed next:

- Reduce the number of suppliers and expand the expectations for the performance of every supplier.
- Expand the areas of responsibility to include purchasing services as well as materials.
- Become a facilitator to help other functional areas develop relationships with suppliers.
- Participate in a cross-functional team to develop effective outsourcing programs.
- Become involved in the strategic planning for the company.

This transition will require a change in buyer attitudes and an upgrade in the skills required of the purchasing employees. Purchasing will become a professional-level job.

Manufacturing

Manufacturing will need to continue to make the transition from a progressive bundle (batch) system to a modular (flow) system.[40] A shift in focus from equipment and labor efficiencies to improving customer service will be needed. The supply chain basics for the manufacturing function are listed next:

- Re-arrange plant layouts to improve the flow of materials.
- Increase worker versatility and empowerment.
- Increase equipment flexibility, such as in reducing setup times.
- Reduce process variation in output and quality.
- Change production performance measures to reflect the new environment.

To facilitate the transition from internal efficiencies to improving customer service, the manufacturing function will need to work more closely with purchasing and marketing to co-ordinate customer orders in reducing response times.

Distribution

The distribution function, which includes the movement of goods from the manufacturer to the distributor, then on to retailer, and from there to the consumer, will also be increasingly concerned with the speed and timing of deliveries. Businesses will be especially concerned with having the right goods and services available to the customer. At the same time, they want to reduce inventories, especially excess and slow-moving inventories. The supply chain basics for the distribution function include the following:

- Retailers and E-commerce sellers will share point-of-sale information with their upstream suppliers.
- The distribution function will increase its capabilities to customize products and services.
- The distribution function will use a greater variety of transportation methods — truck, train, ship, and air — depending on the cost versus response-time balance.
- Distribution costs will receive closer attention in the total cost of ownership analyses, especially concerning offshore outsourcing.
- The distribution function will be increasingly outsourced to private carriers, such as UPS, FedEx, and DHL.

Distribution costs have become a significant portion of the total costs of products and services. The distribution function is also becoming a major factor in customer service considerations. It will become an even more important part of the supply chain in the future as the timing and traceability of deliveries becomes more critical.

Finance and Accounting

The financial function, including accounting, has important roles in improving flow in the supply chain. They must provide the funds to make the physical and structural changes and monitor the results to assure effective utilization of the organization's resources. The supply chain basics for the finance and accounting functions are listed next:

- Help to develop a total cost of ownership analysis, especially for use in evaluating offshore outsourcing.
- Modify reporting systems to provide more meaningful information to operating functions.
- Change the performance measures to evaluate supply chain outcomes, not just local outcomes.
- Work with supply chain partners to achieve an equitable distribution of costs and profits.
- Balance their efforts between external reporting and internal reporting requirements.

Finance and accounting must become participants in the company's efforts to build their supply chains, not just perform the functions of interested bystanders.

Top Management

Top management must become an active participant in building the supply chain. There are many decisions, both strategic and operational, that require their support. The supply chain basics for top management include the following:

- Propose the structural organizational changes that will be needed along with the accompanying responsibilities.
- Analyze the decisions needed to move actively into offshore outsourcing.
- Initiate contacts with other top managers to facilitate developing supply chain partners.
- Propose agreements on changes in performance measures and compensation incentives.
- Serve in the role as the visible motivator of change.

In many companies, the transition to agile and integrated supply chains is a new and complex transition; top management must lead the way.

The configuration change is one of the most visible changes that a company makes in moving toward supply chains. There are many articles written about it because it is often one of the most analyzed and debated areas of change. Although it is a necessary step, it is not sufficient in itself. More is required. Many businesses work to make their chains faster or more cost effective, assuming that those steps are the keys to competitive advantage; however, companies must also build supply chains that are agile, adaptable, and aligned. Great companies create supply chains that adapt when markets or strategies change.[41]

Communication

Communication is the informal and electronic network to communicate between entities, both internal and external. It is technology-driven but people dependent; fraught with a lack of incompatible or local standards; involves various modes — telephone, fax, E-mail, intranet, EDI, Internet; and, above all, is continually evolving.

This phase involves designing and implementing the data exchange systems needed to achieve the instantaneous transfer of data, both within the company and among the entities involved in the supply chain. This network can be expensive and time consuming; in addition, management must be willing to exchange confidential information.

It is insufficient to make the physical changes in the supply chain without upgrades in the communications systems, both within the company and between the company and its suppliers and customers. This may require new technology in the form of hardware and software to enable rapid and reliable communication between employees in a department, between departments within a company, and between companies.

The implementation of new systems can be both expensive and time consuming. In addition to the technology, the employees who will use these new systems require time to accept and learn how to effectively utilize the new technology. A failure to get the employees successfully involved can often lead to poor results, or even failure of the entire effort.

As difficult as it may be to implement the communications system within a company, it is even more difficult to implement inter-company communications. Although communications technology is readily available in the form of EDI or the Internet, the reluctance of companies to adapt between companies often prevents a high level of success. Darren Cooper and Michael Tracey believe that "the strategic aspects of their selection and implementation have for the most part been neglected until recently, and much of the literature fails to address the human and process factors involved in deploying this technology."[42]

IOSs are not a dream. They exist, although not all company relationships have reached the collaboration stage. That will emerge as the technology, infrastructures, and relationships of companies come into alignment.

Culture

Companies often ignore or de-emphasize the need to match the changes necessary with the existing culture of the organization. However, it is often the key to successful implementation and, if ignored, may limit success. It may not cost a lot of money to achieve; however, it requires an understanding of human relations and a willingness and ability to effect major changes in the thinking of individuals and the actions of teams.

Making the Configuration and Communication changes requires a high level of acceptance on the part of employees. In fact, the change from batch to lean requires a major cultural shift in the company. For many in the organization, it may signal the end of an era. No more mass production or local optimization thinking; it is time to shift paradigms from the traditional to the contemporary.

The program to change or adapt the culture of the organization to the program of building supply chains often lacks a systematic approach that can be done in a series of sequential steps. How best to blend or adapt the program to the culture or modify the culture to the program takes insight that involves an understanding of human nature. Failing to account for the impact changes have on the employees can stall or even derail the most desirable of changes.

When the *Harvard Business Review* convened a panel of leading thinkers in the field of supply chain management, technology was not the top concern. People and relationships were the dominant issues of the day. The opportunities and problems created by globalization, for example, are requiring companies to establish relationships with new types of suppliers. The ever-present pressure for speed and cost containment is making it even more important to break down stubbornly high internal barriers and establish more effective cross-functional relationships. The costs of failure have never been higher. The leading supply chain performers are applying new technology, new innovations, and process thinking to far greater advantage than the companies that are lagging. This roundtable gathered many of the leading thinkers and doers in the field of supply chain management. Together, they took a wide-ranging view of such topics as developing talent, the role of the chief executive, and the latest technologies, exploring both the tactical and the strategic in the current state of supply chain management.[43] Cultural differences make a difference in the "fit" among supply chain members. Although trade legislation may influence supply chain combinations, they must consider cultural differences as well.[44]

When successfully implemented, the changes described above will help a company approach a lean production status. However, is a lean company enough? Or will it be necessary to move to a lean and integrated supply chain?

Customization

Mass customization is an extension of agile manufacturing. Lean is being able to produce customized goods or services at optimal efficiency. Agile is being able to move from one product or service to another within the known product lines without noticeable interruption. Mass customization is the ability to make any of the known products at a high volume and to customize that product to a customer's individual specification, at a cost comparable to that of making a standard product.

Is the market-of-one imminent? Can businesses provide a unique product or service to each customer? Although it appears unlikely, a number of companies are working toward that objective. Where it is possible, companies are using modular components to facilitate postponement of the customizing process. They are moving from the "make to stock" (MTS) mode to the "assemble to order" mode. In time, they may be able to move back even further to the make to order mode if they can reach an acceptable compromise between level of customization and response time.

Another interim step is to provide greater variety when it is not possible to use the postponement strategy, such as in grocery items or small tools. Greater variety can be used to appeal to smaller market niches, but it is unlikely to approach the market of one. One of the negatives about greater variety is that it forces consumers to make more decisions about which item they want, a decision-making process that may add stress to their lives.[45] The answer lies in combining product modularity, process flexibility, customer relationships, and employee versatility into a dynamic system. All of these features of a business are necessary.

The Integrated Supply Chain

Everyone talks about an integrated supply chain, but how do you get there? In most cases, the transition passes through three distinct and often overlapping phases: interfacing, interlacing, and integrating. During the interfacing phase, two separate entities begin to pass information back and forth through dissimilar information processing systems. The movement of goods improves, but everyone realizes it could be better. The next move is interlacing, where changes connect the flow of information more closely and directly between the two entities. In the integrating phase, a level of connection is achieved where information flow is efficient and effective for both entities. Few companies have reached the final phase of integration. To achieve even the beginning levels of supply chain integration requires extending communications into cooperation and collaboration. Increasing competition stimulates independent firms to collaborate in a supply chain that allows them to gain mutual benefits. This requires the collective knowledge of the coordination mode, including the ability to synchronize interdependent processes, to integrate information systems, and to cope with distributed learning.[46]

Another study found that the degree of trust, power, continuity, and communication between supply chain partners would enhance commitment and, consequently, the integration of the SCM business process.[47]

Coordination

This means working together to achieve a mutual objective. It suggests that companies begin to understand what their suppliers or customers want and need, and what each participant should do to help achieve the objectives. It can be as simple as sending an order acknowledgment to the customer, or as involved as meeting a one-hour delivery window imposed by the customer.

Coordination provides the infrastructure and opportunities to make it possible to work together. It includes working in an "open-system" environment, and linking entities together both inside and outside the company. It involves policies, procedures, contracts, and other means necessary to work together and is among the early attempts to align the supply chain.

Collaboration

Collaboration implies a greater level of interconnectivity between customer and supplier. Collaboration involves freely sharing information and working together to develop a more meaningful plan for the flow of both information and goods through the supply chain. Several programs enhance collaboration, most recently collaboration, planning, forecasting, and replenishment (CPFR), a program that attempts to integrate the entire stock replenishment process.

Collaboration includes the motivation to work together effectively for the welfare of the entire supply chain. It builds on the coordination infrastructure and requires less tangible qualities such as trust (see Managerial Comment 10-1). It maximizes the benefits to the supply chain, not the individual entity, and provides the ultimate service to the customer.

Managerial Comment 10-1.
Trust

Robert Handfield and Christian Bechtel report that one of the most misunderstood concepts and a ripe area for research in SCM is in the area of trust. Trust (and its cousin, collaboration) seems to be the single most discussed element in making supply chains function effectively and efficiently. They found that four bodies of theory (transaction cost economics, organizational design, relational theory, and network theory) support the premise that effective communication of requirements with appropriate safeguards and approaches is critical to effective customer and supplier relationship management.

They conclude that, although trust is generally considered critical to successful SCM, it is one of the most difficult attributes to measure. There are eight different conceptual paradigms of trust: reliability, competence, goodwill (openness), goodwill (benevolence), vulnerability, loyalty, multiple forms of trust, combining trust with vulnerability, and nonpartisan proactive-based trust. When it is all said and done, integrity in action and thinking seem to be the foundations of trust between companies. Although much of the research to date has focused on analytical approaches to managing supply chains, the area that requires the greatest work is managing supply chain relationships.

Source: Adapted from Handfield, R.B. and Bechtel, C., Trust, power, dependence, and economics: Can SCM research borrow paradigms? *International Journal of Integrated Supply Management*, 1, 1, 3, 2004.[48]

Although coordination and collaboration are necessary, they are difficult to achieve. Internal company objectives may inhibit the attainment of supply chain objectives. The barriers include economic, technology, and cultural matters. Some things change slowly, especially those that may have an unknown effect on the jobs of employees at all levels of the organization.

Once the framework of a lean supply chain has been established, companies must move from a program mode to an institutionalized version in which the key features of the program become part of good management. Specific programs can be sustained for a limited period; however, at some point, the main elements of the program must be assimilated into the management practice of the company, or become lost as the program declines into ineffectiveness or is terminated by management.

Steps in the Change Process

Some of these action steps will be described in the following section.

Investigate

Although it is easy to be carried away with enthusiasm about the possibilities of the change, the first need is to investigate why the change would be beneficial. A company should go through the process of deciding what it wants to accomplish from the change. If the company is satisfied that the change to lean production will provide the desired results, then a commitment can be made to going through with the change.

Initiate

Once management has made the commitment to change, they have to initiate action. But action without a plan is premature. At this point, they need to develop the concept that will carry the program through to a successful conclusion.

Invigorate

Now is the time to get the rest of the organization involved. The participation of the management team and the nonmanagerial employees is required. At this stage, management should anticipate where it will find enthusiastic acceptance of the change plan and where it may encounter resistance. Although acceptance throughout the organization is necessary, it is critical to have support when the program is in the critical beginning stages.

Implement

The actual implementation stage should follow three main areas. First, the configuration of the physical elements of the production or service processes needs to be addressed. Second, communication within the organization and with suppliers and customers needs to be established in which the details of the program are outlined. Finally, the organizational culture of the firm must be changed.

Integrate

Integration comes slowly. It begins with communications and gradually progresses through coordination, cooperation, and collaboration. One of the primary requirements is trust among supply chain partners and trust comes from a series of successful experiences in working together. It begins at the individual level and gradually builds to a collective trust relationship between entities. Integration comes as much from human relationships as from technology and structure changes.

Institutionalize

Specific programs can only be sustained for a limited period; at some point, the main elements of the program must be assimilated into the management practice of the company, or they will be lost as the program declines into ineffectiveness or is terminated by management. The goal, of course, is to make the program part of the normal operating life of the organization. In other words, as the novelty of the new supply chain configuration gives way to familiarity and effectiveness, it gradually becomes institutionalized into the business.

Innovate

Once a company builds a facsimile of an integrated supply chain, it is time to change it. The business world is constantly changing and there is a need to adjust the supply chains to keep pace. We will describe this further in the following section.

A Look Ahead

Suppose that lean production is achieved and then extended to a lean supply chain. Is that the final answer, or should the company move on to agile manufacturing?

Agile manufacturing is first lean, and then agile. It is an extension of lean manufacturing, and viewed by some as a competitive asset for the future. The three inter-related support areas for agile manufacturing are technology, organization, and people. Each of these areas was described more fully in Chapter 7, Chapter 8, and Chapter 9. The following quotes point out the need for collaboration among supply chain members.

> You can be lean by yourself, but you cannot be agile by yourself.[49]

> Virtual companies are "the epitome of the agile enterprise" — a vehicle to tap the knowledge of participating companies.[50]

> In agile manufacturing, the aim is to combine the organization, people and technology into an integrated and coordinated whole.[51]

Is agile manufacturing the final answer? No, there may be a need to move even further, toward mass customization and complexity management.

Up to now, we have described actions that are difficult but reasonably straightforward. In the next two phases, we believe that companies will have to demonstrate significant innovation skills.

Complexity

Managing the complexity of a business in today's global competitive environment is challenging. It is a test for management even greater than achieving mass customization. It goes beyond customizing a known product for an individual customer. It involves dealing with all of the internal and external issues that may exist, at the same time making the product at a competitive cost and lead time. Operating in a complex, or chaotic, environment tests the capabilities of any company.

In Chapter 3, we described open systems and the need for organizations to deal with the complexities of the forces that are external to the organization's control. These forces, especially those resulting from global competition, require that businesses develop global strategies affecting product design (products must be designed to fit the market they serve), processes (to reflect the best blend of resource costs, work rules, and local preferences), and locations (how to blend a myriad of decision criteria in the face of potentially dramatic change).

Complexity is difficult and, even worse, introduces some paradoxes that increase the confusion. Some major paradoxes of complexity include:

- The optimization paradox. Despite the potentially huge economies from designing supply chains from a global view, most manufacturers optimize locally.
- The customer collaboration paradox. Despite the need to be much more responsive to customers, few manufacturers are collaborating closely with them.
- The innovation paradox. Product innovation is continuing to accelerate, yet few manufacturers are preparing their supply chains for faster new product introduction.
- The flexibility paradox. Flexibility is a key priority, but it is being sacrificed in the drive to cut unit cost.
- The risk paradox. Keeping supply chain quality high is critical, yet manufacturers' risk of supply chain failures keeps growing.[52]

Traditional decision making relied on linear thinking — the future will be largely like the past. Today, however, many businesses are finding that using the past as a basis for planning the future may not be very useful; in fact, it may be downright misleading. But how can organizations apply nonlinear thinking? It takes a completely new theory, a theory appropriately called complexity theory. William Frederick describes complexity theory as a variant of chaos theory that explains the organizational and evolutionary dynamics that occur as complex living systems interact with each other and their environments. He suggests that both the corporation and community are natural systems interacting and coevolving in response to environmental factors. The corporation is hypothesized as a complex system that must adapt itself to the environment in which it operates.[53]

Dealing with complexity will be a necessary core competency for any successful organization. Frederick poses the following question: "Is the corporation a self-organized complex adaptive system (CAS) housing an autocatalytic component, operating on a fitness landscape, and exposed to the risk of chaotic change while being held in its niche by a strange attractor?"[54]

Many companies will probably find complexity, or its variant, chaos theory, as too "far out" to be considered seriously. However, a few companies are already trying to decide how to incorporate this theory in their thinking. A full explanation is beyond the scope of this book; we can only suggest that complexity theory will be an important consideration for the future.

Clairvoyance

Clairvoyance is the power to perceive things that are out of the natural range of human senses. Few articles or books combine clairvoyance and business, presumably because businesses align with rationality and logic rather than with a discipline that has been associated with frauds and horoscopes. Up until now, businesses have been dealing with the known world, even though it may be complex and unexpected. In the future, they will have to deal with greater uncertainty. The present forecasting techniques will be increasingly inadequate. Current methods rely heavily on extension, usually linear, from present known positions. At some point, organizations will have to build bridges between disconnected islands of products or services. In other words, they will have to be clairvoyant. Organizations will need people who have the "gift" to foretell the future, and that is a skill not taught in business schools.

Perhaps it would be more palatable to say that businesses need people who possess the wisdom to assimilate fragments of leading indicator information into a holistic and reasonably accurate view of the future. Wisdom is the distillation and organization of knowledge in such a way that it has relevance to the decision at hand. Some individuals seem to possess an uncanny ability to make correct decisions while others struggle in a hopeless maze of contradictory information. For the moment, we will call that ability "clairvoyance." It raises several questions. What does it take to see into the future? Is it a skill that is inherited — a sixth sense — or can it be learned? If so, how? Is clairvoyance something like "thinking outside the box"? Does it result from empowered cross-functional teams? Can it be created through the use of biotechnology? Can it be directed and controlled or is it the result of random insights? We touched on knowledge management in earlier chapters and will address it again in Chapter 12.

Just as with chaos and complexity theory, the detailed discussion of clairvoyance is beyond the scope of this book. We can suggest books that we believe make what we have called clairvoyance a bit more tangible and realistic. In *Blink*, Malcolm Gladwell describes how some people have the knack of making decisions without the need for exhaustive analyses; they see the big picture and just intuitively know! In *A Whole New Mind*, Daniel Pink describes the emergence of "right-brain" thinking as coming of age in the world. Henry Mintzberg also discussed left and right brain implications (see Managerial Comment 10-2). These ideas represent significantly different ways of making decisions from those approaches that have been practiced for a long time in the business world. Although practitioners of these techniques may not be clairvoyants, they will be advanced beyond most of today's managers.

Managerial Comment 10-2.
Implications for the Left Hemisphere:
Left-Brain Thinking

First, I would not like to suggest that planners and management scientists pack up their bags of techniques and leave organizations, or that they take up basket-weaving or meditation in their spare time. (I haven't — at least not yet!) It seems to me that the left hemisphere is alive and well; the analytic community is firmly established, and indispensable, at the operating and middle levels of most organizations. Its real problems occur at the senior levels. Here analysis must coexist with — perhaps even take its lead from — intuition, a fact that many analysts and planners have been slow to accept. To my mind, organizational effectiveness does not lie in that narrow-minded concept called "rationality"; it lies in a blend of clear-headed logic and powerful intuition.

A major thrust of development in our organizations, ever since Frederick Taylor began experimenting in factories late in the last century, has been to shift activities out of the realm of intuition, toward conscious analysis. That trend will continue. But managers, and those who work with them, need to be careful to distinguish that which is best handled analytically from that which must remain in the realm of intuition. That is where we shall have to continue looking for the lost keys to management.

Source: **Mintzberg, H.,** *Mintzberg on Management, Inside Our Strange World of Organizations,* **The Free Press, 1989.**[55]

Conclusions

In an earlier section, we described the "Cs" of change to move from batch flow to a lean and agile supply chain. A few progressive and venturesome companies are well along on their journey. Some have started while others are still thinking about it.

We suggested there are additional "Cs" to be traveled. Some have been discovered and many companies are well beyond the introductory stage, such as with mass customization and partner collaboration. Others, such as complexity management, hold promise, but are not well defined or understood. Some, such as clairvoyance, are still in the untapped potential stage.

We close this chapter with these thoughts:

- Companies understand batch, but is it not time to change?
- From batch to lean is just the beginning.
- Agile and integrated supply chains may be better, but they are difficult to implement.
- Virtual configurations may be best, but it is still a goal of the future.
- Technology is an enabler, but it requires organization and people to make it happen.
- Not to change is an unrealistic alternative.

What do others think about agile and integrated supply chains? Here is a sampling of their thoughts. To obtain the current ideal of SCM practices:

- Collaboration will be the most strategic capability in the extended supply chains.
- Service and support will become as important as the product itself.
- Companies will improve their service capabilities to adapt in turbulent environments.
- Assets and functions not at the core of value delivery are to be divested.
- The success in turbulent business environments depends on delivering new products and services at a faster rate and transforming value chains into customer-focused virtual networks.[56]

In this chapter, we described the myriad of items that companies must accomplish to become participants in an agile and integrated supply chain. In Chapter 11, we will describe a host of services that will be required to support the supply chains.

References

1. Davis, E.W. and Spekman, R.E., *The Extended Enterprise: Gaining Competitive Advantage through Collaborative Supply Chains,* Prentice Hall, Upper Saddle River, NJ, 2004.
2. Poirier, C.C. and Quinn, F.J., Survey of supply chain progress: Still waiting for the breakthrough, *Supply Chain Management Review,* 10, 8, 18, 2006.
3. Poirier, C.C. and Quinn, F.J., How are we doing? A survey of supply chain progress, *Supply Chain Management Review,* 8, 8, 24, 2004.
4. Davis, E.W. and Spekman, R.E., *The Extended Enterprise: Gaining Competitive Advantage through Collaborative Supply Chains,* Prentice Hall, Upper Saddle River, NJ, 2004.
5. Manrodt, K.B., Abott, J., and Vitasek, K., Understanding the Lean Supply Chain: Beginning the Journey, 2005 Report on Lean Practices in the Supply Chain, APICS.
6. Walker, W.T., *Supply Chain Architecture: A Blueprint for Networking the Flow of Material, Information, and Cash,* CRC Press, Boca Raton, FL, 2005.
7. Taylor, F.W., The Principles of Scientific Management, Harper & Row, New York, 1911.
8. Blackstone, J.H. and Cox, J.F., III, *APICS Dictionary* (Eleventh Edition), APICS — The Educational Society for Resource Management, Falls Church, VA, 2005.
9. Manrodt, K.B., Abott, J., and Vitasek, K., Understanding the Lean Supply Chain: Beginning the Journey, 2005 Report on Lean Practices in the Supply Chain, APICS.
10. Lee, H.L., The Triple-A supply chain, *Harvard Business Review,* 82, 10, 102, 2004.
11. Heckmann, P., Shorten, D., and Engel, H., *Supply Chain Management at 21: The Hard Road to Adulthood,* Booz Allen Hamilton Inc., New York, 2003.
12. Kotzab, H. and Otto, A., General process-oriented management principles to manage supply chains: Theoretical identification and discussion, *Business Process Management Journal,* 10, 3, 336, 2004.
13. Haeckel, S.H., *Adaptive Enterprise: Creating and Leading Sense-and-Respond Organizations,* Harvard Business School, Boston, 1999.
14. Kidd, J., Richter, F.J., and Li, X., Learning and trust in supply chain management, *Management Decision,* 41, 7, 603, 2003.
15. Manrodt, K.B., Abott, J., and Vitasek, K., Understanding the Lean Supply Chain: Beginning the Journey, 2005 Report on Lean Practices in the Supply Chain, APICS.
16. Poirier, C.C. and Quinn, F.J., Survey of supply chain progress: Still waiting for the breakthrough, *Supply Chain Management Review,* 10, 8, 18, 2004.
17. Krikke, H., LeBlanc, L., and van de Velde, S., Product modularity and the design of closed-loop supply chains, *California Management Review,* 46, 2, 23, 2004.
18. Anonymous, The challenge of complexity in global manufacturing, Special report, Deloitte Touche Tohmatsu, 2003.
19. Friedman, T.L., *The World Is Flat: A Brief History of the Twenty-First Century,* Farrar, Straus and Giroux, New York, 2005.
20. Beth, S., Burt, D.N., Copacino, W., and Gobal, C., Supply chain challenges: Building relationships, *Harvard Business Review,* 81, 7, 64, 2003.
21. PRTM, Supply Chain Maturity Model, www.prtm.com/services/supply_chain_maturity_model.asp, 2005.

22. Cigolini, R., Cozzi, M., and Perona, M., A new framework for supply chain management: Conceptual model and empirical test, *International Journal of Operations & Production Management*, 24, 1/2, 7, 2004.
23. Simatupang, T.M, Wright, A.C., and Sridharan, R., The knowledge of coordination for supply chain integration, *Business Process Management Journal*, 8, 3, 289, 2002.
24. Kempainen, K. and Vepsalainen, A.P.J., Trends in industrial supply chains and networks, *International Journal of Physical Distribution & Logistics Management*, 33, 8, 701, 2003.
25. Kannan, V.R. and Tan, K.C., Attitudes of US and European managers to supplier selection and assessment and implications for business performance, *Benchmarking*, 10, 5, 472, 2003.
26. Wong, C.Y., Hvolby, H.H., and Johanses, J., Why use loosely coupled supply chains? Report from the Center of Industrial Production, Aalborg University, Denmark, 2004.
27. Poirier, C.C. and Quinn, F.J., Survey of supply chain progress: Still waiting for the breakthrough, *Supply Chain Management Review*, 10, 8, 18, 2006.
28. Handfield, R.B. and Bechtel, C., Trust, power, dependence, and economics: Can SCM research borrow paradigms? *International Journal of Integrated Supply Management*, 1, 1, 3, 2004.
29. Crandall, R.E., Dream or reality? Achieving lean and agile integrated supply chains, *APICS Magazine*, 15, 10, 20, 2005.
30. Ghosh, S., Thorton, J., DeHondt, G., and Faley, R.H., The paradox of overinvestment in enterprise integration, *Align Journal*, 1, 2, 40, 2007.
31. Cigolini, R., Cozzi, M., and Perona, M., A new framework for supply chain management: Conceptual model and empirical test, *International Journal of Operations & Production Management*, 24, 1/2, 7, 2004.
32. Lambert, D.M., Garcia-Dastugue, S.J., and Croxton, K., An evaluation of process-oriented supply chain management, *Journal of Business Logistics*, 26, 1, 25, 2005.
33. Lambert, D.M., Garcia-Dastugue, S.J., and Croxton, K., An evaluation of process-oriented supply chain management, *Journal of Business Logistics*, 26, 1, 25, 2005.
34. Lambert, D.M., Garcia-Dastugue, S.J., and Croxton, K., An evaluation of process-oriented supply chain management, *Journal of Business Logistics*, 26, 1, 25, 2005.
35. PRTM, Supply Chain Maturity Model, www.prtm.com/services/supply_chain_maturity_model.asp, 2005.
36. Poirier, C.C. and Quinn, F.J., How are we doing? A survey of supply chain progress, *Supply Chain Management Review*, 8, 8, 24, 2004.
37. Booz Allen Hamilton, Senior execs: SCM hasn't delivered, *Industrial Engineer*, 35, 9, 12, 2003.
38. Heckmann, P., Shorten, D., and Engel, H., *Supply Chain Management at 21: The Hard Road to Adulthood*, Booz Allen Hamilton Inc., New York, 2003.
39. Haeckel, S.H., *Adaptive Enterprise: Creating and Leading Sense-and-Respond Organizations*, Harvard Business School, Boston, 1999.
40. Abernathy, F.H., Dunlop, J.T., Hammond, J.H., and Weil, D., *A Stitch in Time: Lean Retailing and the Transformation of Manufacturing — Lessons from the Apparel and Textile Industries*, Oxford University Press, New York, 1999.
41. Lee, H.L., The Triple-A supply chain, *Harvard Business Review*, 82, 10, 102, 2004.

42. Cooper, D.P. and Tracey, M., Supply chain integration via information technology: Strategic implications and future trends, *International Journal of Integrated Supply Management*, 1, 3, 237, 2005.
43. Beth, S., Burt, D.N., Copacino, W., and Gobal, C., Supply chain challenges: Building relationships, *Harvard Business Review*, 81, 7, 64, 2003.
44. Kidd, J., Richter, F.J., and Li, X., Learning and trust in supply chain management, *Management Decision*, 41, 7, 603, 2003.
45. Schwartz, B., *The Paradox of Choice, Why More Is Less*, Harper Collins, New York, 2004.
46. Simatupang, T.M, Wright, A.C., and Sridharan, R., The knowledge of coordination for supply chain integration, *Business Process Management Journal*, 8, 3, 289, 2002.
47. Wu, W.Y., Chaing, C.Y., Wu, Y.J., and Tu, H.J., The influencing factors of commitment and business integration on supply chain management, *Industrial Management + Data Systems*, 104, 3/4, 322, 2004
48. Handfield, R.B. and Bechtel, C., Trust, power, dependence, and economics: Can SCM research borrow paradigms? *International Journal of Integrated Supply Management*, 1, 1, 3, 2004.
49. Sheridan, J.H., Agile manufacturing: Stepping beyond lean production, *Industry Week*, 242, 8, 30, 1993.
50. Sheridan, J.H., Agile manufacturing: Stepping beyond lean production, *Industry Week*, 242, 8, 30, 1993.
51. Kidd, P.T., *Agile Manufacturing: Forging New Frontiers*, Addison-Wesley, Reading, MA, 1994
52. Anonymous, The challenge of complexity in global manufacturing, Special report, Deloitte Touche Tohmatsu, 2003.
53. Frederick, W.C., Creatures, corporations, communities, chaos, complexity, *Business and Society*, 37, 3, 358, 1998.
54. Frederick, W.C., Creatures, corporations, communities, chaos, complexity, *Business and Society*, 37, 3, 358, 1998.
55. Mintzberg, H., *Mintzberg on Management, Inside Our Strange World of Organizations*, The Free Press, 1989.
56. Kempainen, K. and Vepsalainen, A.P.J., Trends in industrial supply chains and networks, *International Journal of Physical Distribution & Logistics Management*, 33, 8, 701, 2003.

Chapter 11

The Role of Services to Complement the Supply Chain

John was again dreaming, only this time, he was on a large stage, so he walked over to some people who were working frantically on PCs. As before, John asked the first one, "What are you doing?" Without looking up, the man replied, "We are designing services to go with the agile and integrated supply chain. It's not good enough just to send a product to the other side; we have to send along pre- and post-sales services to supplement the product. We also have to design services to help the manufacturing companies build their products, such as product development assistance, advertising, auditing, custodial services, food services, employee training, instructional manuals, and a host of other services that are essential to any company's success. We even provide services to other service firms, such as legal aid, health care and fitness programs, and insights into the future. In some cases, the services are more important than the product itself. We design services for the consumer also, such as tourism, restaurants, and theme parks."

As John looked at the boxes being loaded on the supply chain, he could see that people were loading Product A with Service A, or Product B with Service B. One product, labeled D, had services A, B, and C attached with it. There were many other combinations of products and services. Who could keep up with all of the combinations? Would the customers ever be satisfied? Or were they really asking for all of these customized packages? Would they be happy with fewer choices?

Not immediately coming up with an answer, John asked a woman at another PC. She replied, "Some customers ask that the products and services be customized for them. Usually they have to wait longer to get their package. If they are in a hurry, we make a large variety of packages that we keep in stock. I think what we would like to do is to be able to 'mass customize' a package to fit each customer, but that is something that is still in the future, although there are a few companies that are coming close to being able to do that.

"If you are interested in learning more about what we do, or if you would like us to help you, take this pamphlet 'Linking Supply Chains to Services'." John dutifully tucked the pamphlet in alongside the others he had collected. With those comments, she returned to her PC.

Introduction

In Chapter 10, we discussed the evolution of localized processes into integrated and agile supply chains. We showed that the final version of the supply chain is a combination of all types of industries. Figure 11-1 portrays the relationships along the supply chain.

In this chapter, we will describe two additional complementary types of industries — producer services and social services. Although these industries are not a direct part of the supply chain, they are complementary and essential to the success of the supply chain. Figure 11-2 shows their relationship to the basic supply chain.

We will also describe the area designated in Figure 11-2 as consumer, or personal, services. They fit in this chapter because they are complementary to the supply chain and, although some occur during the supply chain operation, many occur after the direct sale of the product or service to the consumer. Examples include the service or repair of automobiles, computer help centers, and the processing of extended credit transactions.

Figure 11-1   Model of basic supply chain.

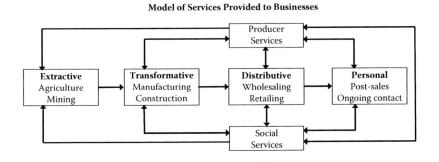

Figure 11-2 Industry classifications.

Producer Services versus Consumer Services

The North American Industry Classification System (NAICS) classification is useful. However, it concentrates primarily on the function of the business and does not distinguish among the types of customers served or the type of institution involved in the distribution of the service. Customers fall into two categories: other producers of goods or services (businesses) and individual consumers.

Gershuny and Miles offer an extended classification, which they credit to Joachim Singelmann and which is reproduced in Table 11-1, that shows agriculture and mining as extractive industries, and construction and manufacturing as transformative industries. The service industries are grouped into four sectors: distributive, producer, social, and personal.[1]

This distinction helps to identify the strong relationship that exists between producer service businesses and the goods-producing industries. If a manufacturing firm goes out of business, that firm will no longer need such services as group insurance, commercial loans, and consulting services. On the other hand, if a service company has been buying computers, office equipment and supplies, and delivery trucks to conduct their business, the goods-producing side of the economy has apparently benefited due to the added sales.

Figure 11-2 shows that the producer and social services provide services to all four of the industry groups along the supply chain – extractive, transformative, distributive, and personal. It also shows that both producer and social service sectors receive outputs from the transformative, distributive services and personal services sectors. There are interdependencies among all of the sectors. Table 11-1 lists the major industries found within each of the sectors shown in Figure 11-2.

Figure 11-3 shows the relationships among the different classifications provided in Table 11-1. Extractive industries provide output primarily to the transformative industries; however, they receive participation from both the producer services group and the social services group. A copper mining company supplies copper ore

Table 11-1 An Extended Sectoral Classification

I. Extractive
(1) Agriculture, fishing, and forestry
(2) Mining
II Transformative
(3) Construction
(4) Food
(5) Textile
(6) Metal
(7) Machinery
(8) Chemical
(9) Miscellaneous manufacturing
(10) Utilities
III. Distributive Services
(11) Transportation and storage
(12) Communication
(13) Wholesale trade
(14) Retail trade (except eating and drinking places)
IV. Producer Services
(15) Banking, credit, and other financial services
(16) Insurance
(17) Real estate
(18) Engineering and architectural services
(19) Accounting and bookkeeping
(20) Miscellaneous business services
(21) Legal services
V. Social Services
(22) Medical and health services
(23) Hospitals
(24) Education
(25) Welfare and religious services
(26) Nonprofit organizations
(27) Postal services
(28) Government
(29) Miscellaneous professional and social services

Table 11-1   An Extended Sectoral Classification (continued)
VI. Personal Services
(30) Domestic services
(31) Hotels and lodging places
(32) Eating and drinking places
(33) Repair services
(34) Laundry and dry cleaning
(35) Barber and beauty shops
(36) Entertainment and recreational services
(37) Miscellaneous personal services

Source: Reproduced from Gershuny, J. I. and Miles, I.D., *The New Service Economy, The Transformation of Employment in Industrial Societies,* Praeger Publishers, 1983.[1]

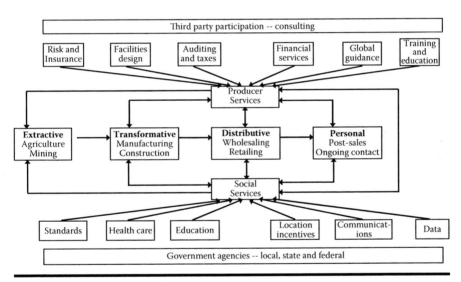

Figure 11-3   Services provided to businesses.

to a refining company and, in return, is a customer of an insurance company or bank (producer services) and the postal service (social services). The transformative industries receive inputs from the extractive industries, the producer service group, and the social services group; they provide output to all of the service groups primarily through the distributive services group. A computer manufacturer (transformative) receives sheet metal from extractive sources, manufactures computers, and sells them to an engineering firm (producer services); to universities (social services); and indirectly, through a distribution network, for example, to hotels that deal with the ultimate consumer.

The distributive services receive inputs from the transformative industries and both producer services and social services; in turn, they provide outputs to all three of the service groups. A wholesale distributor leases its delivery fleet from a truck rental company (producer services) and obtains employee hiring and training services from the local government (social services). It, in turn, distributes office supplies from manufacturers (transformative) to all types of service companies. Producer service industries provide their output to the value-adding industries of agriculture, mining, and manufacturing; in turn, they use inputs from the distributive services group. Banks (producer services) provide a variety of services to all other industries and receive output from all groups except extractive and personal services.

Social services also provide their output to all of the other groups and, in turn, use outputs from the distributive and producer service groups. A hospital (social services) receives services from the transformative and distributive groups and provides services to all other groups. The personal services group receives inputs from distributive services, producer services, and social services; they provide their outputs to the individual consumers. A restaurant (personal services) is primarily a receiver of goods from distributive services, and services from both producer and social service groups to provide output to the consumer.

What are Producer Services?

Much of the early research in service industries was concerned with consumer, or personal, services. Harry I. Greenfield prepared a classic monograph in 1966 entitled "Manpower and the Growth of Producer Services." This was the first attempt to segregate producer services from consumer services and to identify the economic impact of producer services in the general economy. Greenfield defined producer services as "those services which business firms, nonprofit institutions, and governments provide and usually sell to the producer rather than to the consumer."[2] The term "business services" also is used to denote "services provided by businesses for businesses."[3] Managerial Comment 11-1 describes the demand for producer services.

Managerial Comment 11-1. Determinants of Demand for Producer Services

The first determinant to be considered is the desire to produce the existing level of output at a lower cost. The usual explanation of the source of the saving is the greater productivity of the external labor force, which is at once more specialized and more closely supervised. Examples: General maintenance — custodial, laundry service, window cleaning, food service.

A second determinant of the demand for external services, closely related to the first, is the desire to increase the quantity or quality of the output produced by currently employed resources. Example: Use of a market research firm to supplement the internal staff.

A third motive for using external business services is the desire of a firm to eliminate what might be termed troublesome functions, such as staffing low status jobs, dealing with complex jurisdictional or other problems in union-management relations, or coping with irregular work schedules. Example: Use of construction or sub-contractor firms to handle nonroutine maintenance.

A fourth reason for the use of external services is the periodic need for specialized personnel that many small and medium-sized firms cannot afford to employ on a full-time basis. Example; Legal and tax services.

A fifth source of demand for external services arises from the need of companies to adjust to erratic labor requirements. Examples: Hiring part-time or temporary employees for peak loads, such as in taking physical inventory.

A sixth motive for using external services is the desire on the part of many firms to maintain a small, compact, relatively homogeneous labor force. Examples: Concentrate on the core competencies and outsource non-core functions, such as IT maintenance.

Another reason that can be identified as contributory to the increased use of external services is management's desire to enhance the firm's image by being in the forefront of technological or organizational innovations. Example: management consultants, especially IT

> Finally, external services are often called upon to provide guidance with respect to the firm's growth. Examples: economic and financial consultants, management consulting firms, market research firms, investment counselors, and other type of firms specializing in long-range planning.
>
> *Source:* Adapted from Greenfield, H.I., *Manpower and the Growth of Producer Services*, Columbia University Press, New York, 1966.[4]

Greenfield also provides a functional classification of producer services in Table 11-2.

Why Are Producer Services Important?

Producer services are important because they are essential to the operation of any business; yet most businesses are not able to perform all of the services for themselves. They have to depend on other businesses to provide the needed services.

Table 11-3 summarizes the reasons why businesses need outside producer services. The numbers indicate the ranking of the reasons. For example, all categories of the firms surveyed ranked the lack of technical expertise as the top reason for seeking outside producer services. The second most important reason, for all but the large firms, was the lack of size of the business. A consensus third most important reason was the perception that outside firms can perform better. Other reasons listed as a top-four choice by at least two companies included avoidance of risks for in-house service and increased government regulation. Finally, reasons given by at least one group included lack of financial resources to perform services and infrequent need for services. Of the 16 reasons to choose from, only seven were considered among the top four reasons to seek outside producer services.[5]

The Role of Outsourcing

The outsourcing of producer services is common; however, the motivation for outsourcing may vary, depending on the type of producer service. Some producer services, such as institutional food service or custodial service, can usually be outsourced for a lower cost and it reduces the complexity of the workforce and investment requirements. Other producer services, such as legal or architectural services, are outsourced because it would be difficult to maintain the expertise for these services in-house. Other services, such as operating the information technology

Table 11-2 A Functional Classification of Producer Services

Policy Making	
Current Problems	**Longer-Range Planning**
Banking, insurance, and real estate advisors	Business management consulting services
Legal problems	Economic and market research
Human resources and labor relations consultants	Research development laboratories
System analysts	Testing laboratories

Administrative	
Production-Related	**Nonproduction**
Armored car services	Advertising
Detectives, protective services	Bookkeeping, accounting
Equipment rentalFreight forwarding	Coin-operated machine rental
Photofinishing laboratories	Credit bureaus and collection agencies
Safety inspectors	Direct mail advertising
Services to dwellings and other buildings	Duplicating, copying, and stenographic services
Testing laboratories	News syndication
	Private employment agencies
	Routine data processing
	Telephone answering

Source: Adapted from Greenfield, H.I., *Manpower and the Growth of Producer Services*, Columbia University Press, New York, 1966.[4]

function, may be outsourced to acquire a higher level of quality and knowledge in a rapidly changing environment. Whatever the reason, the outsourcing of producer services appears to be increasing.

The Role of Producer Services

The primary role of producer services is to support the supply chain. It should perform a supportive role, but should not be so aggressive or dominant that it influences the strategic direction or effectiveness of the supply chain.

Some producer services involve continuous contact with the supplier. Call centers and banks provide this type of service. Some producer services require only occasional participation in a business. Executive search firms or public accounting

Table 11-3   Reasons Why Businesses Seek Outside Producer Services

Reasons (from perspective of outsourcer)	Large Businesses	Small Businesses	Proprietors	Urban Firms	Rural Firms
Lack of technical expertise	1	1	1	1	1
Size of client too small to produce such service	3	2	2	2	2
Lack financial resources to perform services			4		
Infrequent need for services					3
"Downsizing" or desire to be "lean and mean"					
Perception that outside firms can perform better	3	3	3	3	
Avoidance of risks for in-house service	2			4	
Increased government regulation		4			4
Need for an unbiased, third-party opinion					
Service is not related to a strategic function					
Changes in the nature of the business					
Increasing complexity of managing firm					
Increasing complexity in a global environment					
Need to stay abreast of changing technologies					
Need to stay abreast of economic changes					
Other reasons					

Source: Adapted from Beyers, W.B., Impacts of IT advances and E-commerce on transportation in producer services, Growth and Change, 34, 4, 433, 2003.[5]

audit teams represent this type of service. Even when the service is ongoing, producer services have many of the characteristics of a project and probably should be managed as a project to recognize that it is not a part of the internal operation of a business.

Classes of Producer Services

Producer services can fall into the same categories of functions that we described in Chapter 4 — sustaining operations, problem-solving activities, or continuous improvement programs. An example of sustaining operations is custodial or food services. A problem-solving type of producer service is the use of a crisis management consultant to help during a toy recall triggered by an infant death. A continuous improvement program can use management consultants to provide program direction and training of employees.

Reasons to Acquire Producer Services

In this section, we will look more closely at specific types of producer services and some of the reasons companies elect to acquire these services. We will not try to list all of the producer services activities, but rather will try to provide representative examples of each type.

Gain a Cost Advantage

One of the most obvious reasons to outsource producer services is to gain a cost advantage. If this can be accomplished without a loss of quality and without an increase in risk of causing disruption in normal business operations, it is a popular choice among most businesses.

- Call centers. One of the most publicized forms of producer services is that of call centers. This option is popular, yet controversial. Businesses can have the call center handled for less cost, but they also run the risk of irritating, or even outraging, customers who do not find the service at the call center satisfactory.
- IT programming. IT services, especially in the programming area, are an increasingly active area of outsourcing. Companies are taking advantage of electronic communications capabilities by using businesses in lower-labor cost areas, such as India, to perform routine IT services.

Remove Noncore Types of Activities

These activities are necessary, but are not usually similar to the primary functions of a business.

- Custodial services. Most office locations use a cleaning service to perform custodial services. This service can often be done at the same or even a lower cost, and it removes a dissimilar operation from the company, one that usually has differences in pay grades, employee turnover, special training requirements, and different inventory requirements. This service is usually carried out at night when the regular staff is not there. Even manufacturing operations find it is desirable to outsource the custodial services activity.
- Food service. There is an abundance of food service operations near most manufacturing and service facilities. Consequently, many businesses have abandoned their food service operations. Those that retained them do it more as an added benefit for employees than as a cost saving measure. Even where the food service is on site, it is more often an externally contracted service rather than an internal operation.

Supplement Internal Staffs with Added Expertise

Many businesses want to have representation in a particular activity. However, they do not want to provide capacity in those functions to handle all of the situations that may arise. In many cases, they want to depend on the suppliers of these services to possess the latest knowledge and external contacts necessary to do the complete job. The internal staff will have the general knowledge and the capability to know how best to utilize the external suppliers to provide the most appropriate service to the company.

- Advertising. Most small- to medium-size businesses do not have full-time staff devoted strictly to advertising. Although some may be building their own web pages, advertising through newspapers, magazines, and electronic media is usually handled through agencies that specialize in that form of service. Although large companies may have an internal advertising staff, they usually go through agencies to place their ads.
- Market research. Firms that specialize in market research have two main advantages over in-house research departments. One, they have a vast amount of information already available from their previous work. Second, they have the knowledge and skills to collect additional information through surveys or other research. They can usually perform a better job, and faster, than the in-house research departments.
- Employee selection. Selection of qualified candidates is an increasingly difficult task. It takes personnel who can verify the information on applications,

to screen out individuals who do not fit the position requirements. It takes a combination of hard facts and judgment to eliminate those with undesirable personal characteristics, without violating the rights of that individual. Although HR departments may be able to do this, they often call on employment agencies or search firms to handle the preliminary part of the employee selection process.
- Employee training. All employees, or potential employees, need some level of training. Some of this is conducted before the individual is hired. Learning to drive a tractor-trailer rig, learning to weld, or learning computer literacy, are examples of these skills. Service companies usually perform these basic skills. Once the selection has been made, some additional training may be necessary. Where there are sufficient numbers, or where the training is generic, a business will likely outsource this function.
- Banking. Most businesses, even small ones, have employees who are knowledgeable about the company's finances. However, all businesses have the need to develop a relationship with their bank(s). Businesses need to be able to borrow money, to finance working capital or capital equipment requirements. They cannot accomplish this on the spur of the moment; therefore, they need to establish their relationship with a bank before they need financial resources; and then to continue that relationship to assure they receive those resources and the services the bank can provide.
- Engineering testing. Most engineering departments have some testing capability. However, where specialized testing is required, they usually prefer to outsource this service. They may also use service companies to verify their own testing so that the product can be marketed with an approval rating from the testing company.
- Employee benefits. All human resource departments have some information about employee benefits. Often, they want to supplement their information with other data available from companies specializing in that area. Similar to market research consultants, employee benefits consultants can provide hard information and overall knowledge in this area that is usually superior to the internal HR department.
- Wellness programs. Health care costs are increasing rapidly. This trend is of great concern to businesses because providing health care coverage is a significant cost of conducting business and the potential of it becoming a mandatory obligation is even more threatening. Consequently, many businesses are moving to a preventive strategy of promoting wellness programs. The basic concept is that not only will the costs be lower, but also employees will be more productive and less prone to absenteeism.[6]

Provide Flexible Capacity or Avoid Overload on Key Departments

This reason is similar to the previous one when an internal staff supplements its own capacity with external suppliers. In this case, the reason is not to gain added expertise, but to provide capacity to handle demand variability. The variability may result in predictable demand patterns, such as end-of-fiscal-year activities, or in seasonal buildups or contractions in the number of employees required.

- Inventory-taking service. Internal employees could take the physical inventory and, in many businesses, do. However, a growing number of businesses are outsourcing the inventory-taking activity to external suppliers. In most cases, the external supplier has employees who are more highly trained and efficient in taking inventory. It also avoids the interruptions to the normal operation of a business when employees do not have to take time to do a job for which they lack skills and generally dislike.
- Reservation systems. For a long time, airlines have used travel agencies to handle a part of the ticket reservation workload. Recently, online reservations have become more popular, to the point where some travel agencies have gone out of business. Hotels and other lodging businesses also use this online capability. It will no doubt spread to making reservations with restaurants, doctors, and any place where reservations or appointments are used.
- Registration at colleges and universities. One of the most effective systems to relieve the internal staff of peak loads is the use of computerized registration systems at colleges and universities. Although it may not involve an outside business offering the service, the computerized system acts in much the same way a service provider would act. It handles the registration without requiring the direct attention of the internal staff. The system also relieves the students from the burden of standing in line both to register and to pay their fees.

Acquire Expertise Not Available Internally

Sometimes, it is not prudent to develop internal resources and specialized skills, especially when the requirement for those capabilities may be intermittent or even occasional. Outside service providers can usually fill this void quite well.

- Facilities location. Selecting a site for a new manufacturing plant, distribution center or retail outlet is a daunting task for businesses with little experience in this area. It starts with selecting a country (if considering a foreign location), then a region, and finally, a specific site. Most businesses do not have the data or the knowledge to use the data, in considering demographics, incomes, traffic patterns, local ordinances, and a host of other variables. Facilities location providers can meet this service need.

- Architectural or facilities design. It is rare that companies would have their own architectural or facilities design staffs; they must rely on outside service providers. Exceptions could be retail companies that have hundreds of retail outlets. Even some of these firms may use outside architects and designers to gain experience and expertise with working with external providers.
- Construction. Construction work is also a very specialized service. Even though it is considered as part of the transformative industries, it is a service to most businesses. It involves major project planning capabilities, a capability not available in most other types of businesses.
- Security services. Businesses have purchased security services for a long time. When the concerns were primarily with early identification of fires or screening vendors for purchasing agents, the business could adequately handle this with their own employees. As the needs increased – industrial espionage, computer piracy, or possible terrorism threats – it has become obvious that specialized services, supplemented with specialized equipment, are needed.
- Investment counseling. In addition to the banking relationships described earlier, larger businesses, especially public companies, may need the services of companies that can advise on investment strategies. Public companies need help in developing strategies to issue stock or to buy back stock, to use debt or equity to finance future expansion plans, or to develop strategies to present their best face forward to the investing public.
- Risk and insurance. Historically, business insurance was about fires, floods, thefts and other relatively tangible occurrences. Over time, the risks have increased and now businesses must be protected against less tangible threats as lawsuits from product recalls (children's toys), accidents on the business premises (falls on slippery floors), consequences of inappropriate use of the product (hot coffee spills) and failure to prevent employee misappropriation of funds (Enron). Insurance protection requires ever-increasing knowledge that is difficult to maintain within a business.
- Management consulting. There are management consultants available to meet almost any need. At some point, even small companies can benefit from the use of consultants. Businesses may need help in planning and implementing new software or some management program to reduce costs or improve quality. We described a number of these programs in earlier chapters. It is unwise for a business to attempt these programs without inadequate knowledge — knowledge that can be available from professional consulting firms.

Assist in Strategic Planning

In the preceding examples, most of the supplier services were in the form of a "hands on" arrangement. Suppliers actually did some, or all, of the work that would have been otherwise performed by internal employees. Assistance in strategic planning is different. It is more of an advisory or training role and the internal staff does most of the actual work.

- Global guidance consultants. Almost all businesses, even small ones, are lured by the attraction of global markets. Most need a great deal of help in finding their way through the maze of legal and cultural differences they may encounter in their entry into just one new market. They need help in selecting the market with the most potential and then developing a plan to enter that market. Global guidance consultants are available with these types of services.
- Merger and acquisition specialists. Some industries, such as financial institutions and department stores, are experiencing a wave of mergers as companies move toward economies of scale. There is little doubt that electronic communications capability makes it easier for companies to run larger companies by facilitating information flow. Deciding who and how to buy other companies, or sell a business, is not for novices. Specialists are available for these types of situations.

Comply with Regulatory or Otherwise Mandatory Requirements

Sometimes a business has no choice as to whether or not to use the external service providers. Either the service is required by law or mandated in some form, such as by a major customer.

- Auditing. Audited financial statements are a requirement for public companies. Even private companies may need audited statements if required by their creditors or other stakeholders. Auditing is not only a specialized activity; it implies independence, an independence that requires an outside agency. As companies become more complex in product and service lines, and more global in markets, the skills required to perform a financial audit are challenging to even the most knowledgeable of public accounting firms. As these requirements mount, there is a growing potential for business-auditor conflict in the interpretation and application of various regulatory agency rules and guidelines.
- Environmental compliance. Businesses are experiencing only the tip of the iceberg in environmental compliance issues. What was once a local issue — the dumping of toxic materials into the river — is now a part of the global warming concern. Even when there are knowledgeable employees within the company, many businesses can benefit from outside services to assure they are complying with local, national, and other environmental regulations. This task will become a more demanding requirement in the future.

- Technology compliance. Technology is a major driver of change. An example is the requirement by Wal-Mart that its suppliers use RFID tags on products they purchase in the future. Although this program will be implemented over a period of years, it represents only one of many technology requirements that will change the way businesses operate. Another similar program is the increasing demand for businesses to reduce environmentally or ethically harmful ingredients or processes — hazardous materials in computers or sweat shop conditions in clothing manufacturers.

Manage Major Programs

Most of the foregoing reasons for outsourcing producer services are to sustain the normal operations of a business. A final reason is to assure the business has expertise available to handle programs — both of the problem-solving and continuous improvement kind.

- Crisis management. Even in the most well managed business, a crisis can be fatal. Arthur Andersen was one of the premier public accounting firms in the world; yet the Enron debacle caused its demise. On the other hand, Johnson and Johnson handled the sabotage of its product, Extra Strength Tylenol, so well that it actually gained market share after the crisis was over. Whether it is toys that choke children, canned food that kills pets, slippery floors that cause falls, or computer systems that fail to protect individual identities, crises happen. A number of crisis management consulting firms are available to assist businesses in setting up crisis management plans. These firms can also coach management on the best ways to respond to specific crisis events.
- Continuous improvement programs. In Chapter 4, Chapter 5, and Chapter 6, we described a variety of programs that companies use to make improvements in their operations. Although some companies have the capability to implement these programs on their own, most do not. They need help from outsider service providers, whether it is software vendors or some other form of management consultant.

Steps to Interfacing Business with Producer Services

We have described a number of producer services that businesses may need. In this section, we will outline a general approach to obtaining these services. Buying services is often more demanding than buying products because, although it is possible to see, feel, and even try out the product, this is not possible with services. We believe that a business should take the following steps in planning and implementing the use of producer services in its company.

Recognize the Role of Producer Services in the Company

Most companies do a good job in deciding which products to make and which to buy. They are less diligent in deciding which services to perform internally and which to outsource. They have less formal procedures for evaluating this decision area because, although the cost of doing or outsourcing may be determined, the benefits are often less tangible and more difficult to assign a value. It is important to recognize the need to make the evaluation and to determine the reason for having each of the service activities. Management should understand that the activity area described in Chapter 10 — the integrated supply chain — is the main reason businesses exist. The services described in this chapter supplement the supply chain. Although they are essential and can often provide a competitive advantage, they augment, but do not replace, the activities along the supply chain.

Develop a Strategy for Each Producer Services Area

Each producer service area is different and must be planned as a unique part of a business. The actions taken in the accounting area may not be the same as in the engineering area. In accounting, the emphasis may be on the implementation of an enterprise resources planning (ERP) system; in engineering the focus may be on designing a new quality performance program that involves the purchase of custom test equipment and rigorous training of quality technicians. Once a strategy has been developed, the business must decide whether to perform that service internally or to outsource it.

Select Suppliers to Provide a Continuing Relationship

If the decision is to perform the service internally, management must decide if the internal staff is adequate to handle the service. Are more people required? Is the present staff adequately trained? Are more professionally-trained employees required? Do they need more equipment? If so, what kind? Do they have adequate IT resources? If the decision is to outsource the service, good supplier selection is critical. Many of the same questions listed above must be asked about the supplier. In addition, a total cost comparison should be made. What is the likelihood that the supplier will meet the service level requirements? Can the supplier handle future volume and variety requirements?

Prepare a Plan That Interfaces with the Supply Chain Plan

We reported earlier that the producer services areas are to supplement the integrated supply chain. It takes skill and perseverance to match the capabilities of internal and external producer services suppliers with the needs of their customer

— the supply chain part of the business. The service suppliers must also remember that although they are directly responsible to their company's part of the supply chain, they may also have a responsibility to other members of the supply chain. For example, the IT department may provide sales information for all products and the customers to whom those products were sold. The IT department may also need to provide sales information to suppliers for those products that each supplier sells to the company, but this sales information should not include the customers to whom they were sold. Resolving reporting relationships and personality conflicts is no small part of this matching process.

Implement the Service — Fit It to the Situation

Planning for the services is difficult, but implementation can also be difficult and time-consuming. One of the major obstacles is fitting the people to the processes, or vice versa. Should policies and procedures be the best that can be conceived or should they be modified to fit the people involved? The practical answer is that there is a high likelihood that some adjustments will be made to reflect the situation. People can eventually compensate for system deficiencies, but systems can seldom be designed to compensate for people deficiencies.

Measure the Progress

Developing the producer services package to fit with the integrated supply chain is a major project. Even after the services have been designed and put into place, it is a continuing project to maintain, adjust, and improve the producer services package. The composition of the producer services package is dynamic — there are continuing changes in internal employees, supplier relationships, economic conditions, competitor innovations, and governmental regulations. In addition to the changes in the producer services package itself, there will be changes in the integrated supply chain served. This changes the relationships of the producer services package elements. Somehow, the changes must be identified and their effect on the overall performance of the supply chain evaluated. Performance measurement is one of the critical aspects of any business.

Revise and Assimilate

No matter how well designed and implemented, changes force revisions to the producer services package. Expect this and adjust accordingly. It is a normal part of business. Just as a business must continually update its product and service lines, it must also update its producer services package. It probably helps to view this as a separate part of the business to be planned in much the same way as the integrated supply chain part of the business.

The Future of Producer Services

What changes can we expect in producer services in the future? Several ideas seem to stand out as likely.

There will be an increasing use of IT in producer services. Intra- and inter-organizational systems will become standard in most companies. The need to communicate will drive the use of technology. With it comes a reduction in the personal contact that some employees cherish. The depersonalization of communications will be more acceptable in business-to-business transactions than in business-to-consumer transactions. Even with what appears to be widespread use of IT in business today, it is only the beginning.

With the increased ability to communicate across distances, there will be a continued increase in outsourcing producer services. Most companies do not believe that producer services are part of their core competencies or offer a competitive advantage. Therefore, they will prefer to have these services provided by outside suppliers. At some point, this trend will slow and perhaps reverse itself as companies assign greater value to their information; however, that time still seems to be in the future.

Companies have concentrated their improvement programs primarily along the supply chain. At some point, they will decide that improvement in producer service areas will offer equal, or greater, benefits. In some cases, they will be able to apply the same improvement programs to producer services that they used in supply chain applications. More likely, they will find other types of programs will be needed to achieve the improvement results they desire.

One of the promising areas for improvement comes in the services sciences. This is the application of quantitative techniques to services. Although the core idea is not new, the applications are being packaged into a composite of techniques and methodologies to facilitate the transfer of knowledge from organization to organization.

Producer Services Firms Buy Producer Services

Manufacturing companies are not the only industry group to buy producer services. Producer service firms also need to buy producer services from other firms.[7] In other words, there are no firms that supply a full range of producer services, at least for now. Accounting firms buy HR services; architectural firms buy custodial services; custodial firms buy tax services; and so on. Although it appears likely that producer services firms will continue to be specialists, that may change over time.

Producer services have been fragmented parts of a business operation in the past. In the future, they will be part of an integrated package that will play a more important role in the operation of any type of business or organization.

What Are Social Services?

Social services are similar to producer services in that they are services provided to a business. However, they differ in that they are usually provided by nonprofit organizations, or government agencies, as opposed to the profit-oriented organizations that provide producer services. The bottom half of Figure 11-3 shows the categories of social services.

Why Are Social Services Important?

We saw that producer services are usually necessary in running any type of business. Some social services may be necessary, such as the U.S. Postal Service or intellectual property protection. Some may be optional, such as the use of census data or other business-related information. Some may be welcome as an indirect benefit, such as education systems or protection of assets. Finally, some may be mandatory, such as waste disposal regulations or financial reporting requirements. Whatever the reason, social services play an important role in the lives of most organizations.

The Role of Social Services

In theory, businesses could operate without social services; in practice, it would be difficult. In some cases, social services provide benefits to businesses, benefits that might not otherwise be available to individual businesses or other organizations. In some cases, social services act as a constraint on the operations of a business, although they may provide a benefit to society in general.

A primary role of social services is to provide direct services to businesses that would be beyond the capability of individual businesses to provide, regardless of size. Although some of these services are mandated by federal law (Bureau of Labor statistics and interstate highways), many are provided within the jurisdiction of state (university school system) and local governments (landfill operations). Although it is conceivable that private companies could operate some of the government agencies, the privatization of government agencies is a slow, but not impossible, process.

Another role of social services is to provide indirect services to both businesses and individuals. Police departments and fire departments represent this type of social services. Although private businesses may maintain their own staff to provide security protection and have designated employees available for fire protection, such measures are usually minimal and rely heavily on the local agencies to provide the much needed support in case of a major occurrence.

Sometimes social services promote social movements that motivate businesses and other organizations to begin practices that provide for the general welfare of the population. Recycling programs are an example of this. Although private businesses are participants in the recycling process, the driver is government at all levels. Where necessary, the government can move this activity from a voluntary recommendation to a mandatory requirement by enacting legislation that prescribes what businesses must do or face the penalties.

One of the more valuable social services is the work done by the federal government to provide a vast repository of information. Agencies such as the Census Bureau or the Bureau of Labor Statistics compile and make available, a vast amount of data and information, usually free of charge, to everyone. Individual firms would not have access to the information in these sources, much less have the financial or physical resources needed to set up and maintain these information sources.

Classes of Social Services

We will discuss three major classes of social service organizations — governments, nonprofit organizations and quasi-private organizations. Each group includes a variety of entities that can provide social services to businesses.

Government

Federal, state, and local governments operate throughout the business world. They provide a variety of services that are described in detail in a later section. In some cases, businesses welcome the service; in other cases, businesses consider the services provided by government as unnecessary or detrimental to the business.

Nonprofit Organizations

These include professional organizations that provide education courses, reference materials and conferences where employees of a business can enhance their individual and collective knowledge. This class includes organizations that focus their efforts on improving some situation or group of subjects — endangered species, sweatshops, or global warming. In some cases, the business community may endorse these programs; in others, they may resist.

Quasi-Private Firms

These organizations may have government sponsorship or operate as a nonprofit entity; however, they compete with private businesses. A hospital financially supported by the local government may compete directly with private hospitals. A state-supported university competes with privately funded universities. The U.S. Postal Service competes with Federal Express and United Parcel Service.

Business organizations may tend to have a mixed view of these classes of organizations that provide social services. Do they really help? Or are their services more negative than positive? We will try to sort out some of these issues in the next section.

Specific Social Services

In this section, we will describe a few of the specific social services provided to businesses. We will concentrate on those that most businesses view as positive.

Information

Government agencies, especially at the federal level, collect, organize and make available a vast amount of information, especially economic news. It is largely free and easily available with today's electronic communications systems.

Communications Systems

In addition to providing information, the federal government provides a variety of communication networks that make it possible for businesses to communicate with one another. Most notably is the World Wide Web and Internet system that most businesses would be lost without in today's global trade environment.

Standards

The business community could not operate without standards. It may be units of measure such as for weights or distances. It could be standards for radio frequencies or bar code symbology. It could even have to do with standards of conduct in boardrooms. Without standards, the results would be chaotic. Standards are also necessary to protect society and businesses from harmful forces such as environmental pollution, potential automobile safety issues, and workplace discrimination problems. Usually, these standards are brought forth by the government after a considerable amount of public outcry dictates that something "should be done."[8]

Education and Training

There is little question that we are entering the information age. Companies increasingly depend on knowledge to provide them with a competitive advantage. One way to attain knowledge is to combine information and education. Educational systems — public school plus universities — have been a major factor in the evolution of the business world to its present state. In addition to the broader vista of education, social service providers also provide employee training to meet specific job requirements.

Location Incentives

One of the newer services that appeals to many businesses is that of location incentives. Governments, often a combination of state and local agencies, work together to attract new businesses to communities, in an effort to help those that have suffered job losses from outsourcing or global competition. Although this service is often a subject of debate among rival politicians, it is a social service to businesses, nevertheless.

Support Infrastructure

Businesses are dependent on the infrastructure provided by governments. Businesses need roads with signs and stop lights, water purification systems, waste disposal systems, and zoning regulations to assure a comfortable relationship between businesses and individuals. This is an area where the business community and government must work together to achieve realistic solutions to the needs of a variety of groups.

Protection

Businesses have large investments in resources — facilities, equipment and, most important, people. These assets need protection and police and fire departments provide at least some of the protection. Businesses also need protection from other impediments to their success — predatory pricing, intellectual property theft, trivial lawsuits, and unfair competitive practices.[9] Although the individual organization may have to solicit the help, governments at all levels have an obligation to help protect a business.

Health Care

There is another area of social services that is looming larger in each generation — the subject of health care. Some of the services described above, such as information, education, and protection, contribute to the physical well-being of the population. However, businesses have a special interest in the health of their employees. Healthy employees, both the physical and mental welfare of individuals, are of great value to the organization. Businesses are reorienting their focus to wellness, a more positive perspective than simply taking care of the ill employee.[10] This appears to require a collaborative effort between the business community and the social service providers.

Steps to Interfacing Business with Social Services

The approach outlined above to integrate producer services into a business will apply to social services, but with some qualifications or cautions:

- Businesses may have less choice in selecting the supplier. Although a business will be able to select the location of a new facility, the business will have to accept the service of the local government, once the new location has been settled. This will be especially important in considering foreign locations. In many cases, the business will have a "sole source" situation and must maximize the benefits from that situation, rather than having the flexibility to move to another supplier.
- The relationships with the supplier may be more permanent. As a corollary to the "sole source" situation, relationships with social service providers will last longer. It will be up to the business to work out the best arrangement with its provider. The form of government may not change, the site of most universities does not change, and the campaigners for purer water may span generations.
- Political climates affect social supplier actions. Although the form of the government agency may not change, it is likely that the political climates may change, sometimes frequently or abruptly. When this happens, the relationships with businesses may move in a more positive or a more negative direction. For example, taxes may go up or regulations may increase. Businesses have to adjust to this change and make the best of it, or move, which is often a less desirable choice.
- Consequences may be more than economic. When businesses work collaboratively with social service providers, there may be positive economic results. On the other hand, if the relationship deteriorates, the consequences to the business may be more than just economic. They may face fines for failure to follow waste disposal regulations and this could hurt their public image. If the violations are more serious — for example, criminal acts are committed such as falsifying financial results — then the managers involved could face imprisonment. Sometimes the effect of these acts causes the demise of the business, such as for Enron.

- Social services may not be as good. Government and nonprofit agencies are not thought of as being as effective or efficient as private organizations. Although this may be undeserved, it is something to consider. These agencies exist to provide services to a mass group of customers; they cannot choose their markets. An individual business will gain the most benefit by building a long-term relationship with their providers and working with them through the disruptions or transitions.

The Future of Social Services

How will social services change in the future? Some of the major issues that will have an effect on business are discussed next.

Level of Regulation

In recent years, several industries, such as in transportation, have seen a reduction in the amount of regulation. On the other hand, there is a growing concern about the level of regulation in personal health related issues, such as the food supply or the quality of medicines. There is also likely to be more emphasis on environmental issues. Increased regulations may pose additional constraints on businesses that require a transition to a new operating environment. Although making these changes in processes, businesses may find ways to make improvements that will increase their competitive advantage in the global marketplace. At least one industry has seen both a relaxing of some regulations and a tightening of others. The airline industry was deregulated in terms of route selections during the 1980s, but after 911, safety and passenger regulations have been understandably increased. Such events make this industry one of the most difficult ones to operate in.

Level of Privatization

Governments and academic institutions are slow to change. However, there are indications that the outsourcing movement will eventually reach even the most staid of organizations. Local governments are finding that privatization works in a variety of public service areas. During his tenure as mayor of Indianapolis, Stephen Goldsmith spearheaded a number of successful privatizing projects, in such varied areas of city government as highway litter pickup, printing and copying, courier service, wastewater treatment, construction permits issue, construction and operation of correctional facilities and welfare administration.[11] More recently, Emily Thornton describes a number of cases where governments have sold major assets such as bridges, airports, parking garages and shipping ports to private investors to recover cash to "solve short-term fiscal problems." She believes that "the conditions are ripe for an unprecedented burst of buying and selling. All told, some $100 billion worth of public property could change hands in the next two years."[12]

Universities are facing more competition from distance learning institutions and corporate colleges. It will be interesting to see if state and private universities develop their own in-house distance learning capabilities, or outsource this to private institutions. It appears likely that there will be more privatization of social services, not less, in the future.

Level of Effectiveness and Efficiency

The cost and quality of social services is attracting increased attention from the business community, as well as the individual citizen. It is not only logical, but also politically perceptive for government and government-supported institutions to begin a program to improve the effectiveness of their services (provide the right services) and efficiency (do the services right). This will provide a wonderful opportunity for business and nonprofit organizations to work together in optimizing their mutual offerings.

Emphasis on Business-Related Topics

If business and nonprofit organizations work together more closely, it follows that government agencies will become more business savvy. If they do, they will be better able to identify the right issues to work on and come up with a better balance of solutions. Although this is rational, it may never reach a satisfactory resolution because of continuing diversions into international conflicts and needs.

Specific Issues to Resolve

There is no lack of issues or problems to work on. Most of these will benefit from the collaboration of business and nonprofit organizations. Some of the most obvious issues include health care, global warming, ethical behavior, terrorism, and secular vs. religious disputes.

What Are Consumer Services?

Consumer services are those services that provide outputs directly to the consumer, or business-to-consumer (B2C). Personal services differ from the Distributive Services shown in Table 11-1. Distributive Services include retail and other selling outlets that deliver products and services to the consumer also, although their emphasis is usually on the product complemented with services. Personal services tend to be service businesses that may also deliver products as part of the service. This is the ultimate purpose of business — to service the customer. Most businesses tend to concentrate on how to supply services to the consumer; only a few are consciously considering the services that the consumer provides to the business.

Roles of Consumer Services

Consumer services have two major functions. First, they are the way that many products and services are delivered to the ultimate customer — the individual consumer. The second role of consumer services is to accumulate feedback from the consumer about their impressions, both good and bad, about the products and services they have received. The first role is obvious and is the culmination of the efforts described in Chapter 10 to develop an integrated and agile supply chain to make and deliver the products and services.

The second role is not so obvious. Retail organizations, both those who sell through storefronts and those who sell through websites, are developing very sophisticated ways of collecting data about what the customer buys and, in some cases, what the customer would like to buy or expresses an interest in. See Chapter 10 for a fuller discussion of the work being done in this area. The process for collecting data in the consumer services businesses is less formal. A hair stylist may remember how an individual customer wants her hair styled and a local restaurant may remember which table you prefer when you come to dine. However, much of this knowledge is in the minds of individuals, known as tacit knowledge, and is not compiled as collective information available for the organization — explicit knowledge. There is some evidence that this is changing. An automobile dealer may keep a file of individual customers and a record of the services performed on their vehicles. A motel chain may retain a record of a customer's stays, regardless of location. In most cases, however, the information is inadequate to develop decisions that will influence product or service design.

Future of Consumer Services

Consumer services will continue to grow as the overall economic prosperity of a country increases. Many of the personal services might be considered luxuries, not necessities. Because most businesses providing personal services are small, they have not developed the management information systems that larger businesses have. The exception is, of course, those in the hospitality industries, particularly resorts and higher end hotels, where their IT capabilities are equal to, or better, than in some of the other industries.

Another trend that is already evident is for larger companies, especially those in the retail industry, to add select personal services to their portfolio of services. Department stores provide beauty stylists, many malls house movie theaters, and building supply retailers sponsor pet adoption Saturdays.

Integrated Service Package

In the previous sections, we described a number of services that a business, whether for-profit or nonprofit, needs to have. In the past, most organizations viewed these services as individual functional areas that were a necessary cost to conducting business. In the future, more organizations will view these services as complementary services that make the supply chain part of the business better. Many companies will begin to view this set of services as an integrated service package, not as simply additional business fragments.

The Need for an Integrated Service Package

Why does a company need an integrated service package? Why can't it continue to operate with service islands throughout the company? Do the benefits outweigh the costs?

We do not know with certainty the answers to the above questions, because the premise that an integrated service package is better remains to be tested. We can only speculate that it seems logical. Integration is defined as "to make into a harmonious whole by bringing all parts together."[13] That not only sounds logical; it sounds desirable and financially feasible. Based on the results from other types of integrating efforts, including the one we described in Chapter 10, it may not be easy. In the next section, we will outline how we think organizations can achieve an integrated service package.

Steps in Developing the Integrated Service Package

How does a company go about developing its integrated service package? The steps that we outline below follow closely the steps we outlined in the strategic planning process in Chapter 2.

Recognize the Need to Change

Is the present combination of services adequate? Could they be improved? Just as management ponders its need to improve its supply chain arrangement, it must also recognize the need to make improvements in the company's service package. They must also recognize that there are no quick and easy fixes. The transition program will take time and perseverance; unfortunately, most companies have competitors who are also just beginning to recognize the need to make this transition.

Identify the Service Package Components

Once companies have decided that they need to develop a service package, the next step is to decide which services they should include in the package. The combination of services, those that work in synchronization with each other, is especially important to consider. They should try to eliminate any services that they decide are unnecessary for their future operation. They should look especially for duplication among different functional areas. Is there redundancy in data collection or reporting? If so, organizations should try to decide where and how best to perform the work. If the service cannot be eliminated, there may be a way to simplify it. This is a good time to convert a marginally running process into an excellent process. After simplification, it may be possible to combine some of the services into a new process that spans previously separated functional areas. Companies should not be bound by the old functional arrangements; the new combination of services may offer a way to streamline the organization.

Develop Objectives for the Service Package

In this phase, management should decide what they want the service package to accomplish. Is the objective to increase the effectiveness of the supply chain part of the business? Is it to reduce the cost of providing the services? Is it to increase the effectiveness and efficiency of the service package components? It is possible to develop very specific objectives for the supply chain part of the business. Such traditional measures as sales volume, profit, return on investment and inventory turns are just a few of the useful tangible measures. It is more difficult to measure the effect of the service package on the effectiveness of the supply chain, or even the efficiency of the service package. Nevertheless, it is important that there be some way to measure the performance of the service package. For example, a university that is seeking to add an online program to its academic offerings may set a goal to attain a certain level of enrollment, while staying within a specified budget. Developing performance measure in this area will be an interesting challenge for both practitioners and academics.

Develop Strategies for the Service Package

Strategies are the action plans to achieve the goals. Companies should first develop systemwide or cross-functional strategies that involve integration, collaboration and all the other steps necessary for success. Initially, this step will be difficult because it is not something that has been done in most companies. Rather than frustrate all of the individual functions with a plethora of business strategies, it is probably a good idea to go slowly in developing overall service package strategies. Let them come naturally, one at a time. Implementing one good strategy that gains acceptance throughout the organization is better than failing to implement multiple strategies. One of the first strategies will need to be the organization structure for the service package. This may require some changes in the basic functions of the business. The traditional structure may no longer be suitable for the integrated service package approach. For example, the university seeking to develop its online and nontraditional education programs may set up a new department to oversee this endeavor.

Develop Objectives for Each Component

Once management decides on a strategy or strategies for the total integrated service package, each component, or functional area, should develop objectives for their part of the total service package. These objectives should support the overall strategies. Managers of the functional areas need to recognize that their objectives must fit with the total strategies and not necessarily the best for their functional areas — a global optimum, not a local optimum.

Develop Strategies for Each Component

Once the objectives are identified, strategies to meet the objectives should be developed. These action plans will usually be of a project nature and should have schedule dates, costs, and outcomes well defined, along with the responsibilities assigned for each part of the plan. It may not be realistic to plan every strategy at the same level of detail. The most critical should be carefully planned; other plans of less importance may depend on more "hands-on" management.

Implement the Strategic Plan

After the planning phase, it follows that the plans must be implemented. Although implementing an integrated service package may not carry the excitement of implementing the supply chain part of the business, it is important and should be done with the professionalism deserved. It will be easier to see the costs of this program faster than its benefits; therefore, top management must have faith that it will eventually pay off in tangible benefits.

Evaluate the Results

Just as with any type of management program, the progress of the implementation and the value of the outcomes need to be carefully assessed. This step does not mean that the monitoring has to be in real-time; however, it should be timely, accurate, and with the full awareness of not only the participants, but also the other stakeholders in the company.

Revise as Needed

Any plan should be dynamic and be revised to reflect changing conditions. The service package program is supplementary; therefore, it will also be reactive to the supply chain part of the business, not proactive in its own right. Revision to reflect changing needs is appropriate; this type of planning is after all, a fluid endeavor.

Summary

In Chapter 10, we described how a company could move toward an integrated and agile supply chain. Regardless of the company's role in the supply chain — fabricator, assembler, distributor, or retailer — it is an essential part of the total supply chain. The power of the supply chain is that it integrates a number of activities that were previously performed independently or, at best, with localized interfacing in a limited number of companies.

In this chapter, we have described the services that a business must provide, either internally or through outside suppliers, to be successful. There can be no disputing the fact that services are essential to complement the supply chain functions. We believe that organizations will be served best if they view these services as being a part of an integrated package of services that integrate smoothly into the supply chain part of a company's business.

References

1. Gershuny, J.I. and Miles, I.D., *The New Service Economy, The Transformation of Employment in Industrial Societies,* Praeger Publishers, 1983, from Singelmann, Joachim, *From Agriculture to Services, The Transformation of Industrial Employment,* Sage Publications, Beverly Hills, CA, 1978.
2. Greenfield, H.I., *Manpower and the Growth of Producer Services,* Columbia University Press, New York, 1966.
3. Sako, M., Outsourcing and offshoring: Implications for productivity of business services, *Oxford Review of Economic Policy,* 22, 4, 499, 2006.
4. Greenfield, H.I., *Manpower and the Growth of Producer Services,* Columbia University Press, New York, 1966.
5. Beyers, W.B., Impacts of IT advances and E-commerce on transportation in producer services, *Growth and Change,* 34, 4, 433, 2003.
6. Collins, J., Workplace wellness, *Business and Economic Review,* 51, 1, 3, 2004.
7. Juleff-Tranter, L.E., Advanced producer services: Just a service to manufacturing? *The Service Industries Journal,* 16, 3, 389, 1996.
8. Hartley, R., *Business Ethics: Violations of the Public Trust,* John Wiley & Sons, Inc, New York, 1993.
9. Steiner, G. and Steiner, J., *Business, Government, and Society: A Managerial Perspective* (Eleventh Edition), McGraw-Hill Irwin, New York, 2005.

10. Robbins, S. and Coulter, M., *Management* (Ninth Edition), Pearson-Prentice Hall, Upper Saddle River, NJ, 2007.
11. Goldsmith, S., Can business really do business with government? *Harvard Business Review,* 75, 3, 110, 1997.
12. Thornton, E., Roads to riches, *Business Week,* May 7, 2007, p. 50.
13. *The American College Dictionary,* Second College Edition, Houghton-Mifflin Company, Boston, 1982

Chapter 12

The Future of Improvement Programs

John was tired, but as he looked down the corridor, he could see that there was only one more stage. "I might as well see what is there," he thought. "Then I will see what I can figure out from all of the things I have seen." As John came closer to the final stage, he heard sounds. He could hear clinking sounds, whirling sounds, hammering sounds, roaring sounds, and a number of other sounds that he couldn't distinguish because of the complexity of it all. "What in the world is going on? It must be alive. Inanimate objects couldn't make all of those different sounds."

He walked up to the stage expectantly. All he saw was a series of doors. He could hear activity behind each door; that was where the sounds were coming from. He also noticed that there were signs on each door. They looked temporary, but John could see that they could be made permanent without too much trouble. He read some of the names on the doors. They included Inter-Organizational Systems (IOS), Knowledge Transfer Systems (KTS), Virtual Reality, Service-Oriented Architecture (SOA), Left Brain–Right Brain (that was eye-catching), Service Sciences, and Chaos Theory. There were more, but John had to catch his breath. What in the world were all of these things? They sounded exciting, but what did they have to do with running a business?

"Oh, well," he thought. "As long as I'm here, I may as well find out about them." He walked up to the first door, knocked, and woke up. "That was some dream," he thought, "and I had to wake up just when it was getting interesting." As he got out of bed, he saw a bag with what looked like a number of pamphlets in it.

Introduction

In Chapter 1, we presented a traditional approach to improvement programs — logical and sequential. In this chapter, we will look ahead to describe how improvement programs in the future will enter a murkier environment — more complex and chaotic.[1] Although improvement programs will still be essential to the success of a business, they will become more closely intertwined with the sustaining and problem-solving parts of the business, so that it will be more difficult to separate one mode of operating from another.

The Background to Improvement Programs

To understand the future of improvement programs, it would be good to take a brief look at where we have been in this book. Each chapter presented a topic area that serves as a building block to initiating improvement programs. We also showed how there is a vanishing boundary between manufacturing and service industries; in other words, improvements from manufacturing can be used in services and vice versa.

The Vanishing Manufacturing/Services Boundary

Chapter 1, Introduction to the Manufacturing/Services Boundary, described some of the ways in which the manufacturing/services boundary is being eliminated, or bridged over. We organized this material to tie in with the rest of the book.

The Foundation Topics

In Chapter 2, Chapter 3, and Chapter 4, we described some of the key topics that related to improvement programs.

> Chapter 2. Critical Success Factors and Strategic Planning: In this chapter, we presented the increasing array of critical success factors (CSFs). We identified CSFs as those items that determine what it takes to compete. In other words, these are things a company "must" do to compete effectively. This chapter explained that the manufacturing/services boundaries are eroding. Rapidly

increasing global competition is forcing companies to compete more effectively to survive. They do this by identifying the CSFs needed and building them into their strategic and tactical planning.

Chapter 3. The ITO Model: The ITO Model is the DNA of the supply chain. This chapter explained that the manufacturing/service boundaries are disappearing because we "can," that is, we have the capability to build relationships, both physical and personal, that are meaningful and productive. We are learning how to connect between entities in a supply chain and build the product–service bundle to be both effective and efficient.

Chapter 4. The Role of Management Programs in Continuous Improvement: Business innovation and improvement involves the blending and packaging of management theories into programs. Businesses have found that it is often convenient to use "programs" to focus attention on one, or a few, CSFs. To reduce costs, we have a program of continuous improvement. To improve quality, we have a total quality management program. Programs involve packaging basic management ideas into a package that addresses a specific need — from management theory to a systems approach using contingency theory to customize it.

Knowledge Transfer across the Manufacturing/Services Boundary

In Chapter 5 and Chapter 6, we showed some of the ways in which knowledge is being transferred across the manufacturing-services boundary. In Chapter 5, we described how improvement programs that originated in manufacturing have been extended to the services sector. In Chapter 6, we described how improvement programs that originated in the services sector have been extended to the manufacturing sector. In so doing, we tried to convey the impression that the lines between manufacturing and services are blurring so much that it is becoming more difficult to identify where programs originate and how they move from one sector to the other. Some programs, such as supply chain management, share in the knowledge from both sectors.

Chapter 5. How Manufacturing Techniques Are Being Adapted to Service Operations: This chapter described some of the major programs that have been used to move manufacturing concepts and techniques into service applications. Cost reduction and quality improvement have been primary objectives, such as in JIT and TQM, although there have been a number of variations in a program's appearance. Much of the effort has emphasized the use of equipment and automated processes to achieve these objectives.

Chapter 6. How Service Techniques Are Being Extended to Manufacturing: This chapter described some of the concepts and techniques that originated in service industries and extended into the manufacturing areas. These efforts have centered largely on customer service programs, such as CRM and Quick Response programs. In services, much of the attention is on the proper use of employees, as well as equipment and facilities.

Agents of Change

In Chapter 7, Chapter 8, and Chapter 9, we described three change agents that affect the success of any improvement program — technology, structure, and culture. These disparate elements must blend as a synchronized force in making it possible for improvement programs to reach their intended goals.

> Chapter 7. The Role of Technology in Continuous Improvement: This chapter described the progression in using technology to develop more sophisticated, and complex, processes that facilitate the application of resources, such as people, equipment, and facilities. It described technologies that drive the processes, such as CNC and FMS. It also described the major increases in the application of information processing technology, such as ERP systems, RFID, SOA, and the like.
>
> Chapter 8. The Role of Infrastructure in Continuous Improvement: Structure is how companies organize and how they support the input–transformation–output (ITO) process. In many cases, companies have stressed the application of technology, but have not changed their structures. One of the most significant needs is to move from a functional orientation to a process or horizontal orientation. The need for accurate, complete, and rapid communication necessitates a different way of organizing. Key issues include outsourcing, teams, and project management.
>
> Chapter 9. Understanding Organizational Culture — The Elusive Key to Change: Just as with structure, companies have often forgotten to proactively consider the implications of change for employees. Often, employees are expected to understand intuitively, accept, and participate willingly in the change process, even though in many cases the change may be detrimental to their personal welfare — loss of earnings or even their job. As a result, it is important for managers to understand the fundamentals of organizational culture. This chapter presented those fundamentals, and outlined how to go about changing the company culture.

Integration of Related Entities

In Chapter 10 and Chapter 11, we blended the segments described in earlier chapters. In Chapter 10, we showed how the main stream of a business flows along the supply chain and builds from batch operations to an agile and integrated system.

In Chapter 11, we showed how the support services necessary for a business to be successful could be organized into an integrated services package to support the supply chain described in Chapter 10.

> Chapter 10. Integrated Supply Chains — From Dream to Reality: In this chapter, we described the core supply chain from raw materials to end consumer. We overviewed the phases that a company goes through in developing its supply chain and the way the pieces fit together. Building an agile and integrated supply chain is a complex process that requires a blending of technology, structure, and cultures. It is not as sequential and well orchestrated as companies would like; it often involves false starts and the need to reorient as circumstances change. However, some companies persevere and eventually achieve a workable version of a supply chain. We also described some of the current issues in supply chains, including how to achieve equity among participants, how to use the ITO process to move from a gathering of independent entities into a closely linked unit, and how to make the supply chain relevant to today's needs, but flexible enough to change to meet tomorrow's needs.
>
> Chapter 11. The Role of Services to Complement the Supply Chain: This chapter dealt with the links from support services to business, such as providing group health care plans, equipment financing, product liability insurance, and management consulting. It also presented support services provided to the individual consumer, such as hospital services, home mortgages, automobile insurance, road maintenance, and preparation of tax returns.

In the first 11 chapters, we describe a rapidly changing business world. No doubt, it will change even more in the future. In each chapter, we tried to highlight a good "takeaway" for the reader. We also tried to link the chapters together in a logical manner. Our goal is to present the book as an ITO model in which an expectant reader is transformed into an enlightened reader.

In this last chapter, we offer some ideas about changes that could face the business community in the years ahead. We have organized this chapter to project trends in the topics we have covered in earlier chapters, such as critical success factors (Chapter 3), management programs (Chapter 4), technology (Chapter 7), structure (Chapter 8), and cultures (Chapter 9). One of the ideas we should stress is the need to discover how to move from functional silos to integrated entities, from left-brain to right-brain thinking, from local to global optimization, and from repetitive perfection to dynamic "satisficing."[2]

Future Areas of Emphasis

The future will most likely not be a linear extension of the past. The future will have twists and turns that will appear logical after they occur, but may not be apparent ahead of time. This follows the reasoning offered in chaos theory, a subject we will describe briefly in a later section. Therefore, we offer the following as our opinion about future areas of emphasis for organizations.

Services Will Continue to Increase as a Critical Success Factor (CSF) in Business

Several factors will force companies to increase their dependence on services as CSFs in the future. To compete successfully, they will need to keep increasing the level of the product/service package they offer to consumers. In many cases, it will be easier to add services than it will be to add product features.

Continuing Need to Integrate the Product and Service Bundle

As businesses add services to their product offerings, there will be a greater need to add these services as part of an integrated product/service package. A piecemeal approach will not have the appeal of one that has a holistic appearance.

Continuing Increase in Globalization

Globalization will continue to expand, for two main reasons. One will be the result of a search for lower costs. Perhaps even more important will be the desire to locate operations closer to the market. This will necessitate the need to be able to manage a diverse and widespread organization structure effectively. Of particular interest is the movement of nearly every major auto maker to China. Partnerships with Chinese auto makers have allowed auto makers to market their cars in China, and take advantage of low labor costs as well. The future configuration of the auto industry will be intriguing to watch. Is it possible that U.S. auto makers will manufacture vehicles in China, only to export them back to the United States? This scenario is worrisome to many, but only time will tell what the U.S. auto industry may look like in the future.

Outsourcing Will Become a More Focused Activity

As part of the globalization expansion, outsourcing will continue to be a major consideration. Although the past motivation to outsource has been largely the quest for lower costs, this will begin to change in the future. When businesses begin to analyze outsourcing from a total cost of ownership perspective and consider such indirect costs as longer lead times, increased inventories, transportation costs of products back to the home country, and flexibility constraints, they will begin to be more selective in their outsourcing efforts. Although they will continue to outsource, they will view it from a strategic objective.

Increased Need for Project Management Competencies

The requirements for good project management will grow. The dynamic nature of business will increase the need to view product development, outsourcing, and improvement programs as projects that need careful management not only to achieve desired results, but also to assure that projects are completed and do not interfere with the normal operation of the business.

Decision Making Will Deal with a Blend of Hard and Soft Variables

Managers strive to make good decisions. Most of the time, they try to accomplish this task by analyzing facts — hard numbers about costs, sales, productivity, response times — or a variety of other bits of data. Scholars have built an impressive array of methodologies to help in the decision-making process — time value of money, linear programming, expected value analysis, and economic value added, just to name a few. These analyses have had difficulty in dealing with soft numbers, or bits of information that are more difficult to quantify — things like customer satisfaction, consumer optimism, propensity to buy, and job satisfaction. Several noted authors warn that decision making in the future will require a blend of hard and soft variables.[3-5] Consequently, we will have to rethink our dependence on numbers and rely more heavily on the judgment or wisdom of the decision makers.

Decisions Will Become More Complex

Future decisions will be more complex because they will have further-reaching consequences. They will require more advanced decision-making methods. Will the answer be in more quantitative solution techniques? Will they come from greater involvement of the consumers — through surveys, focus groups, and blogs? Will companies have to be satisfied with less than optimum decisions because of the insurmountable complexity, or, in the terms of Simon, satisficing?[6]

Future of Improvement Programs

In the preceding section, we outlined some of the areas of emphasis we see ahead for organizations, especially business organizations. Where will the improvement programs that we have described throughout the book fit in? In this section, we will describe the changes that we expect in this area.

Improvement Programs Will Become Increasingly Important

Businesses will have to continue to improve. Much of the improvement will come from implementing improvement programs such as those described in this book. Improvements will come in various forms; without them, many businesses will ultimately fail.

Improvement Programs Will Be Customized to Specific Applications

As we pointed out in Chapter 4, many of the improvement programs have failed to realize the outcomes that companies desired because the program was not adequately adapted to fit the particular business that undertook the implementation. This will change in the future as companies become more proficient in selecting and adapting the improvement program to their specific need and situation.

Improvement Programs Will Be Better Managed

In addition to adapting improvement programs to fit their specific situation, businesses will need to do a better job of managing the implementation of the improvement program. They will plan and prepare, introduce, implement, monitor, and measure each step of the program's life so that it will be accepted and used for the benefit of the company, the employees, the customers, and other stakeholders. Project management will become a core competency for all successful companies.

Improvement Programs Will Become a Regular Part of Business Operations

By definition, an improvement program will have a finite life. At the end of its implementation, it will be important that the business assimilate the positive features of the program into its sustaining operations. Bolton and Heap call this the "lock-in" phase in which the business takes time to consolidate its gains before going on to the next improvement program.[7] Juran calls this the control phase of a company's life before moving on to a "breakthrough" or the next improvement program.[8]

Improvement Programs Will Lose Their Manufacturing or Service Heritage

We described programs in Chapter 5 and Chapter 6 as originating in either manufacturing or services. Some of these distinctions were somewhat arbitrary to illustrate the point that the manufacturing and services boundary is disappearing. We believe this will be even more evident in the future, where improvement programs will be a blend of manufacturing and services joint efforts, not a product of one or the other.

The Drivers of Change

As we described in Chapter 7, Chapter 8, and Chapter 9, there are three major drivers of change — technology, structure, and culture. In the past, they have been separate entities, sometimes at odds with one another as businesses attempt to introduce improvement programs. Consequently, improvement programs have been less effective for stakeholders, more disruptive and controversial among participants, and less durable than desired by management. To say that these conflicts will have to change in the future is saying the obvious, but how will it happen? We will use an analogy that may seem far-fetched in a business setting — children, families, and communities — but we will see if it makes sense.

Technology

Technology is like children. It is spontaneous, sometimes radical, often disruptive, and seldom traditional. It is a driver of change. It wants to go places, see things, pose questions, and rebel against the status quo. Technology wants to deviate from the traditional ways of doing things, but is often constrained by the structure of the organization, just as children are constrained by the structure of the family.

In the past, children were more influenced by their parents than by their peers or society in general. This is changing as children have greater exposure to the outside world. In a sense, technology is going through the same transition. At one time, technology grew from focused efforts within an organization. Today, technology advances come from knowledge transfer across multiple disciplines, fostered by electronic communication systems and more open information accessibility.

Structure

The structure of an organization is like the family. It provides vision, mission, strategies, programs, and processes to carry on the daily routines of operation. Although it may sometimes be a barrier to change introduced by technology, eventually it accepts good ideas and squashes bad ideas.

Structures of today's organizations are changing, some rapidly. The technology of the Internet, the openness of outsourcing, and the kinetic energy of employee empowerment are examples of the need for technology and structure to accommodate each other if there is to be harmony in the operation of the business.

Culture

The culture of an organization is like the community. It responds to the antics of technology, often with disapproval or rejection. Only over time will it adjust to, or accept, the change that technology offers. Organizations with deeply embedded cultures will find it difficult to implement any kind of improvement or change program. To improve, they must find a way to change the culture of the organization, a formidable task.

Sometimes communities take the initiative in change, perhaps as the result of a crisis or a reaction to some anticipated action with which they can mount organized opposition. When this happens, the culture can become a driver of change, although the momentum often is short-lived and somewhat unpredictable. It is not likely that cultures will change dramatically in the future. Companies that implement improvement programs will need to manage their organizational culture change as well.

Most Likely Future Methodologies

What are some of the methodologies, or programs, that will become popular in the future? Some are already well along the program life cycle while others are just appearing in the early stages of discovery. We describe them below, in no particular order of importance or timing.

Integrated Supply Chains

In Chapter 10, we described how to build an agile and integrated supply chain. To our knowledge, no company is there yet. Although some have moved well along the path we described, most businesses are still in the early stages. This journey will consume a large part of a company's effort in the future. They will probably move fast in setting up the physical parts of the supply chain — the customers, the products, and the suppliers. It will take much longer to develop the collaborative relationships that will make the supply chains most effective. Unfortunately, once a specific supply chain is achieved, it will become obsolete and a new supply chain will be required. Companies will have to be agile in creating new supply chains, perhaps using the virtual organization idea described in a later section.

Outsourcing

Outsourcing, especially offshore outsourcing, is currently in the growth part of its life cycle. From its humble beginnings as a "make or buy" decision area in manufacturing, outsourcing has grown to become a strategic weapon to gain a competitive advantage in both manufacturing and service activities. As it matures as an improvement program, outsourcing will be refined and adapted to fit specific business needs best. Some of the expectations for the future include the following:

- The outsourcing trend will continue in the United States; however, many companies are being more cautious and some are recalling outsourced work because of unsatisfactory results.[9]
- Extensive outsourcing will require a major restructuring of the purchasing, or procurement, function. Purchased services will increase as a percentage of the total costs and their composition will increase in complexity. It will take a multi-functional team to manage the outsourcing programs effectively.[10]
- Project managers, and project management skills, will become increasingly important. Although some in-house projects can be nursed along informally, outsourcing requires the formal coordination of functions and tasks, both internal and external.
- Although politicians lament the trend, it appears unlikely that the federal government will do anything substantial to stop the outsourcing movement.[11] Although there may be short-term hurdles, there will not be permanent barriers.
- U.S. companies will continue to outsource, but at a more deliberate pace as they move up the learning curve. Companies will view outsourcing as a management function and will learn to analyze, plan, manage, evaluate, and control the process. When appropriate, outsourcing will become a core function of a business.

Developing a successful outsourcing program is a little like making a soufflé. When done correctly, it is a masterpiece to enjoy. When done incorrectly, it is a mess to clean up.[12]

Total Cost of Ownership

One of the outgrowths of the outsourcing movement will be a more comprehensive method of determining the total cost of ownership of products and services. At the present time, the direct savings in labor costs is often enough to justify an outsourcing program. The major negative is increased lead time, although there are other "inconveniences" such as variances in quality and communication barriers. At some point in the future, these direct savings will decrease and indirect benefits and costs will become a more important consideration in the decision-making process. Such factors as value will assume greater importance. What is the value of more or less customer service? What is the value of proprietary knowledge that has been transferred to a supplier? What is the value of being a global market participant? Answering such questions will require a blend of known quantitative techniques coupled with new approaches to be discovered and refined.

Performance Measurement

Just as decision making will involve a blend of quantitative and qualitative variables, so will performance measurement. The traditional performance measurement systems have been focused on overall company financial performance and sometimes the specific measures used at lower levels of management were difficult to relate to overall performance. Consequently, depending on the use of the measures, they may or may not have been meaningful. To make performance measurement systems more effective in the future, several things have to happen:

- Performance measures should be selected to meet specific needs. It may not be possible to use the same measure for multiple tasks. For example, inventory turns may be a useful measure for inventory management purposes, but not necessarily for measuring the performance of a supervisor of a parts fabrication shop. Some measures work better for groups than for individuals.
- Performance measures should reflect the strategic and operational plans throughout the organization. If the plan is to minimize the level of work-in-process inventory, then machine utilization is probably not a fair measure to evaluate the effectiveness of a department.
- Discourage local optimization and encourage global objectives. Most businesses are trying to optimize the performance of their supply chains. This is not necessarily conducive to optimizing the performance of local functions or departments. It is difficult to encourage and facilitate inter-departmental or inter-company collaboration; using performance measures that introduce conflicts only makes it more difficult.
- Include both quantitative and qualitative measures. As we have discussed in earlier sections, there is a need to consider both quantitative and qualitative measures in decision-making analyses. It follows that businesses should learn to use qualitative variables as performance measures. This will encourage the acceptance of such concepts as satisficing[13] or "it is better to be roughly right than precisely wrong."[14]

- Performance measures should be accepted by all who use them. Although sometimes difficult and always time-consuming, it is a good idea to get acceptance of the performance measures by all of those being measured. Most employees do not like to have their performance measured; it is tolerable if they believe the measure being used is fair and appropriate.
- Performance measures should be usable and available. There must be a way to collect the data to be used in performance measurement. If there is not, then another measure should be found. This does not mean that it will not take time to set up a method of collecting the data, especially in the case of qualitative measures. Data collection should be routinized; converting data to useful information for analysis and action is where the human resources should be focused.

Project Management

Project management is an underutilized methodology. It has been around for several decades and is used in the military and in major construction projects. However, it can be useful within organizations to plan and monitor new product development, improvement programs, outsourcing projects, and other unique activities. As company operations become more volatile and complex, there will be greater opportunities to use project management techniques. Businesses are changing from a repetitive transaction orientation to a customizing process orientation.

Mass Customization

Like outsourcing, mass customization is in the growth part of its life cycle. The objective of mass customization is to provide a customized product or service to an individual customer — a market of one. Examples include Dell's customizing a computer, Subway's customizing a sandwich, and a physical therapist designing a treatment program for a patient. The objective is to provide high-volume customized products or services at prices competitive with standard products and services. Mass customization is closely allied with the concept of postponement where the customizing is done on standard modules.

One approach that may sometimes be thought of as mass customization is really not. Having 36 varieties of canned tomatoes at a grocery store is not mass customization; it is niche marketing. Introducing greater variety may be more of an attempt by a company to coax out extra sales than a response to consumer needs or wants. Some writers are even suggesting that having so many choices is causing stress among consumers because of the time and anxiety of making the decisions.[15] Needless to say, increased variety in manufacturing or service operations introduces another form of variance that is difficult to manage.

An interesting conflict arises between mass customization and offshore outsourcing. Mass customization needs short response times; offshore outsourcing requires longer response times. It will take a careful balancing of the two programs to arrive at an acceptable blend of the two concepts. One obvious approach is to outsource the standard modules and do the customizing as a domestic operation.

Virtual Organizations

Virtual organizations make it possible to adapt the organization to the specific need, whether it is to launch a new product — the Boeing 787 — or to assemble assistance for responding to a natural disaster — a hurricane. Virtual organizations may be for varying lengths of time — enough to get the job done. It is no longer feasible, and often does not make economic sense, for a company to design a product, build facilities to produce it, design and launch a marketing program, arrange financing, and do all of the other things necessary to introduce a new product. This approach made sense when the product life expectancy was years or even decades and changes in the succeeding products were minor. Today the product life cycles are in months or, hopefully, a few years, and the changes in succeeding products may be substantial. Today, the answer is to fit together a group of participants who can work together to accomplish the overall objectives — the virtual organization. Although the concept is still in the early stages, it makes sense and will become a more prominent factor in the future.

Information Technology

Information technology (IT) has enabled organizations to do things that previous generations only dreamed of when reading Jules Verne or Dick Tracy. Yet, most authorities agree that IT is only in the early stages of its potential. To accomplish supply chain integration, offshore outsourcing, mass customization, and virtual organizations, IT is necessary. Inter-organizational systems (IOS) to facilitate greater cooperation and coordination among businesses on projects will become standard. Advancement in such technologies as service-oriented architecture (SOA) will make it possible for "my computer to talk with your computer." As John Teresko, veteran columnist with *Industry Week,* puts it: "Remember when the factory floor and enterprise IT were two separate, unconnected worlds? Unfortunately, integrating these two disparate departments still remains a challenge for much of manufacturing, according to Cisco Systems and Rockwell Automation. Their advice: Think of integrating them as a way of future-proofing your business."[16] IT will be essential for companies of all sizes, both manufacturing and service.

Environmental Design

Business, society and government do not always work well together. One exception in the future may be in a program to protect the global environment. Society is becoming increasingly aware of and alarmed at the prospect of global warming consequences. Governments are noticing that society is concerned, and are taking steps to announce their intentions to act. Businesses are looking for ways to accommodate the need to change their ways and, if possible, to make money in doing it. This mixture of concerns will likely lead to some positive actions in the future. Right now, the efforts are piecemeal and cautious; however, there seems to be enough momentum to mount a significant effort. This will necessitate the use of supply chains, knowledge transfer, cross-functional teams, virtual organizations, IOSs, and a host of other tools and techniques to make such an effort successful. In a sense, it is a massive improvement program that incorporates the ITO model introduced in Chapter 3, with the objective to transform an input of an unprotected environment into an output of a protected environment. Incorporated in this transformation is the concept of reverse logistics.

Service Sciences

In Chapter 6, we described some of the improvement programs that originated in services and how they could be extended into manufacturing and other service organizations. In Chapter 10, we described how distributive services could be combined with extractive and transformative industries to form integrated and agile supply chains. In Chapter 11, we described how producer and social services form an integrated package to supplement the supply chains. Services are a vital and essential part of the world's economy.

However, most of the positive results from services have been achieved without a formal science or theory of services. Today, there is a growing interest in the field of services sciences. This is a program to enrich the contributions of services by adding a scientific approach — in the form of quantitative analysis theories and increased IT applications — to the already significant intuitive and ad hoc approach to service applications. Services sciences is a field that is gaining recognition among scholars and practitioners alike.

Chaos and Complexity

Most have probably heard numerous variations of the butterfly illustration used to describe events in chaos theory. The illustrations go something like this: a butterfly flapping its wings in Brazil can create tiny air currents that eventually lead to a hurricane in the Gulf of Mexico. The butterfly example illustrates a fundamental concept in chaos theory called "sensitive dependence to initial conditions."[17] This idea maintains that small differences in initial conditions can have vastly different outcomes. This concept becomes important to business researchers as they attempt to understand why similar companies in the same industry may have very different success outcomes. Chaos theory has also had applications in the area of understanding crisis events in the life of an organization.[18]

Perhaps the most promising aspect of chaos theory is its use in understanding complex phenomena, or the type of environment in which most organizations find themselves. Although technically chaos theory is rooted in mathematics and science, its use in understanding complex business phenomena is mainly metaphorical. Although some may write this area off as being too abstract, applications of chaos theory have been made in a number of areas, including the fields of psychology,[19] social work,[20] strategic management,[21] health care management,[22] public relations,[23] and public management.[24] Its future use looks promising.

Most Likely Improvement Programs

What kinds of improvement programs will we see in the future? What will be their goals? Probably much the same as today. We will look first at the types of improvement programs that will continue to receive attention. In a later section, we will look at the industries in which these improvement programs will have the greatest potential.

Cost Reduction

The pressure to reduce costs is almost a given in considering improvement program objectives. No matter how much businesses do, customers expect them to do more. Cost reduction programs almost seem to have inertia in them that keeps them moving along. Although there will be continued interest in reducing costs, the area of emphasis may shift somewhat in the future, to include greater efforts on reducing capital costs as well as the cost of the product or service provided.

- Product or service costs. Historically, companies have focused on reducing the costs of making products or providing services. With the help of engineers and empowered employees, companies have reduced direct labor costs to a very low percentage of the total cost. Many companies then switched their emphasis to the direct purchase side of the costs and worked to reduce direct material and other direct purchases through a variety of improvement programs and more recently, outsourcing. In addition, some companies are working to reduce overhead costs. This is proving more difficult because it is easy to identify the components of direct materials and direct labor costs; it is much more difficult to identify the components of overhead costs because most accounting systems conceal their true identity. However, overhead costs, as a group, are now usually two to four times as large as direct labor costs in manufacturing plants, and cost reduction programs must find ways to address these areas. Cost reduction programs of the future will continue to spend some time on direct labor costs, but the emphasis will also shift to direct purchases and overhead expense areas.

- Working capital costs. Working capital consists of accounts receivable, inventories, and accounts payable. Although the initial reaction may be that businesses do not want to reduce accounts receivable, that may not be the case. Consider this. One of the reasons that customers do not pay for products or services received is that there may be an unresolved problem and they are withholding payment until the problem is resolved. It may be because of a partial delivery, defective units, or errors in the invoice. Whatever the reason, it means that the business has excess accounts receivable because of unresolved issues. This could go unnoticed if the management information system does not report it to someone who can take action. As a result, there is a cost associated with having excess accounts receivable.

  Businesses pay a lot of attention to inventory, but often the focus is on having enough of individual items versus having too much in total. Inventory has two primary costs — having inventory (carrying costs) and the cost of being out of inventory (stockout costs). Any program to reduce costs must recognize the need to find the right level of inventory not just in total, but also for individual items. That is a huge task because many businesses, especially those in wholesale and retail, have thousands of items to manage. Although computer systems can do a lot of the work on staple items, they need considerable human input for seasonal and style items. Inventory will remain a key area for cost reduction; however, it requires finesse, not brute force.

  Accounts payable has also been an area where many companies have taken the approach that they just want concessions from their suppliers. Large companies have used their clout to force smaller suppliers to grant extended terms, or they have just not paid their suppliers on time. As building collaborative relationships becomes more important in future supply chains, it appears that greater consideration will be given to the total purchase package so that accounts payable will become part of a holistic solution.

- Capital equipment and facilities. Companies spend time and effort to justify their acquisition of capital equipment and facilities; they spend less time in considering the costs they incur in operating these investments. Even small companies perform time value of money analyses and consider buy versus lease alternatives. However, just as with overhead costs, it is more difficult to determine the cost of operating capital investments and the cost reduction potential in that area. There are isolated instances such as when the cost of fuel attracts attention for a large truck fleet, or when seasonal surges overload the capacity of a distribution center. However, the availability of such programs as outsourcing makes it possible to consider their impact on reducing the costs of capital equipment and facilities. However, such analyses require that the analysts assemble reliable operating cost information.

Response Time Reduction

As we described in Chapter 2, response times are rapidly becoming a key CSF for many businesses, manufacturing and service alike. James Womack and Daniel Jones reported that it takes an average of 309 days to move from aluminum oxide to an aluminum soft drink can in an individual's home. Although that seems like a long time, the amazing part is that, during the 309 days, only about three hours worth of work is performed.[25] That suggests that the total response time could be reduced by reducing inventories and move times. Of course, consumers do not experience that length of wait because they can buy whatever they want off the shelf at their local grocery store. However, if the business community is to move to mass customization, they must find ways to reduce response times.

- Existing products or services. There are several variations of response time. One is the time to respond to a customer's order for an existing product or service — a can of peas at a grocery store or a book from an online seller. A variation of that is the time to deliver a product or service that requires some slight modification — a sub at a deli counter or a computer with selected accessories added. An even longer response time results when the supplier must order the raw materials or assemble the service package — a wedding cake or public stock offering. Finally, the longest time is when the product or service must be designed — a house or an entire wedding. The above examples follow a hierarchy of approaches long known in manufacturing as "make to stock" (MTS), "assemble to order" (ATO), "make to order" (MTO) and "engineer to order" (ETO). MTS has the shortest lead time, but the least customization. Conversely, ETO has the greatest customization, but the longest lead times. Although the concepts are not new, the emphasis is now on how to increase customization while reducing lead times.

At some point in the above series of examples, the product or service may cease to be an existing model, but become a new model. This moves the discussion into the realm of response times to design new products or services. The automotive industry has moved rapidly in reducing the time it takes to design and move new models into production. What used to take five to ten years now takes 12 to 24 months. Techniques such as concurrent engineering are being used to achieve these reductions. The reductions in time have come largely because of reducing wait times caused by sequential processing where one functional area waits for upstream participants to finish their work. Another reduction is because there is less recycling of designs as participants resolve questions. These reductions are similar to the reduction of wait and move times described above. The actual design may be reduced some, but the greatest reductions have been in reducing nonvalue-added activities.

Response times are important and yet receive little attention in still another area. That is in the response time to implement improvement programs. This book is filled with ideas about improvement programs. If these programs can provide a benefit to companies, it follows that the sooner the programs are implemented, the sooner the company can begin receiving the benefits. In the future, companies will need to pay more attention to planning and implementing improvement programs faster.

Quality

The quality of products and services is a basic CSF, as are cost reduction and response time reduction. Both TQM and Six Sigma are improvement programs aimed directly at improving quality. Although they both originated in manufacturing, they are being successfully implemented in a number of service applications, as described in Chapter 5. Unfortunately, there are also many examples of only partial successes or even failures.

Product quality lends itself to measurement of tangible attributes. Does the product conform to its specifications? It is usually easy to know when a product is defective, a sure measure of quality. Product failure can often be corrected, hopefully before there is major damage or injury. Companies will continue to emphasize their efforts to improve product quality. Another aspect to product quality is reliability. Historically, most consumers were more concerned with quality at the time of their purchase; today many are also concerned with the expected reliability of the product over the time they will own it. This enters into their estimate of value, measured as the lifetime cost of a product. A growing number of consumers will pay more initially for a product if they believe it will last longer and have lower operating costs. Automobiles and appliances fall into this product category. As with overhead and stockout costs, it is still difficult to determine the lifetime costs of a product.

The quality of services is even more difficult to evaluate because a service may conform to specifications, but be unsatisfactory to a consumer. In evaluating service quality, a consumer looks at both the service and the process of how the service was provided. Did the cashier smile when she scanned your groceries? Was the waiter pleasant when he brought you the menus? Did the doctor spend enough time explaining the surgical process he was going to perform? Did the technician ask if you were comfortable as she started the ultrasound to clean your teeth? Did the help desk attendant adequately understand your problem and offer an answer that not only solved your problem, but also matched your level of computer literacy? As the world moves to a greater dependence on services to provide competitive advantage, it will become more important to understand service quality and how to improve it.

Customer Service Level

Customer service has moved into the forefront in recent years, as the movement to be responsive to customer needs becomes the foundation on which companies build their strategies. It could be thought of as a part of service quality, but it is probably better to view it as a separate CSF. We discussed the role of the customer extensively in Chapter 6. To paraphrase Drucker: Businesses only exist to serve the customer.[26] Customer service has many facets. It requires that the order taker quote a realistic delivery date, the inventory manager has the item in stock, the distribution center sends the correct item, the delivery person delivers the item to the correct address, the invoice is correct, and the check from the customer is credited to the correct account. These actions, and many more, all go into customer service. It is evident that the service side of customer service is equal in importance to the product side of the transaction.

Some of the items included in customer service can be measured, such as the fill rate on orders or the number of invoice errors. It may even be possible to trace the number of choices telephone callers have to make before being connected with a person, and then how many transfers they must wait through before they connect with someone that can help them. No matter how proficient companies become at measuring the measureables, they will still have to deal with the intangibles of individual customer differences. When we can identify and understand the DNA of humans and what makes them different, maybe we will be able to anticipate and prevent a customer from becoming dissatisfied.

Flexibility

In recent years, the increased emphasis on becoming lean and, with that, reducing inventories has required companies to become more flexible. Flexibility, in this case, means handling volume and mix fluctuations within the present system. Manufacturing and distributive companies handled volume fluctuation by building inventories in the slow periods and selling them down during the busy periods. This worked, but at the expense of higher average inventories. It also worked best with stable product lines in which new products were introduced at a pace that enabled companies to phase out the older products in an orderly fashion. To operate with lower inventories and with greater turnover in products from year to year, companies have to formulate better forecasts. To accomplish this, companies need to share information. Flexibility is as much about information sharing as it is about processes that can adjust to volume and mix variances.

Companies have worked to design flexible capacities to match flexible demands. This is especially true in scheduling employees. Service companies have become very good at scheduling employees to match demand in such businesses as restaurants, grocery stores, and clothing stores. They have volume fluctuations daily, weekly, and seasonally. In the future, they will need to become even better at this type of scheduling. One of the constraints will be moving from the traditional five-day, eight-hour work week that has been the norm. The business community and employees will have to find a way to accommodate this transition.

Flexibility also implies the capability to handle product and service mix variations. The emphasis has been to reduce the lot size so that the existing process can respond to changes in mix by making a wider variety of products during a single day. To do this, manufacturers have had to reduce setup times so that there is a less severe time penalty in switching from one product to another. This will continue to be the emphasis, and, in addition, manufacturing processes will be designed to make setups easier.

A number of service businesses already deal with customer mix variation. Employees have to have the versatility to handle what amounts to a lot size of one. A call center attendant can expect that the next call will be different from the last call. An order taker at L.L. Bean or Land's End can expect that no two callers will want the same product. The process provides information that, in combination with the ability of the employee, makes it possible for the employee to handle a variety of calls seamlessly. Where the process is automated, such as with online check writing, the computer is programmed to handle the variety. Flexibility will probably join cost, response time, and quality as a primary CSF for most businesses.

Agility

Agility is an extension of flexibility. It is the ability to change from one system of manufacturing or service to a new system to remain competitive. In manufacturing, it is the ability of an automobile manufacturer to move from a substantial part of its production capacity from SUVs to hybrids. In retailing, it could be for a retail bookstore chain to launch an online book-selling service (Barnes and Noble), the conversion of a "mom and pop" local chain to a "big box" national chain (Lowe's), or the conversion of a clothing manufacturer to a chain of retail clothing stores (Jos. A. Bank). In banking, it is the introduction of investment advisory services. In health care, it is the addition of a wellness center to help prevent health problems.

The key to being agile is in designing processes that can be converted or added to without being detrimental to the existing process. An even greater requirement is the ability of management to recognize the need and to introduce a program that will successfully achieve the transition. Agility is a strategic and long-term consideration. Only a few companies are doing it; some are thinking about it; many have not seriously considered it.

Compatibility

As supply chains mature, there will be an increasing emphasis on the compatibility of supply chain members. They will have to work together and this means increased communication, cooperation, coordination, and collaboration. This will require new technologies in communication, but it will also require large quantities of human interaction, which will be a challenge that will take years to effect. One of the essential ingredients is trust, and trust is not well understood in the business community. A great deal of research is being done in trying to understand trust and how it is developed not only between individuals, but also between organizations.[27] Although this area holds great promise, its realization will take time and patience.

Integration

Compatibility implies separate and distinct organizations working together. A stage even beyond that is integration of organizations until they become almost inseparable. This arrangement could mean the integration of entities into the supply chain until they are so completely coordinated that they become an integrated entity. In addition to supply chains, virtual organizations also hold promise of a way to develop integrated entities, if only for a short period of time. This concept could be extended beyond just business organizations to include social and political organizations to blend their interests and capabilities. The effort in global environmental initiatives would seem to be a worthy objective for an integrated system of organizations.

Industries Most Likely to Stress Continuous Improvement

Which industries will be the most active in implementing improvement programs? We believe that the emphasis will continue to shift to the services areas of business. Some of the specific industries that should be the most involved in continuous improvement programs are described below.

Health Care: Hospitals and Wellness Centers

One of the industries that offers tremendous opportunities for improvement is health care. There are governmental and social pressures as well as business pressures, to effect change. They are concerned with reducing costs, improving quality by reducing errors in patient medication and treatments, reducing response times (lengths of stay), improving flexibility in both volume and mix, increasing agility in handling a greater variety of illnesses and surgical procedures, becoming more compatible (patient friendly), and providing smooth transitions involving a host of independent operators (personal physician, specialist physician, surgeon, physical therapist, and life-style counselor). In other words, they are expected to become active in all types of improvement programs. At the same time, they will have to continue their sustaining and problem-solving parts of their business. They will have a host of new technologies, archaic organization structures (most have to outsource their primary transformation process of surgery), and a culture that is under attack from a number of influential outside sources (insurance companies, governments at all levels, privacy advocates, and trial lawyers, to name a few).

The traditional role of the health care profession has been to treat the sick. That is no longer good enough. The goal is to extend life expectancies and to improve the quality of life during that extended period. "Prevent" is the watchword, and there has been an explosion of wellness centers, personal trainers, weight reduction systems and procedures, physical and psychological trainers, assisted living centers, extended care centers, and life extension centers.

If that were NOT enough, there is the financial consideration of medical insurance that is taxing the best minds of government and industry. What a wonderful opportunity for a few good improvement programs!

Pharmaceuticals

Intimately connected with the health care industries is the pharmaceutical industry. Medications abound; just watch the six o'clock news on your local station. It is difficult to imagine a single person who is not taking medicine for some perceived or real ailment. It starts at infancy and continues forever. If there is an industry that views people as life-long customers, it is the pharmaceutical industry. There is a medication for every part of the body, from the hair on the top of one's head to the nails on one's toes. These comments should not be interpreted as criticism. Because of this intense search and marketing of medications, humans are living longer and, in some cases, such as infantile paralysis or measles, diseases have been essentially eradicated where vaccines are available.

There has been vast improvement. However, the approach has been similar to that in manufacturing — mass production or group solutions. As a result, just as with clothing, some sizes fit some people better than other people. There are side effects from some medications, either because of being the wrong medication for an ailment or for being incompatible with some biological configuration. Some side effects are minor and not worth worrying about; however, some side effects can kill. The profession struggles with trying to achieve a Six Sigma level of effectiveness.

Some research, such as in DNA, offers exciting promise. If the medication could be adapted to an individual's DNA, then the right medication could be prescribed for each individual. That is, if we could link up the supply chain that includes the manufacturer, the physician, the pharmacy, and a foolproof system to get a willing and totally compliant patient to take the correct medication. Of course, this means that the pharmaceutical manufacturer would have to enter the mass customization arena to match their medication with the patient. They might also have to make their medication compatible with a competitor's medication and share some of their proprietary information.

Local Government

If it were not enough that government agencies are expected to be effective — do the right things — they are also expected to be efficient. This challenge means they also need improvement programs. It may mean computerized processes to reduce the response time to getting automobile license renewals. On the other hand, it could mean scheduling night-time operating hours to accommodate the working public. How about simplified income tax legislation to make it easier for individuals to file tax returns? There are cost, quality, response time, and flexibility opportunities in government, just as in industry.

One idea that has had some success is privatization. Perhaps some governments, especially at the local level, are too complex and they need to outsource some of their functions to private industry, such as janitorial services or transportation systems. What are the core competencies of local governments?

Retailing

Although retailing is one of the oldest industries, it continues to evolve as it attempts to keep pace with the changing demands of customers. There is no doubt that the advent of the Internet has forced retailers to seek continuous improvement. The Internet has made retailing a 24/7/365 option. Even when their physical stores are closed, the retailer's online Website can continue to generate sales in "virtually" any part of the world where one has Internet access. The Internet has also forced retailers to compete on the basis of time. Deliveries are expected to be prompt and at a competitive price. Customers are usually given a choice of delivery options, depending on how fast they want their product and what they are willing to pay.

Retailers are also candidates for improvement programs related to quality and inventory control practices. Quality of service is expected from all retailers in this day and time, as consumers have many options to choose from and have high expectations on how they want to be treated. Inventory control is still important for all of the traditional reasons, the product must be available to the customer, and yet holding costs must not be too prohibitive.

Education

Most U.S. citizens believe that the educational system is one of the best in the world. Although that has been true, there is reason to believe that improvement is necessary, especially in science and technology. We include snippets of information from a joint report from The National Academy of Sciences, The National Academy of Engineering, and The Institute of Medicine in Managerial Comment 12-1.

Managerial Comment 12-1.
The Future of Education in the United States

Charge to the Committee

The National Academies was asked by Senator Lamar Alexander and Senator Jeff Bingaman of the Committee on Energy and Natural Resources, with endorsement by Representative Sherwood Boehlert and Representative Bart Gordon of the House Committee on Science, to respond to the following questions:

What are the top 10 actions, in priority order, that federal policymakers could take to enhance the science and technology enterprise so that the United States can successfully compete, prosper, and be secure in the global community of the 21st century?

What strategy, with several concrete steps, could be used to implement each of those actions?

Findings

Although the U.S. economy is doing well today, current trends in each of those criteria indicate that the United States may not fare as well in the future without government intervention. The nation must prepare with great urgency to preserve its strategic and economic security. Because other nations have, and probably will continue to have, the competitive advantage of a low wage structure, the United States must compete by optimizing its knowledge-based resources, particularly in science and technology, and by sustaining the most fertile environment for new and revitalized industries and the well-paying jobs they bring. We have already seen that capital, factories, and laboratories readily move wherever they are thought to have the greatest promise of return to investors.

Recommendations
The committee identified two key challenges that are tightly coupled to scientific and engineering prowess: creating high-quality jobs for Americans, and responding to the nation's need for clean, affordable, and reliable energy. To address those challenges, the committee structured its ideas according to four basic recommendations that focus on the human, financial, and knowledge capital necessary for U.S. prosperity.

The four recommendations focus on actions in K-12 education (10,000 Teachers, 10 Million Minds), research (Sowing the Seeds), higher education (Best and Brightest), and economic policy (Incentives for Innovation) that are set forth in the following sections.

Some actions involve changes in the law. Others require financial support that would come from reallocation of existing funds or, if necessary, from new funds. Overall, the committee believes that the investments are modest relative to the magnitude of the return the nation can expect in the creation of new high-quality jobs and in responding to its energy needs.

The committee notes that the nation is unlikely to receive some sudden "wakeup" call: rather, the problem is one that is likely to evidence itself gradually over a surprisingly short period.

Recommendation A: Increase America's talent pool by vastly improving K-12 science and mathematics education.

Recommendation B: Sustain and strengthen the nation's traditional commitment to long-term basic research that has the potential to be transformational to maintain the flow of new ideas that fuel the economy, provide security, and enhance the quality of life.

Recommendation C: Make the United States the most attractive setting in which to study and perform research so that we can develop, recruit, and retain the best and brightest students, scientists, and engineers from within the United States and throughout the world.

Recommendation D: Ensure that the United States is the premier place in the world to innovate; invest in downstream activities such as manufacturing and marketing; and create high-paying jobs based on innovation by such actions as modernizing the patent system, realigning tax policies to encourage innovation, and ensuring affordable broadband access.

Conclusion
We have led the world for decades, and we continue to do so in many research fields today, but the world is changing rapidly, and our advantages are no longer unique.... Without a renewed effort to bolster the foundations of our competitiveness, we can expect to lose our privileged position.... We owe our current prosperity, security, and good health to the investments of past generations, and we are obliged to renew those commitments in education, research, and innovation policies to ensure that the American people continue to benefit from the remarkable opportunities provided by the rapid development of the global economy and its not inconsiderable underpinnings in science and technology.

Source: Excerpts from Rising Above the Gathering Storm: Energizing and Employing America for a Brighter Future, The National Academies Press, 2006.[28]

The improvement program outlined in Managerial Comment 12-1 is comprehensive in its scope. It implies the need for improvements in practically all of the areas that we have described in earlier sections of this chapter.

Knowledge Management: Where Does It Fit?

As organizations consider their need to improve, it becomes obvious to most that one of the keys to success is in their ability to manage knowledge. The emphasis in improvement programs will shift from the physical orientation toward products and services to information flow and knowledge management. We will discuss some of the following questions in this section, using Figure 12-1 as a reference diagram.

- How does knowledge flow along the knowledge corridor from data to information to knowledge to wisdom?
- Can we use the supply chain analogy to show the flow of information?
- Will increased knowledge some day replace some of the need for "things"?
- How does knowledge management relate to the educational system?

Figure 12-1 shows the flow along what we are describing as the knowledge corridor.

From Data to Information

We will describe Figure 12-1 as a series of transformation processes moving from left to right. The first transformation is in converting data to information. In the top section, along the horizontal axis, we begin with data. It is transformed into information, largely by organizing bits of data into meaningful clusters of information. An example is to take individual daily sales by item, by customer and by store, and summarize this into stock replenishment orders by item, buying trends by customer, and revenue performance by store. It is usually collected transaction by transaction and, in today's environment, computers, in the form of point-of-sale terminals, do most of the data collection. Computers are also programmed to do most of the data-organizing activities. At the present time, most organizations are faced with the problem of having too much data, not too little. Their problem is how to make it more useful, e.g., convert it to information.

Most organizations are still learning to use the information that has accumulated effectively. The intent is to use it in their sustaining operations to make routine decisions in inventory management, customer relationship management (CRM), and resource utilization. To do this, they have to have an effective system for disseminating the right information to the right people to make the right decisions. They want to send enough information to users so that they can make better decisions; however, they do not want to overburden users with excess information. It is a complex and never-ending process to tailor the collection and organization processes so that there is an effective flow from the data collection point to the decision-making point. At the present time, the effectiveness of information systems ranges from leaders with smooth and comprehensive flows to laggards who have undirected data collection and erratic information flow. We have labeled the overall situation as fragmented.

Figure 12-1 also shows converging lines that move from a widely separated state at the data stage of the knowledge chain to a narrowly separated state on the right-hand side of the diagram. This is to show that there must be a selection process that separates and preserves the most important data into more concise elements of information.

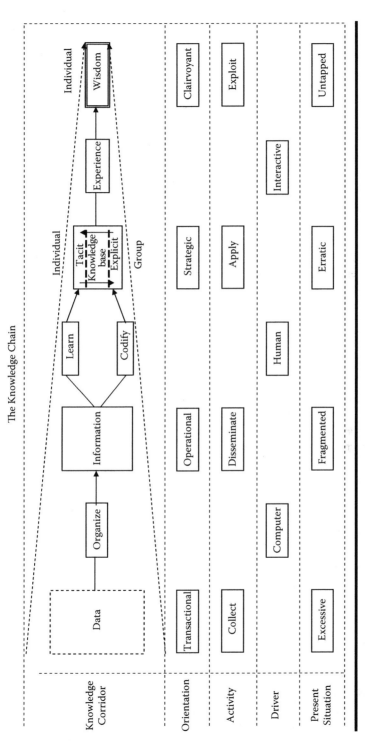

Figure 12-1 The flow of information along the knowledge corridor.

From Information to Knowledge

The next major transformation process is in converting information into knowledge. Although the conversion of data to information can be handled largely by computers, most authorities agree that people are needed to convert information to knowledge. This conversion has two major paths — learning and codifying. Learning is a process in which individuals convert information to tacit knowledge, or knowledge that is lodged within their own minds. Tacit knowledge remains with an individual until it is shared with another individual or codified to make it available to groups. From an organization's perspective, tacit knowledge is only valuable as long as that individual stays with the organization. It is to the organization's benefit to convert tacit knowledge to explicit knowledge.

Codifying information involves documenting in some formal process the rules, policies, and procedures of an organization; the result is known as explicit knowledge. Explicit knowledge is often the result of a group effort and is available to a wide cross-section of the organization. It also extends beyond the tenure of any individual or groups of individuals; it is part of the organization's knowledge base.

Knowledge is used in strategic planning and making decisions. The goal is to use the knowledge that has been derived from the collecting and disseminating processes described earlier. At the present time, it appears that there is even greater dispersion among organizations as regards the use of knowledge than there is with respect to the use of information. Information comes from systems and there is great consistency among information systems because of the efforts of many stakeholders to standardize information systems to facilitate intercompany communications. Knowledge, however, comes from individuals and there is great variation among people. Companies are often willing to share information; they are less willing to share knowledge. At the present time, it appears that the most composite description of the handling of knowledge is erratic, reflecting the even wider range of use by organizations.

As with the movement from data to information, Figure 12-1 continues the converging lines to convey the concept that knowledge is extracted from information into more concentrated and focused resources.

From Knowledge to Wisdom

The final stage shown in Figure 12-1 is the transformation of knowledge into wisdom. This is unexplored territory for most companies and for the individuals within those companies. Wisdom implies a level of understanding even beyond that shown for knowledge. Wisdom is largely individually oriented at this point. It is hard to conceive as wisdom emanating from an inanimate organization. Individuals gain wisdom in a variety of ways, but largely through their own experiences. Some individuals gain wisdom while others, with similar experiences, do not. The expression "20 years of experience" versus "one year of experience twenty times" seems to capture this distinction. One of the anomalies is that some people have it, but most do not. Individuals with wisdom may almost appear surreal because they seem to have an uncanny knack of predicting or spotting problems or opportunities before most others do. They may even appear clairvoyant, a characteristic that we suggested in Chapter 10 would be a desirable capability for businesses to have in the future. When an organization discovers wisdom in an individual, they should exploit it for the good of both the organization and the individual.

Gaining wisdom is a learning process; exactly how that process works is still largely unknown. In fact, some individuals may have wisdom, but are not aware of it because there is no way to recognize and document it. As a result, wisdom is an untapped resource in most organizations because, although some individuals within the organization may have wisdom, it goes unrecognized. It takes a rare combination of circumstances and individuals to expose wisdom. We believe that wisdom is largely undiscovered and even untapped in most organizations. The converging lines previously described continue to narrow as the diagram reaches wisdom.

The knowledge chain is a useful concept for continuous improvement. It implies that improvement programs can help in the development along the knowledge chain. The concept also suggests that the selection and implementation of improvement programs will gain effectiveness and efficiency through the application of information, knowledge, and wisdom.

We described the ITO Model in Chapter 3 as the DNA of the supply chain. It is also a model for other kinds of transformations. As we described in the previous section, the flow of knowledge can also be viewed as a form of supply chain with the ultimate consumer being the individual with wisdom and the fragments of data that enter the information chain being the inputs. Hopefully, this book will aid those who read it in their own quest for knowledge and wisdom.

Notes

1. Emblemsvag, J. and Bras, B., Process thinking — A new paradigm for science and engineering, *Futures*, 32, 7, 635, 2000.
2. Simon, H.A., *Administrative Behavior: A Study of Decision Making Processes in Administrative Organizations* (Fourth Edition), The Free Press, New York, 1997.
3. Emblemsvag, J. and Bras, B., Process thinking — A new paradigm for science and engineering, *Futures*, 32, 7, 635, 2000.
4. Mintzberg, H., *Mintzberg on Management, The Free Press, Inside Our Strange World of Organizations,* The Free Press, New York, 1989.
5. Pink, D.H., *A Whole New Mind: Moving from the Information Age to the Conceptual Age*, Riverhead Books, New York, 2005.
6. Simon, H.A., *Administrative Behavior: A Study of Decision Making Processes in Administrative Organizations* (Fourth Edition), The Free Press, New York, 1997.

7. Bolton, M. and Heap, J., The myth of continuous improvement, *Work Study*, 51, 6/7, 309, 2002.
8. Juran, J.M., *Managerial Breakthrough, A New Concept of the Manager's Job*, McGraw-Hill, Inc., New York, 1964.
9. Landis, K.M., Mishra, S., and Porrello, K., Calling a change in the outsourcing market: The realities for the world's largest organizations, Deloitte Consulting, Email: callingachange@deloitte.com, April 2005.
10. Venkatesan, R., Strategic Sourcing: To make or not to make, *Harvard Business Review*, 70, 2, 98, 1992.
11. Gibson, S., Outsourcing Is Growing Up, www.eweek.com, March 21, 2005.
12. Crandall, R.E., Device or strategy? Exploring which role outsourcing should plan, *APICS Magazine*, 15, 7, 21, 2005.
13. Simon, H.A., *Administrative Behavior: A Study of Decision Making Processes in Administrative Organizations* (Fourth Edition), The Free Press, New York, 1997.
14. This quote has been attributed to several economists — Greenspan and Keynes, to name two. We expect that the quote goes back a long way, but we haven't found the sources yet.
15. Schwartz, B., *The Paradox of Choice, Why More Is Less*, Harper Collins, New York, 2004.
16. Teresko, J., Integrating IT with manufacturing, www.industryweek.com/PrintArticle.aspx?ArticleID=14151, June 1, 2007.
17. "The butterfly effect" was coined and discovered by Edward Lorenz. See Lorenz, E., *The Essence of Chaos*, The University of Washington Press, Seattle, 1993. Interested readers should also consult James Gleick's popular book for a less technical explanation of chaos theory. See Gleick, J., *Chaos: Making a New Science*, Penguin Books, New York, 1987.
18. Crandall, W., Crisis, chaos, and creative destruction: Getting better from bad, in *Re-Discovering Schumpeter Four Score Years Later: Creative Destruction Evolving into "Mode 3,"* Carayannis, E.G. and Ziemnowicz, C., Eds., MacMillan Palgrave Press, 2007.
19. See Barton, S., Chaos, self-organization, and psychology, *American Psychologist*, 49(1), 5–14, 1994; Carver, C., Dynamical social psychology: Chaos and catastrophe for all, *Psychological Inquiry*, 8(2), 110–119, 1997.
20. See Bolland, K. and Atherton, C., Chaos theory: An alternative approach to social work practice and research, *Families in Society: The Journal of Contemporary Human Services*, 80(4), 367–373, 1999; Hudson, C., At the edge of chaos: A new paradigm for social work? *Journal of Social Work Education*, 36(2), 215–230, 2000.
21. See Dervitsiotis, K., Navigating in turbulent environmental conditions for sustainable business excellence, *Total Quality Management*, 15(5–6), 807–827, 2004; Edgar, D. and Nisbet, L., A matter of chaos — Some issues for hospitality businesses, *International Journal of Contemporary Hospitality Management*, 8(2), 6–9, 1996.
22. McDaniel, R., Jordan, M., and Fleeman, B., Surprise, surprise, surprise! A complexity view of the unexpected, *Health Care Management Review*, 28(3), 266–278, 2003.
23. Murphy, P., Chaos theory as a model for managing issues and crises, *Public Relations Review*, 22(2), 95–113, 1996.
24. Farazmand, A., Chaos and transformation theories: A theoretical analysis with implications for organization theory and public management, *Public Organization Review*, 3, 339–372, 2003.

25. Womack, J.P. and Jones, D.T., *Lean Thinking: Banish Waste and Create Wealth in Your Corporation,* Simon & Schuster, New York, 1996.
26. Drucker, P.F., *Management: Task, Responsibilities, Practices,* Harper & Row, New York, 1974.
27. Magsood, T., Walker, D., and Finegan, A., Extending the "knowledge advantage": Creating learning chains, *The Learning Organization,* 14, 2, 123, 2007.
28. Committee on Prospering in the Global Economy of the 21st Century: An Agenda for American Science and Technology, National Academy of Sciences, National Academy of Engineering, Institute of Medicine, Rising Above the Gathering Storm: Energizing and Employing America for a Brighter Economic Future, The National Academies Press, 2006; also see Fahey, L. and Prusak, L., The eleven deadliest sins of knowledge management, *California Management Review,* 40(3), 265–276, 1998; Girad, J., Where is the knowledge we have lost in managers? *Journal of Knowledge Management,* 10(6), 22–38, 2006; Lester, D. and Parnell, J., *Organizational Theory: A Strategic Perspective,* Atomic Dog Publishing, Mason, OH, 2007.

Index

A

Administrative management, 100–102
Agriculture, future technology, 206–209
Alternate organizational structures, 251–254
    horizontal organization, 252–253
    matrix organization, 251–252
    virtual organization, 253–254
Anthony, Scott, 6, 18, 37, 49, 145
Artifacts, in organizational culture, 283–291
Automation, 140–141, 210

B

Basic ITO model, 58–65. *See also* ITO model
    basic model extended, 61–62
    for both manufacturing, services, 58
    components of, 62–65
    customers, 62–63
    general systems model, 58–61
    inputs, 64–65
    outputs, 63–64
    supply chain configurations, extending into, 65–75
    transformation process, 64
    resources, 60
Batch flow, transition from, to integrated and agile supply chain, 321
Behavioral issues, 226–227
Benefits of using models, 57
Boundary between manufacturing/services, disappearance of, 1–18, 380
    critical success factors, 4–5
    customer input, 9
    customer relationship management, 9
    customer service improvements, 17
    economy, evolution of, 4
    efficiency, 5–7
    financing of purchases, 8–9
    health care, 6
    industry focus, *vs.* functional focus, 13
    integrated product, service functions, 17–18
    internal improvements, costs, quality, 15–17
    investment banking, 6–7
    manufacturing, 7–9
    service, differences between, 2–4
    mass customization, 13–14
    online purchasing, 9
    outsourcing, 11–12
    personal, organizational relationships, changes in, 12
    post-sales services, 8
    process perspective, movement towards, 11
    retail industry, 7
    separated disciplines, 15
    services, 13
    supply chains
        expansion of, 5
        integrating companies into, 12–13
    tasks, resources, techniques, blending into programs, 5
    technology development, 9–10
Bowen, John, 3, 13, 18, 118, 134, 141, 143–145, 188, 194
Burgess, Thomas, 45, 49
Business intelligence, use of, 185

Business planning process, 32–33
Business services. *See* Producer services

C

Cadmium, EU market restrictions, 77
Ceremonies, in organizational culture, 286–287
Change, drivers of, 387–388
    culture, 388
    in improvement program, 10
    structure, 387–388
    technology, 387
Chaos complexity, 393–394
Chase model, 122–123
Christensen, Clayton, 6, 18, 145
Classical management functions, 245–248
    controlling, 248
    directing, 247–248
    organizing, 246–247
    planning, 246
Classical management thinkers, 70–71
Closed ITO model. *See also* ITO model
    with multiple customer roles, 62
Closed loop supply chain, 77
Closed systems, 67–73
    strategy, 70–71
CLSC. *See* Closed loop supply chain
Collaboration, 171–172, 225
Complexity of marketplace, 151
Comprehensive supply chain model, 320–322
Conceptual technologies, 208
Configuration of integration supply chain, 327–331
    distribution, 329
    finance, accounting, 329–330
    manufacturing, 328–329
    marketing, 327–328
    purchasing, 328
    top management, 330–331
Construction technology, 206–209
Consumer services
    defined, 371–372
    future of, 372
    roles of, 372
Contingency theory, 103–105
Continuous improvement, 200–205
Continuous replenishment programs, 170
Control, as management function, 248
Core competency, 72

Corporate life, defined, 20
Costing, activity-based, 139
Costs, benefits of added technology, 227–228
Creativity, fostering, 280
Critical success factors, 2, 4–5, 19–50, 53, 64, 84, 128, 148, 380, 383–384, 396–399, critical success factors27–29
    business planning process, 32–33
    characteristics of, 37–39
    defined, 21–22
    effect on operations deployments, 44
    evolution in U.S., 22–26
    growth, 40–41
    hierarchy of, 33–34
    interrelated, examples, 34
    manufacturing flows, 44–45
    manufacturing strategies, 45
    maturity, 41
    in operational planning, 41–45
    performance measurement, 47
    planning process, 33
        hierarchy, 30–33
    positioning strategy, 43–44
    project and program plans, 33
    recommendations, 42–43
    role, 34–41
    in selecting management programs, 45–47
    service increase, 384
    start-up, 40
    strategic planning
        characteristics of, 37–39
        defined, 35–36
    in strategic planning, 31–32, 34–41
        characteristics of, 37–39
    temporal nature of, 39–41
    use at IBM, 46
Cross-functional teams, 258–259
CRP. *See* Continuous replenishment programs
CSFs. *See* Critical success factors
Culture
    as driver of change, 388
    slogans, strategy, interplay of, 301
Culture of organization, 271–306
    artifacts, 283–291
    behavioral norms, 287–288
    ceremonies, 286–287
    changing, steps in, 293–303
    commitment to company, 279–281

components of, 282–291
identity, 276–277
importance of, 276–282
language, 284–286
organizational rewards, 290–291
physical surroundings, 288–290
rituals, 286–287
stability, 281–282
stories, 284
strategy, culture, aligning, 292–293
symbols, 286–287
types of, 291–293
value system, 283, 287–288
Customer
  acceptance, 232–233
  encounter, managing, 185
  input, 9, 184
  relationship management, 9, 164–169
    supply chain, 168
  service improvements, 17
  as shopper, maturity of, 151
Customer-centric strategy, product-centric strategy, compared, 154
Customer-engaged organization, steps to, 153–155
Customization
  evolution to, 14
  mass, 174–175
    operational effectiveness in, 176

D

Day-to-day operations management, 87–88
Decentralization, 254–255
Decision-making criteria, 225–228
Demand for producer services, determinants of, 351–352
Demand management, 178–179
Directing, as classical management function, 247–248
Discontinuous, continuous incremental change, contrasted, 89–90
Distributive services, 161

E

E-commerce, 212–214
  definitions, concepts, 212
  implementation, 215
Economic life cycle, changes during, 26–29
Economy, evolution of, 4

Education, 402–404
  in U.S., future of, 403–405
Employee management theories, 98
Employee turnover data, for managers, 264
Empowered employees, 255–256
Enterprise resource planning, 140, 224
Environmental design, 393
ERP. *See* Enterprise resource planning
Evolution of management programs, 97–105
Extended ITO model. *See also* ITO model
  with customers, suppliers added, 60

F

Fayol's list of management principles, 100–101
Financing of purchases, 8–9
Flexibility, planning for, 174–180, 259–260, 358
Flexible organization forms, 176–177
Ford, Henry, 25, 132
Ford, Robert, 3, 188
Forward logistics supply chain, 79
Functional classification, producer services, 353
Functional organization structure, 249
Future developments, 379–412
  agents of change, 382
  areas of emphases, 383–385
  boundary between manufacturing/services, disappearance of, 380
  change drivers, 387–388
    culture, 388
    structure, 387–388
    technology, 387
  chaos complexity, 393–394
  education, 402–404
  environmental design, 393
  health care industry, 400–401
  improvement programs, 385–387, 394–400
    agility, 399
    compatibility, 400
    cost reduction, 394–396
    customer service level, 398
    flexibility, 398–399
    integration, 400
    quality, 397
    response time reduction, 396–397
  information technology, 392
  integrated supply chains, 388–389

knowledge management, 405–409
  data, to information, 406–407
  information, to knowledge, 408
  knowledge, to wisdom, 408–409
knowledge transfer, 381–382
local government, 402
management programs, 111–113
mass customization, 391–392
methodologies of future, 388–394
outsourcing, 384
performance measurement, 390–391
pharmaceuticals, 401–402
project management, 391
related entities, integration of, 382–383
retailing, 402
service sciences, 393
in technology, 231–235, 237
total cost of ownership, 389–390
virtual organizations, 392

G

Gates, Bill, view of future technology, 234–235
Geographic organization structure, 250–251
Globalization, 23–26, 384
Government social services, 366

H

Health care industry, 6, 43, 368–369, 400–401
Hexavalent chromium, EU market restrictions, 77
Horizontal communication, 175–178
Horizontal organization, 252–253
Horizontal structures, 255
Hospital industry. *See* Health care industry
Human relations management, 99–100
Human resources, 214–219

I

IBM, critical success factors, 46
Identity, organizational culture and, 276–277
Improvement programs, 394–400
  agility, 399
  compatibility, 400
  cost reduction, 394–396

customer service level, 398
flexibility, 398–399
focus of, 164
future of, 385–387
integration, 400
quality, 397
response time reduction, 396–397
Industrial Revolution, 23
Industry classifications, 158, 347
Industry focus, *vs.* functional focus, 13
Information
  flow along knowledge corridor, 407
  replacement of inventory with, 180
Information technology, 222–223, 392
  matching to business, issues, 223
Infrastructure, 241–270
  alternate organizational structures, 251–254
    horizontal organization, 252–253
    matrix organization, 251–252
    virtual organization, 253–254
  change in, 267–268
  classical management functions, 245–248
    controlling, 248
    directing, 247–248
    organizing, 246–247
    planning, 246
  cross-functional teams, 258–259
  decentralization, 254–255
  defined, 242–243
  empowered employees, 255–256
  flexibility, 259–260
  higher skill variety, 256–257
  horizontal structures, 255
  Internet, role in changing organizational structure, 261–262
  knowledge management integration, 263–267
  multi-directional communications, 259
  organic structures, 260–261
  organization structure, 248–251
    functional, 249
    geographic, 250–251
    product, 249–250
  self-directed work teams, 257–258
  strategies, 243–245
    business strategy, 244–245
    corporate strategy, 243–244
  trends, 254–261
Innovation, 285–286, 289

Input-Transformation-Output Model. *See* Basic ITO model; ITO model
Integrated product, service functions, 17–18
Integrated service package, 373
   need for, 373
   steps in developing, 373–376
Integrated supply chains, 307–344, 388–389
   agile, lean supply chain, steps to achieve, 325–335
      change process, steps, 335–340
      clairvoyance, 339–340
      complexity, 338–339
      implementation, 336
      initiation, 336
      innovation, 337
      institutionalization, 336–337
      integration, 336
      investigation, 335–336
      invigoration, 336
   collaboration, 334–335
   commitment, 325–326
   communication, 331
   complexity, 338–339
   configuration, 327–331
      distribution, 329
      finance, accounting, 329–330
      manufacturing, 328–329
      marketing, 327–328
      purchasing, 328
      top management, 330–331
   coordination, 334
   culture, 331–332
   customization, 332–333
   delivery process, 324–325
   demand variation, 325
   implementation, 336
   initiation, 336
   innovation, 337
   institutionalization, 336–337
   integration, 336
   investigation, 335–336
   invigoration, 336
   material flows, 210
   organization hierarchy, 310–311
   present status, 308–310
   production process, 324
   purchasing process, 324
   role of supply chain, 312–313
   setting stage, 313–317
   supply chain models, 317–325
      comprehensive, 320–322
      evolution of, 317–320
      lean production, 322–325
      systems theory, 312
Inter-organizational communications, 46, 180–183
Inter-organizational virtual organization, 253
Interactions among plans, 31
Interdisciplinary design methodology, 219
Interfacing business, producer services, steps, 361–364
Internal improvements, costs, quality, 15–17
Internet, role in changing organizational structure, 261–262
Internet EDI, 181–183
Interrelated critical success factors, examples, 34
Inventories, 125–126
Investment banking, 6–7
IOS, inter–organizational systems
IT. *See* Information technology
ITO model, 51–83
   basic ITO model, 58–65
      basic model extended, 61–62
      components of model, 62–65
      customers, 62–63
      extending into supply chain configurations, 65–75
      general systems model, 58–61
      inputs, 64–65
      outputs, 63–64
      transformation process, 64
   benefits of using models, 57
   closed systems, 67–73
   examples, 66
   feedback, 73
   models, use of, 56–57
   open systems, 67–73
   pizza chain example, 247
   relationship-building, 73–75
   reverse logistics, 76–80
      barriers to, 78
      benefits of, 77–78
      system design, implementation, 79–80
   role in business relationships building, 73
   supply chain, 67
   types of models, 55–56
ITO modules, examples, 59

J

JIT, 132–134

Total Quality Management, compared, 104

K

Key performance indicators, 37, 47
King, Stephen, 45, 49, 191
Knowledge corridor, flow of information, 407
Knowledge management, 405–409
    data to information, 406–407
    information to knowledge, 408
    integration, 263–267
    knowledge to wisdom, 408–409
    strategies, 105
Knowledge transfer, 381–382
    from services to manufacturing, 162–163
KPI. *See* Key performance indicators

L

Lead, EU market restrictions, 77
Lean production, 322–325
Lean supply chains
    implementation, 309
    progress in developing, 312
Left-brain thinking, 340
Life cycles, management program, 92–96
    at end of life cycle, 94
    implications for management, 94–96
    life-cycle stages, 92–94
Local government, 402
Location near market, 179

M

Make-to-stock, to make-to-order, 157
Management operations, relationship of programs to, 87
Management program life cycle, 93
Management programs, 83–116
    administrative management, 100–102
    contingency theory, 103–105
    day-to-day operations, sustaining, 87–88
    defined, 85–92
    employee management theories, 98
    evaluation of, 110–111
    evolution of, 97–105
    examples, 86
    future of, 111–113
    human relations management, 99–100
    implementation, 109–111
    importance of, 96–97
    improvement programs, 88–92
    life cycles, 92–96
        at end of life cycle, 94
        implications for management, 94–96
        life-cycle stages, 92–94
        matching program with need, 107–109
    preparation, 109–110
    problem-solving activities, 88
    scientific management, 98–99
    systems theory, 102–103
Management thought, programs, evolution of, 104
Manufacturing, 7–9
    areas of expertise, 162–163
    flows, 44–45
    objectives, 125–130
    process types, 119
    services, differences between, 2–4, 186–187
    techniques, 117–146
        activity-based costing, 139
        automation, 140–141
        description of types, 120–127
        Enterprise Resources Planning, 140
        materials requirements planning, 139–140
        objectives, 125–130
            inventories, 125–126
            product costs, 125
            product development time, 129–130
            quality, 127–128
            resource utilization, 126–127
            response time, 128–129
        performance measures, 140
        product costing, 139
        product-process relationship, 122
        resource utilization, 141
        service industry classifications, 122–123
            Chase model, 122–123
            Schmenner model, 123
        service objectives, 130–131
        services
            manufacturing, comparison of, 123–125
            programs, 131–138
        types, description, 120–127

technology, 206–209
Manufacturing techniques and services programs
    JIT, 132–134
    lean, 132–134
    Six Sigma, 135–138
    Total Quality Management, 135–138
Mass customization, 13–14, 174–175, 391–392
    evolution to, 14
    movement toward, 157–158
    operational effectiveness in, 176
Materials requirements planning, 139–140
Matrix, product, service process, 124
Matrix organization, 251–252
Mechanistic, organic organization, compared, 261
Mercury, EU market restrictions, 77
Methodologies of future, 388–394
Mining, future technology, 206–209
Models
    nature of, 54
    use of, 56–57
MTS. *See* Make-to-stock
Multi-directional communications, 259

N

National economic life cycles, changes during, 26–28
New conditions, adaptation of programs to, 108
Nonprofit organizations, social services, 366
Nonquantitative performance measurement, 185
Normal transformation process of company, transitions in, 91

O

Online purchasing, 9
Open system ITO model, 69. *See also* ITO model
Open systems, 67–73
    environment, 184
    strategy, 71–73
Organic structure, 260–261
    mechanistic organization, compared, 261
Organization hierarchy, 310–311
Organization structure, 248–251

    functional, 249
    geographic, 250–251
    product, 249–250
Organizational culture, 202–204, 271–306
    artifacts, 283–291
    behavioral norms, 287–288
    ceremonies, 286–287
    changing
        focus areas, 294
        steps in, 293–303
    commitment to company, 279–281
    components of, 282–291
    identity, 276–277
    importance of, 276–282
    language, 284–286
    organizational rewards, 290–291
    physical surroundings, 288–290
    rituals, 286–287
    stability, 281–282
    stories, 284
    strategy, culture, aligning, 292–293
    symbols, 286–287
    types of, 291–293
    value systems, 283, 287–288
Organizational rewards, 290–291
Organizational structures, 251–254
    horizontal organization, 252–253
    matrix organization, 251–252
    virtual organization, 253–254
Organizing, as classical management function, 246–247
Outside producer services, reasons why businesses seek, 354
Outsourcing, 11–12, 352–353, 384

P

PBDE. *See* Polybrominated diphenyl ether
Performance measurement, 47, 140, 390–391
Performance of equipment, enhancing, 219
Personal, organizational relationships, changes in, 12
Personal services, 161
Pharmaceuticals, 401–402
Planning, as classical management function, 246
Planning process, hierarchy of, 30–33
Polybrominated diphenyl ether, EU market restrictions, 77
Positioning strategy, 43–44

Post-sales services, 8
Privatization, 370–371
Problem-solving activities, 88
Process perspective, movement towards, 11
Process technologies in manufacturing, 207
Producer services, 161
    classes of, 355
    in company, recognition of, 362
    defined, 350–365
    future of, 364
    importance of, 352
    reasons to acquire, 355–361
    role of, 353–355
    vs. consumer services, 347–350
Producer services firms, purchase of producer services, 364–365
Product-centric, customer-centric strategies, compared, 154
Product costing, 125, 139
Product development time, 129–130
Product-process relationship, manufacturing, 122
Professor, hiring, potential scenarios, 296
Program extensions, success rates, 106
Programs developed in services, examples, 163–185
Publication life cycle, 93

Q

Quality issues, 127–128, 184
Quasi-private firms, social services, 367

R

Regulatory compliance, 360–361, 370
Relationship-building, 73–75
Resource enhancement, technology for, 214–225
Resource utilization, 126–127, 141
Response systems, 169–170
Response time, 128–129, 169–174
Restriction on Hazardous Substances Directive, 77
Retail industry, 7, 402
    swing of power to, 150–151
Reverse logistics, 76–80
    barriers to, 78
    benefits of, 77–78
    supply chain, 79
    system design, implementation, 79–80

Rituals, in organizational culture, 286–287
Robots, categories of, 211
RoHS Directive. *See* Restriction on Hazardous Substances Directive
Role of supply chains, 312–313
Roth, Erik, 6, 18, 145

S

Sales, operations planning, 86, 112, 164, 171–172, 179, 315
Schmenner model, 123
Scientific management, 98–99
Sectoral classification, extended, 348–349
Self-directed work teams, 180, 257–258
Self-service, 161, 211–212
Service, manufacturing, differences between, 2–4
Service continuum, product, 8
Service industry classifications, 122–123
    Chase model, 122–123
    Schmenner model, 123
Service objectives, 130–131
Service-oriented architecture, 224–225
Service sciences, 393
Service techniques, 147–195
    business intelligence, use of, 185
    collaborative, 171–172, 188–189
    complexity of marketplace, 151
    consumer response, 170
    continuous replenishment programs, 170
    customer as shopper, maturity of, 151
    customer encounter, managing, 185
    customer inputs, 184
    customer relationship management, 164–169
    supply chain, 168
    demand management, 178–179
    distributive services, 161
    flexibility, 174–180
    horizontal communication, 175–178
    information, replacement of inventory with, 180
    inter-organizational communications, 180–183
    Internet EDI, 181–183
    knowledge transfer, from services to manufacturing, 162–163
    location near market, 179
    make-to-stock, to make-to-order, 157

manufacturing, services, differences, 186–187
manufacturing expertise, areas of, 162–163
mass customization, 174–175
movement toward, 157–158
need for added services, 156
nonquantitative performance measurement, 185
open-system environment, 184
personal services, 161
producer services, 161
product-centric, to customer-centric, 153–155
programs developed in services, examples, 163–185
quality, 184
response systems, 169–170
response time, 169–174
retail, swing of power to, 150–151
rise of services, 150
sales, operations planning, 171
self-directed teams, 180
self-service, 161
service expertise, areas of, 163
services
defined, 158–162
as separate business segment, 156
social services, 161–162
supply chain management, 172
vendor-managed inventory, 171
Services, 13
defined, 158–162
manufacturing, comparison of, 123–125
programs, 131–138
JIT, 132–134
lean, 132–134
Six Sigma, 135–138
Total Quality Management, 135–138
Services complementing supply chain, 345–378
added expertise, 356–357
consumer services
defined, 371–372
future of, 372
roles of, 372
cost advantage, 355
expertise, 358–359
flexibility, 358
integrated service package, 373
need for, 373
steps in developing, 373–376
interfacing business, producer services, steps, 361–364
outsourcing, role of, 352–353
plan, supply chain interface, 362–363
privatization, 370–371
producer services
classes of, 355
defined, 350–365
future of, 364
importance of, 352
reasons to acquire, 355–361
role in company, recognition of, 362
role of, 353–355
vs. consumer services, 347–350
producer services firms, purchase of producer services, 364–365
progress measurement, 363
regulatory compliance, 360–361, 370
removal of noncore activities, 356
revisions, assimilation, 363–364
social services, 367–369
classes of, 366–367
government, 366
nonprofit organizations, 366
quasi-private firms, 367
communications systems, 367
defined, 365–371
education, 368
future of, 370–371
health care, 368–369
importance of, 365
information, 367
interfacing business with, steps, 369–370
location incentives, 368
protection, 368
role of, 365–366
standards, 367
support infrastructure, 368
training, 368
strategic planning, 359–360
strategy for producer services area, 362
suppliers, 362
Six Sigma, 135–138
Slogans, 285–286
culture, strategy, interplay of, 301
SOA. *See* Service-oriented architecture
Social services, 161–162, 367–369
classes of, 366–367
government, 366

nonprofit organizations, 366
quasi-private firms, 367
communications systems, 367
defined, 365–371
education, 368
future of, 370–371
health care, 368–369
importance of, 365
information, 367
interfacing business with, steps, 369–370
location incentives, 368
protection, 368
role of, 365–366
standards, 367
support infrastructure, 368
training, 368
Soft technologies, 208
Source of performance information, providing, 219–220
Stories, in organizational culture, 284
Strategic needs, vs. short-term needs, 225–226
Strategic planning, 19–50
business planning process, 32–33
critical success factors
characteristics of, 37–39
defined, 21–22
evolution in U.S., 22–26
hierarchy of, 33–34
in operational planning, 41–45
role, 34–41
in selecting management programs, 45–47
in strategic planning, 36–37
temporal nature of, 39–41
critical success factors plan, 33
decline, 41
defined, 35–36
economic life cycle, U.S., changes during, 26–29
essence of, 35
growth, 40–41
manufacturing flows, 44–45
manufacturing strategies, 45
maturity, 41
national economic life cycles, changes during, 26–28
performance measurement, 47
planning process, hierarchy of, 30–33
positioning strategy, 43–44
process, 31–32

project and program plans, 33
start-up, 40
strategic planning, defined, 35–36
Strategies, 243–245
business strategy, 244–245
corporate strategy, 243–244
Strategy-culture linkage, 298
Supply chain, 12–13, 67. *See also* Integrated supply chains
agile, lean, achievement of, 321, 325–335
customer relationship management, 168
evolution of, 74, 311
expansion of, 5
management, 172
model of, 346
services complementing, 345–378
Symbols, in organizational culture, 286–287
Systems theory, 102–103, 312

T

Tasks, resources, techniques, blending into programs, 5
Teams, self-directed, 180, 257–258
Technology, 195–240
agriculture, 206–209
behavioral vs. scientific management issues, 226–227
construction, 206–209
costs vs. benefits of added technology, 227–228
customer acceptance, 232–233
decision making criteria, 225–228
development, 9–10
as driver of change, 387
E-business, 212–214
economic feasibility, 233–234
equipment, 219–220
performance of, 219
source of performance information, 219–220
facilities, 220–222
design, 220–221
layout, 222
location, 221–222
future developments, 231–235, 237
human resources, 214–219
as aid to employee, 215–216
as integral part of process, 218–219
as substitute for employee, 216–217
information technology, 222–223

infrastructure, 201–202
integrated systems, 224–225
   enterprise resource planning systems, 224
   inter-organizational systems, 224
   service-oriented architecture, 224–225
   manufacturing, 206–209
   mining, 206–209
   organizational culture, 202–204
   process, adding technology to, steps, 228–231
   for resource enhancement, 214–225
   role in continuous improvement, 200–205
   self-service, 211–212
   services, 209–211
   strategic needs *vs.* short-term needs, 225–226
   technical feasibility, 234
   transfer, 205
   trust, 225
   work force acceptance, 233
Thinking, left-brain, 340
Time line, management programs, 112
Total cost of ownership, 389–390
Total Quality Management, 6, 17, 23, 85–86, 88–89, 94–96, 103–105, 112, 118, 132, 135–138, 142, 144, 155, 164, 186, 312, 381, 397, 410

Transition to integrated society, phases in, 28
Trust, 225, 334–335

V

Values, in culture of organization, 283
Variation, effect on process, 73
Vendor-managed inventory, 171
Virtual organization, 253–254, 392
VMI. *See* Vendor-managed inventory

W

Wal-Mart, 7, 15, 25, 142, 150, 182, 193, 244, 268, 288, 313, 361
Waste Electrical and Electronic Equipment Directive, 77
Web-based supply chain management systems, 46
WEEE Directive. *See* Waste Electrical and Electronic Equipment Directive
Work force acceptance, 233
Work teams, 180, 257–258
World War II, 23, 25
WSCMS. *See* Web-based supply chain management systems

HD31 .C69 2008

Crandall, Richard E.,

New methods of competing
 in the global
      c2008.

2008 09 18

0 1341 1117120 0

Due Date	Date Returned
MAR 2 0 2010	APR 7 2010

www.library.humber.ca

RECEIVED

OCT 1 4 2008

GUELPH HUMBER LIBRARY
205 Humber College Blvd
Toronto, ON M9W 5L7